D1258085

RECENT DEVELOPMENTS IN PARTICLE SYMMETRIES

CONTRIBUTORS:

J. S. BELL
N. CABIBBO
L. N. COOPER
S. FOCARDI
P. FRANZINI
U. MEYER-BERKHOUT
A. PAIS
J. PRENTKI
L. A. RADICATI
D. SHARP
G. A. SNOW
J. STEINBERGER
V. F. WEISSKOPF

RECENT DEVELOPMENTS IN PARTICLE SYMMETRIES

1965 International School of Physics "Ettore Majorana," a CERN-MPI-NATO Advanced Study Institute

Editor: A. ZICHICHI

1966

ACADEMIC PRESS New York and London

ACADEMIC PRESS INC.
111 Fifth Avenue, New York, New York 10003

United Kingdom Edition published by
ACADEMIC PRESS INC. (LONDON) LTD.
Berkeley Square House, London W.1

LIBRARY OF CONGRESS CATALOG CARD NUMBER: *66-20676*

PRINTED IN THE UNITED STATES OF AMERICA.

FOREWORD

During two weeks in September-October 1965, 90 physicists from twenty-six countries met in ERICE to attend the third course of the International School of Physics "Ettore Majorana", the proceedings of which are contained in this book.

The countries represented at the school were: Australia, Austria, Belgium, Canada, Czechoslovakia, Democratic Republic of Germany, Federal Republic of Germany, Finland, France, Hong Kong, Hungary, India, Israel, Italy, Japan, Netherlands, Nigeria, Norway, Pakistan, Poland, Rumania, South Africa, Sweden, Switzerland, United Kingdom, and the United States.

The school was sponsored by the European Organization for Nuclear Research (CERN), the Italian Ministry of Public Education (MPI) and the North Atlantic Treaty Organization (NATO). Many organizations gave their moral and financial support; we would like to acknowledge in a particular way the National Science Foundation of the United States of America and the Federazione degli Industriali Siciliani. The personal scientific sponsorship of Professors Gilberto Bernardini, Abraham Pais and Victor F. Weisskopf was of vital importance for the realization and the great success of the school. Our gratitude to them is very deep and sincere. The school has among its most distinguished supporters the Italian Minister of State H.E. Dr. Bernardo Mattarella and the Italian Ambassador H.E. the Count Giusto Giusti del Giardino, to whom we wish to express our most sincere and grateful appreciation.

The programme of the school was greatly influenced by the remarkable successes of the unitary symmetry approach to particle physics. These successes are testified by the following facts: the baryon and meson super multiplicities together with the observed relation between spin-multiplicities and SU_3-multiplicities; the equality between the coefficients of the Gell Mann-Okubo mass formulae for the baryon octet and decuplet, and for the pseudoscalar and vector meson octets respectively; the ratio of neutron to proton magnetic moments; the D/F ratio for the effective coupling of the pseudoscalar octet to the baryons; the pseudoscalar coupling constant between pions and nucleons; the initial degeneracy of the ω_1 and ω_8

the amount of φ-ω mixing; the forbiddenness of $\varphi \nrightarrow \rho\pi$; the absence of appreciable mixing between the η and the X^{o} (960); the relations between decay amplitudes in non-leptonic strange particle decays; the forbiddenness of $\bar{p}p \rightarrow$ strange particles and of $\bar{p}p \nrightarrow \varphi + n\pi$; and the Johnson-Treiman relation for (πp) and (Kp) total cross-sections.

Unless we want to believe that all the above-mentioned facts are in accidental agreement with theory, we have to conclude that the unitary symmetries are expected to be of extreme relevance for the understanding of elementary particles. Nevertheless, many problems await solutions, as for instance how to incorporate relativistic invariance in the theory and how to understand the symmetry breaking. It is the regularity of this symmetry breaking which is, in fact, even more extraordinary than the symmetry itself.

All these aspects of our present-day problems in elementary particle physics, together with some related experimental investigations, were taught and discussed in ERICE. I hope the reader will enjoy the book as much as the students enjoyed attending the lectures and the discussions. Thanks to the work of the scientific secretaries the discussions have been reproduced as faithfully as possible.

At various stages of my work, I enjoyed the collaboration of many friends whose contributions have been extremely important for the School and are highly appreciated. I would like to thank most warmly: Dr. T.Massam, who acted as deputy director; Dr. G.B. Cvijanovitch, who acted as administrator of the school; Drs. G. Altarelli, W. Blum, G. Cicogna, H.C. Dehne, V.P. Henri, K. Lassila, N. Lipshutz, F.N. Ndili, C. Noack, E. Remiddi, A. Saeed, F. Strocchi, R. Van Royen and M. Zulauf, who acted as scientific secretaries; and Miss C. Mason, who was responsible for the final typing of all the proceedings in Geneva. A final word of acknowledgement to all those who, in Erice and in Geneva, helped me on so many occasions and to whom I feel very much indebted, in particular to Mrs. Mary Bell.

TABLE OF CONTENTS

SEMINARS

DISCUSSIONS

OPENING CEREMONY

Alla cerimonia inaugurale di apertura del corso, che ha avuto luogo nel salone dell'Albergo Jolly in ERICE, Lunedi 27 Settembre, hanno partecipato: l'On.le Vincenzo Occhipinti in rappresentanza dell'Assembla Regionale Siciliana, S.E. il Prefetto di Trapani Dr. G. Napoletano, il Sindaco di ERICE Prof. A.Savalli, il Sindaco di Trapani Prof. A. Calcara, l'Assessore Provinciale Dr. S. Garofalo, in rappresentanza del Presidente della Provincia, il Presidente della Federazione degli Industriali Siciliani Avv.to G. Messina, il Provveditore agli Studi Dr. G. Purpi, il Questore di Trapani Dr. F. Lo Cascio ed altre Autorità civili e militari.

Il Direttore della Scuola ha tenuto il discorso inaugurale di apertura del corso. Hanno quindi preso la parola il Sindaco di Erice ed il Chiarissimo Professore A. Pais dell'Istituto Rockeffeler di New York.

* * *

DISCORSO INAUGURALE DEL CHIARISSIMO PROFESSORE A. ZICHICHI, DIRETTORE DELLA SCUOLA

Eccellenze, Onorevoli, Signore e Signori,

siamo qui riuniti stasera per celebrare l'apertura del terzo corso della Scuola Internazionale di Fisica "Ettore Majorana" che, come Loro sanno, è sotto gli auspici del Centro Europeo per le Ricerche Nucleari, della NATO e del Ministero della Pubblica Istruzione. A questo corso partecipano 91 fisici provenienti da 50 laboratori di 26 Nazioni. Figurano, tra i nomi degli studenti, scienziati di affermata reputazione internazionale, e poichè sarebbe troppo lungo elencarli, mi limitero soltantò a citare i nomi delle Nazioni cui essi appartengono: Australia, Austria, Belgio, Canada, Cecoslovachia, Finlandia, Francia, Giappone, Repubblica Federale Tedesca, Repubblica Democratica Tedesca, Hong Kong, India, Inghilterra, Israele, Italia, Nigeria, Norvegia, Olanda, Pakistan, Polonia, Romania, Stati Uniti, Svezia, Svizzera, Sud-Africa ed Ungheria.

Come molti di Loro non sanno, S.E. il Ministro Bernardo Mattarella è uno dei maggiori sostenitori di questa nostra iniziativa e noi dobbiamo al Suo autorevole appoggio se è possibile oggi avere in ERICE questa attività culturale scientifica altamente qualificata. E' con profonda gratitudine che io desidero ringraziarlo anche a nome di tutti coloro cui stanno a cuore questi nostri tentativi di attrarre in ERICE, quindi in Sicilia, quelle correnti del pensiero scientifico più avanzato, da cui non ci si può isolare senza cadere in un progressivo immobilismo intellettuale.

All'On.le Enzo Occhipinti, che ha sostenuto e che continua a sostenere validamente presso gli organi competenti dell'Assemblea Regionale Siciliana la nostra Scuola, vada l'espressione della nostra più sincera riconoscenza.

Vorrei rivolgere un vivo ringraziamento a S.E. il Ministro per la Pubblica Istruzione, On.le Prof. Luigi Gui, a S.E. l'Ambasciatore Manlio Brosio, Segretario Generale della NATO, ed al Direttore Generale del Centro Europeo per le Ricerche Nucleari, Chiarissimo Professore Victor F. Weisskopf, per l'appoggio che essi hanno sempre voluto concedere a questa nostra Scuola; vorrei anche ringraziare calorosamente tutti i membri del Comitato Scientifico della Scuola ed in particolare modo il suo Presidente, Chiarissimo Professore Gilberto Bernardini. Ci hanno telegrafato parole di augurio e di vivo incoraggiamento Autorità politiche e scientifiche, tra cui vorrei citare: il Presidente del Governo Regionale, On.le F. Coniglio; l'Assessore Regionale alla Pubblica Istruzione On.le D. Giacalone; il Rappresentante Permanente dell'Italia presso le Organizzazioni Internazionali di Ginevra, S.E. l'Ambasciatore Conte Giusto Giusti del Giardino; il Presidente dello Istituto Nazionale di Fisica Nucleare, Chiarissimo Professore Edoardo Amaldi, il Presidente del Comitato Regionale per le Ricerche Nucleari, Chiarissimo Professore Sebastiano Sciuti, ed infine il Direttore del Centro Siciliano di Fisica Nucleare, Chiarissimo Professore Italo Federico Quercia, che ha telegrafato anche a nome delle tre Università Siciliane.

Questa sera abbiamo l'onore di avere con noi il Professore Schall della Segreteria Generale per gli Affari Scientifici della NATO, qui venuto tanto gentilmente a visitarci, per testimoniare l'interesse e l'appoggio che la NATO rivolge alla Scuola "Ettore Majorana". Sono qui con noi alcuni tra i nomi più prestigiosi della fisica mondiale, che hanno tanto gentilmente accolto l'invito ad insegnare in questo corso: il Professore John Bell di Ginevra, il Professore Nicola Cabibbo di Ginevra, il Professore Leon Cooper della Brown University, il Professore Paolo Franzini della Columbia University, il Professore Abraham Pais del Rockefeller Institute di New York, il Professore Meyer-Berkhout dell'Università di Amburgo, il Professore Jack Prentki del "Collège de France" di Parigi, il Professore Luigi Radicati dei Principi di Brozolo della Scuola Normale Superiore di Pisa, il Professore Sharp del California Institute of Technology di Pasadena, il Professore George Snow dell'Università di Maryland, il Professore Jack Steinberger dell'Università di Columbia ed il Professore Victor F. Weisskopf, Direttore Generale del Centro Europeo per le Ricerche Nucleari.

Vorrei adesso illustrare brevemente il tema del corso, cercando soprattutto di giustificare il motivo del nostro interesse in questo settore della ricerca scientifica.

Il potenziamento della fisica delle particelle elementari ha indubbiamento rappresentato uno degli aspetti più coerenti della politica scientifica seguita negli ultimi decenni dalle Nazioni più avanzate nel campo tecnico ed industriale, come l'America, la Russia, l'Inghilterra, la Germania e la Francia. Questo interesse che le Nazioni più progredite dal punto di vista tecnico e scientifico dedicano alla cosiddetta fisica delle particelle elementari è determinato dalla importanza che questo tipo di ricerca ha rispetto alla comprensione di quei fenomeni che investono direttamente la nostra esistenza. Non bisogna dimenticare che è la soluzione di questi problemi a determinare il progresso tecnico delle società moderne. I lavori dei grandi scienziati degli ultimi secoli hanno infatti determinato la trasformazione industriale della nostra comunità. E basterebbe forse ricordare l'ultimo di questi rivolgimenti, quello nucleare, iniziato in modo violento attraverso l'esplosione di bombe atomiche per poi svilupparsi nelle nuove sorgenti di energia che diverranno indubbiamente entro la fine di questo secolo di vitale importanza per l'uomo. Mentre è difficile prevedere quale particolare linea di ricerca sia suscettibile di applicazione più o meno immediata, è indubbiamente certo che la comprensione dei problemi del mondo che ci circonda è l'unica via verso il progresso della nostra specie. Una nazione che si isoli da questa ricerca fondamentale, la quale rappresenta la continuazione di quello che fu il pensiero scientifico di uomini come Galilei, Newton, Maxwell, Rutherford, Einstein e Fermi, è destinata a ridurre la sua influenza nel mondo scientifico ad un ruolo di marginale importanza con le naturali conseguenze politiche e sociali. Ed infatti le distanze in questo rapido progresso del pensiero scientifico moderno sono in continuo aumento; è come nel mondo economico in cui chi è più ricco diventa sempre più ricco, e chi è più povero diventa sempre più povero. Potenziare quindi le attività correlate con lo studio dei fenomeni fondamentali che servono a farci capire la natura che ci circonda è di importanza troppo fondamentale per essere oggetto di dubbio. C'é però il guaio che lo studio di quello che noi fisici chiamiamo "fenomeni fondamentali" appare ai non iniziati uno studio avulso dalla realtà quotidiana. Potrei dilungarmi sui nostri grandi predecessori come Galilei, Newton, Einstein etc., i cui lavori apparivano ai loro contemporanei come opere completamente avulse dalla realtà quotidiana, per convincervi di quanto errato sia un tale giudizio sui fisici. Mi limiterò invece a citarvi un solo esempio, quello del sole. L'uomo ha vissuto su questa terra per migliaia di anni cercando di capire cosa fosse il sole e fino a poche decine di anni fà il dilemma sole era difficile distinguerlo da un problema divino; infatti non si poteva certo capire quale fosse l'origine di questa enorme sorgente di luce. Ebbene, dopo tante migliaia di anni, è solo grazie alla fisica moderna che adesso l'uomo può dire di avere capito come funziona il sole: esso non è altro che una grande candela a combustione nucleare d'idrogeno. E questa candela non è una "bomba" grazie al fatto che esistono le cosiddette "interazioni deboli". Se le interazioni deboli fossero "forti", l'uomo non sarebbe mai esistito su questa nostra terra.

La problematica dei fisici moderni non è certo meno affascinante, come adesso vedremo. Io ho recentemente letto nella stampa nazionale degli articoli in cui si diceva che la fisica è arrivata ad un punto morto; questo non potrebbe essere più falso; viceversa la fisica è oggigiorno ad un punto che potrebbe risultare veramente rivoluzionario.

Fino a circa vent'anni fà si pensava che l'uomo avesse capito quasi tutto di quello che era la struttura della materia. La situazione era infatti la seguente : si sapeva che esistevano quattro tipi di interazioni fondamentali ed un certo numero di particelle "elementari", tutte con funzioni essenziali per la nostra stessa esistenza.

Incominciamo dalle interazioni:

i) le interazioni gravitazionali sono responsabili dei moti celesti e in particolare, ci tengono legati alla terra;

ii) le interazioni deboli, come abbiamo accennato prima, tengono sotto controllo la lenta combustione nucleare della nostra essenziale sorgente di energia: il sole;

iii) le interazioni elettromagnetiche sono responsabili della formazione degli atomi, delle molecole e quindi di tutta l'enorme varietà del mondo che ci circonda. Sono di natura elettromagnetica non solo la vista, ma anche il tatto, e l'olfatto e possiamo senza altro dire che tutta la nostra esistenza è basata sui fenomeni eletromagnetici;

iv) le interazioni nucleari servono a tenere insieme protoni e neutroni in un nucleo e sono quindi responsabili della formazione dei nuclei atomici.

Passiamo adesso alle particelle. Le particelle conosciute fino al 1947 erano: il protone, il neutrone, il pione, l'elettrone, il muone, il fotone ed il neutrino. Sembrava allora che il capitolo della fisica delle particelle elementari stesse per chiudersi, in quanto a ciascuna particella era assegnata una funzione e non c'era proprio bisogno di nient'altro. Infatti i protoni ed i neutroni unendosi insieme, grazie alle interazioni nucleari, formano i nuclei, l'ente base di queste interazioni nucleari essendo il pione che appunto agisce come una specie di colla nucleare che permette ai protoni ed ai neutroni di stare legati in un nucleo; gli elettroni unendosi ai nuclei, grazie alle interazioni elettromagnetiche, formano gli atomi. I fotoni sono i quanti di luce emessa quando un atomo passa da uno stato eccitato ad un altro meno eccitato; mentre i neutrini ed i muoni*) intervengono in quei processi "deboli" di cui abbiamo parlato prima.

Ma vediamo cosa doveva succedere dal 1947 in poi: con l'avvento delle nuove macchine acceleratrici, di energia sempre più elevata, il numero di particelle che hanno lo stesso diritto ad essere considerate altrettanto "elementari" quanto il protone, il neutrone, il pione, il fotone, l'elettrone, il muone ed il neutrino è andato sempre crescendo. Oggi se ne contano più di 100. È difficile però pensare che possano essere elementari cento oggetti e questo ci fa ritornare ad un dilemma analogo a quello che fu della chimica ai tempi della famosa tavola di Mendeléeff. Si pensava allora che quegli elementi di cui se ne contavano quasi un centinaio, fossero i costituenti ultimi della materia; era però difficile accettare che la Natura avesse bisogno di qualcosa come cento oggetti, tutti elementari, e questo sospetto era corroborato dalle sorprendenti regolarità osservate nella tavola di Mendeléeff. Come è stato poi scoperto, all'origine di queste regolarità stava e sta il fatto che gli atomi sono tutti fatti di protoni, neutroni ed elettroni; quindi queste diverse decine di elementi sono in verità costituiti da tre soli oggetti "elementari". Come ho accennato prima, la cosa straordinaria nella fisica di questi ultimi anni è l'enorme numero di particelle "elementari" scoperte. Si ritorna quindi a perdere la base del concetto di elementarità, che presuppone un numero quanto più piccolo possibile ed è certo incompatibile con una multitudine di enti "elementari". Ma anche adesso, così come fu per la tavola di Mendeléeff, questa moltitudine di particelle elementari mostra delle notevoli regolarità. Tali regolarità potrebbero essere spiegate se si supponesse che in verità le particelle fin'ora considerate "elementari" non sono tali, ma constano di altri enti. E basterebbero tre oggetti "super-elementari",

---

*) In verità il muone era già un mistero. Questo mistero non è stato ancora chiarito dalla fisica moderna, in quanto il muone risulta, in tutte le osservazioni sperimentali sinora eseguite, essere nient'altro che un "elettrone pesante", e non si può quindi capire cosa ci stia a fare in natura.

veri costituenti ultimi della materia, per spiegare le regolarità osservate nella classificazione delle particelle fin'ora considerate "elementari". E' bene dire subito che molti problemi teorici restano ancora da risolvere prima che una formulazione rigorosa della teoria delle particelle elementari, basata sulla esistenza di questi tre enti "super-elementari", sia possibile. Le difficoltà teoriche passerebbero ovviamente in seconda linea qualora questi costituenti ultimi della materia venissero scoperti; il che rivoluzionerebbe la nostra attuale comprensione della struttura intima della materia. Una cosa molto interessante in questo contesto è il fatto che se tali particelle esistessero, dovrebbero essere legate tra di loro con energie di legame circa mille volte più forti di quelle proprie alle forze nucleari. Il che significa che se mai ci fosse una qualsiasi applicazione pratica, questa implicherebbe una sorgente di energia mille volte più potente di qualunque altra fin'ora conosciuta. Noi ovviamente non studiamo problemi applicativi; noi studiamo la natura in sè per il piacere profondo di svelarne i più reconditi misteri. E' però certo che la fisica moderna si trova oggi ad affrontare problemi di estremo interesse che dovrebbero affascinare l'uomo della strada non meno della conquista della luna. La differenza sta forse nel fatto che tutti riescono a capire cosa significa mettere i piedi sulla luna, mentre così facile non è riuscire ad immaginare cosa significa lo studio di fenomeni che avvengono a distanze inferiori ad un millesimo di miliardesimo di millimetro.

Spero di avere detto abbastanza per giustificare l'interesse e l'entusiasmo che una grande parte dei fisici di tutto il mondo dedicano a questa appassionante problematica della fisica moderna, cui il corso è dedicato. Saranno infatti trattati e discussi quei problemi che si pongono quando si voglia passare alla formulazione rigorosa di una teoria delle particelle elementari, basata sulle "regolarità" o "simmetrie" osservate. Un problema interessante, al quale sarà dedicato un gruppo di lezioni è ad esempio il seguente: si osservano, come abbiamo accennato sopra, certe regolarità, o proprietà di simmetria nella fisica delle particelle elementari. Ma queste proprietà di simmetria non sono rigorosamente valide; esse vengono violate; però il modo in cui questo avviene è perfettamente controllato; esso non può avvenire a caso, ma sempre in un modo ben determinato. E' la regolarità di queste violazioni che è forse più sorprendente delle simmetrie stesse. Un altro problema che sarà discusso in questo corso è quello che riguarda il modo di incorporare in queste teorie l'invarianza relativistica. Ma questi non sono che due tra i tanti problemi che verranno trattati.

Il corso durerà due settimane e si articolerà in trenta lezioni, dieci discussioni ed alcuni seminari che io prevedo tutti interessantissimi e che mi auguro pieni di vivacità,come è già tradizione in questa nostra giovanissima istituzione.

PAROLE DI BENVENUTO DEL SINDACO DI ERICE

Professore A. Savalli

Eccellenze, Onorevoli, Signore e Signori,

come Sindaco di questa città, mi è assai gradito rivolgere Loro a nome mio personale, del Consiglio Comunale che ho l'onore di presiedere, e della Cittadina tutta, il più cordiale saluto di benvenuto in questa mitica vetta ed è con vivo piacere che desidero esprimere un sentito ringraziamento al Centro Europeo Ricerche Nucleari, al Ministero della Pubblica Istruzione, alla NATO, alla Direzione della Scuola Internazionale di Fisica "Ettore Majorana", e particolarmente al suo Direttore, Chiarissimo Professore Antonino Zichichi, i quali hanno voluto che Erice accogliesse ora per la terza volta il corso internazionale di fisica nucleare. Questo monte, famoso fin dai tempi più remoti per le civiltà che qui si sono succedute, è orgoglioso di accogliere fisici di rinomanza mondiale che qui, in questo silenzio claustrale, in questa cittadina dove ogni pietra parla dell'antica gloria, potranno e sapranno impartire ai giovani convenuti i dettami del nuovo scibile che in poco volger d'anni sta cambiando il volto del mondo, nella visione certa di un vivere migliore, lontano dai bisogni e rivolto alla conquista dell'universo. Nella speranza certa, che gli organi competenti nazionali e regionali, sensibili come sono ai nuovi bisogni della vita sociale, faranno sì che venga potenziata in Erice l'attività della Scuola "Ettore Majorana", auguro che questo terzo corso possa essere coronato dal più valido successo e formulo a tutti gli auguri di buono studio, di ottima permanenza in questa nostra Erice ed ancora un caloroso invito a voi tutti di ritornare qui tra di noi.

INVITED SPEECH BY PROFESSOR PAIS

Eccellenze, Onorevoli, Signore e Signori,

Well, a previous speaker has apologised for the fact that something had to be said in Italian. I think it is only proper that I have to apologise for the fact that something has to be said in English. I regret that I cannot speak your beautiful language as I would like to, although I begin to understand a little bit of what I have heard before.

I stand here with very mixed feelings, because unlike Professor Zichichi, who has many notes, I have nothing, not even in my pocket, because I did not know I would stand here until about one minute before the gentleman asked me to join on the podium. So, let me then just make a few general remarks. In the first place, I had the pleasure of arriving in Sicily last Friday. It seems ages ago. I have been travelling over your island at great speed, because tomorrow I have to start work, and I will not have time to look at the beautiful things, but I have seen the mosaics in the Casa Palatina, and I have seen your famous ruins at Agrigento and Selinunte, and I begin to understand more about the great heritage of your island of Sicily. And then I came to Erice. It is nice and hot in the daytime and it is nice and cold in the night time. I personally like that because I was born in a climate like that and, tomorrow morning, I can assure you I will be much more comfortable, because first of all I will not wear a tie, and secondly I will not wear a jacket, and I will be able to talk about something I have thought about longer than just these few moments.

It is also a great privilege to talk to such an international audience of students as is presented here. This is a great privilege. Youth in the sciences is of the essence. You have to be young to do the work and to start the work, that is for sure, and I may perhaps remind you of the story about the great Professor Pauli, who is no longer with us, and who once had a visit from a young student who had, he thought, made a very interesting discovery. And so Pauli destroyed his idea in a very few minutes, and when the young man left, Pauli shook his head, and said "so young, and already so unknown!"

Now, the subject of particle physics, the subject of physics in general, to the outsider may seem to be a narrow path that is followed logically step by step, and from new insight to new insight, but those of us who are actively engaged in this field know that sometimes you think you understand something and you do not, and sometimes you understand something and you do not know why, and it is about such subjects that I will be talking during the coming days.

To conclude my few brief remarks I may perhaps tell you the spirit in which many of these discussions are held in science, and in order to tell you about it I want to tell you about a visit which one famous physicist made to the laboratory of a colleague of his. He was shown around, and the guest and the host discussed beautiful and very clever experiments, they discussed all kinds of apparatus, and then when they had finally seen everything in the laboratory together, they came to the door to leave the laboratory, and as they were leaving the guest saw above the door a horseshoe. So he said to his host: "You, a scientist, do you believe in that?". So his host said: "Of course I do not, but they say it works even if you do not believe in it".

Thank you.

HIGHER SYMMETRIES

A. Pais,
Rockefeller Institute,
New York.

I. INTRODUCTION ABOUT UNITARY SYMMETRY

The unitary group U(N) is the group of linear transformations of a complex vector x^α:

$$x^\alpha \equiv (x^1, x^2, \ldots x^N), \tag{1}$$

$$x^\alpha \to x'^{,\alpha} = a^\alpha_\beta x^\beta, \tag{2}$$

such that

$$x^*_\alpha{}' x'^{,\alpha} = x^*_\alpha x^\alpha, \tag{3}$$

where x^*_α is the vector whose components are the complex conjugate of those of x^α. Equation (3) implies

$$a^\dagger a = 1 \tag{4}$$

i.e. the matrix a^α_β is unitary. If we further impose

$$\det a = 1 \tag{5}$$

we obtain the special unitary (unimodular) group SU(N). We shall mean by y_α (lower index) a quantity which transforms like x^*_α.

---

*) Notes for first three lectures taken by G. Altarelli and F. Strocchi. Notes for fourth lecture taken by K. Lassila and C. Noack.

Starting from the vector x^α, by application of all the matrices a, we generate a complex vector space of dimensions N. In this space we may define a basis x_i^α, i = 1,2, ... N. In physical applications to each vector of the basis is associated a physical state.

On the vector space x^α a representation is defined, denoted by $\underline{N}$, which is called the defining representation. $\underline{N}^*$, the "conjugate representation", is defined on the vector space x^*_α. In physical applications the generators of a group are associated to conserved quantities. The validity of these conservation laws varies in different cases. This is illustrated by the following well-known examples.

a) Strict kinematical group SU(2): angular momentum

In this case, strict means that it applies without exception, kinematical means that the physical quantities associated to the generators of the group, i.e. the angular momentum operators, J^2 and J_z are conserved independently of the dynamics. In this case the defining representation has two dimensions: $\underline{N}$ = 2, the "spinor".

b) Approximate kinematical group SU(2): isospin

Here, the invariance is not a strict one as the isospin symmetry is valid only in the limit e = 0. However, it is still of a kinematical character because the electric charge e is a non-dynamical parameter. $\underline{N}$ = 2 is the "nucleon".

c) Approximate dynamical group SU(2): normal coupling in atoms (Russell Saunders)

The Hamiltonian describing the interaction between the nucleus and the electrons may be split into three parts:

$$H_{int} = H_{\text{central nucleus field}} + H_{\text{electrons' static Coulomb field}} + H'_{\text{the rest}}.$$

If H' can be neglected in the first approximation, then not only $\vec{J}$ is conserved, but also $\vec{L}$ and $\vec{S}$. When we are in this situation, we speak of normal coupling. For a fixed j, the multiplets are labelled by $^{2s+1}L_j$ (normal multiplet structure). $\underline{N} = 2$: the electron with spin. Wherever H' may be neglected, v/c effects such as spin orbit couplings are effectively small. Unlike the previous case, the parameters defining the domain of validity are now of dynamical nature. The SU(2) in this case is said to correspond to an approximate dynamical group.

d) Approximate group SU(3): particle symmetry

We do not know whether this is an approximate kinematical group or a dynamical one. If we assume that the Hamiltonian may be split into two parts

$$H = H_1 + gH' ,$$

where H_1 is SU(3) symmetric, H' is not, and g is a "medium strong" coupling constant, then we have an approximate kinematical symmetry. ($\underline{N} = 3$, the quarks or triplets.)

e) Approximate dynamical group SU(4): Wigner theory

This is the nuclear analogue of case c). The group works well for low-lying nuclear states, where spin-orbit effects are small.

f) Approximate dynamical group SU(6)

This is the analogue of case e), for particle physics. The neglect of spin-orbit coupling corresponds to a situation in which the super multiplets would be degenerate. They split into SU(3) multiplets because of spin-orbit interactions. ($\underline{N} = 6$, the quarks or triplets with spin ½ = sextet.)

The pseudo-unitary group U(m,n) is defined as the set of all linear transformations acting on a complex vector space

$$x^{\alpha} \equiv (x^1, x^2, \ldots x^N) , \quad N = m + n ,$$

which leave the form

$$x^+ \Gamma x = x^*_{\alpha} \Gamma^{\alpha}_{\beta} x^{\beta}$$

invariant. Here, Γ is the following diagonal matrix

$$\Gamma = \begin{pmatrix} \overbrace{1 \; 1 \cdots 1}^{m} & \\ & \overbrace{-1 \; -1 \cdots -1}^{n} \end{pmatrix}$$

Then if

$$x^{\alpha} \to x'^{\alpha} = a^{\alpha}_{\beta} x^{\beta} ,$$

we require that

$$x'^{\dagger} \Gamma x' = x^{\dagger} a^{\dagger} \Gamma a x = x^{\dagger} \Gamma x$$

that is,

$$\bar{a} a = \Gamma a^{\dagger} \Gamma a = 1 .$$

If in addition

$$\det a = 1 ,$$

then we have SU(m,n).

II. USEFUL TOOLS

A tensor

$$T^{\alpha_1 \ldots \alpha_n}_{\beta_1 \ldots \beta_m}$$

is an object which transforms as

$$x^{\alpha_1}(1)\, x^{\alpha_2}(2) \ldots x^{\alpha_n}(n)\, y^{*\beta_1}(1') \ldots y^{*\beta_m}(m')$$

under the group. Operating on

$$T^{\alpha_1 \ldots \alpha_n}_{\beta_1 \ldots \beta_m}$$

with all the transformations of the group, we get a vector space. The representation of the group defined on this vector space is, in general, reducible for two following reasons:

i) The contraction of a certain number of upper indices with the same number of lower indices is an invariant operation. Then, if the result is not zero, we obtain a lower dimensional invariant vector space and, correspondingly, a lower dimensional representation. So irreducible tensors must be traceless.

ii) Consider the particular case of a second order tensor $T^{\alpha\beta}$. Its symmetrical and antisymmetrical parts transform separately one from the other, so that the corresponding vector spaces are invariant and the representation is reducible. In general, irreducible tensors must have a definite permutational symmetry in both upper and lower indices separately.

<u>Exercise</u>: Prove that permutational symmetries are respected by the group. (The proof is found for example in Hamermesh's book.)

* * *

In SU(N) the tracelessness and permutational symmetries are connected. In fact, for a tensor like T^{α}_{β}, the condition

$$T^{\alpha}_{\alpha} = 0$$

may be written in the following way. Define

$$T^{\alpha_1 \cdots \alpha_N} = \epsilon^{\beta\alpha_1 \cdots \alpha_{N-1}} T^{\alpha_N}_{\beta} ,$$

where $\epsilon^{A_1 \cdots A_N}$ is totally antisymmetric in all indices and $\epsilon^{12 \cdots N} = 1$ (Levi-Civita tensor). In terms of $T^{\alpha_1 \cdots \alpha_N}$, the trace condition becomes

$$T^{\alpha_1 \cdots \alpha_N} \pm T^{\alpha_2 \cdots \alpha_N \alpha_1} + \ldots \pm T^{\alpha_N \alpha_1 \cdots \alpha_{N-1}} = 0 ,$$

where the upper (lower) signs hold for N = odd (even). So each tensor in SU(N) may be written with all indices up.

The problem of getting all the irreducible representations of SU(N) can be treated by the general method of Young. These rules are valid for any unitary unimodular group as well as for finite dimensional representations of pseudo-unitary unimodular groups. These rules are:

i) Fix an integer $n \geq 0$ and make a partition $(n_1, n_2 \ldots n_N)$ such that

$$\Sigma n_i = n ,$$

$$n_1 \geq n_2 \geq n_3 \ldots \geq n_N \geq 0 .$$

To each partition may be associated a diagram (Young tableau) which is illustrated in the figure below

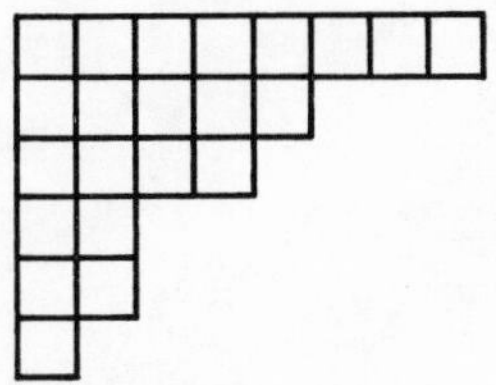

where the number of the boxes in the first row is n_1, in the second row n_2, ... and in the last n_N. Young tableaux are in a one to one correspondence to the irreducible representations of the group. A given representation will be labelled by the symbol $(n_1, n_2, \dots n_N)$. For brevity's sake the following corrections will be used

$$(n_1, 0, 0, \dots 0) = (n_1)$$
$$(n_1, 1, 1, 1, 0, \dots 0) = (n_1, 1^3) \text{ etc.}$$

ii) The dimension of a given representation is obtained by the following formula

$$\begin{aligned} D_N(n_1, n_2, \dots n_N) = \frac{(n_1 - n_2 + 1)}{1!} &\times (n_1 - n_3 + 2) \times \dots \times (n_1 - n_N + N - 1) \times \\ &\times \frac{(n_2 - n_3 + 1)}{2!} \times \dots \times (n_2 - n_N + N - 2) \times \\ &\quad \ddots \qquad \vdots \\ &\qquad \times \frac{(n_{N-1} - n_N + 1)}{(N-1)!} . \end{aligned}$$

For N = 2, use the first column only, for N = 3 use the first and second columns, etc. From this formula we see that

$$D_N(n_1, n_2, \dots n_N) = D_N(n_1 - n_N,\ n_1 - n_{N-1},\ \dots\ n_1 - n_2,\ 0)$$
$$= D_N(n_1 - n_N,\ n_2 - n_N,\ \dots\ n_{N-1} - n_N,\ 0).$$

Exercises: 1) Show that

$$D_N(1) = D_N(1^{N-1}) = N\ ,$$
$$D_N(2, 1^{N-2}) = N^2 - 1$$

[$D_N(2,1^{N-2})$ is the adjoint representation]. Example: $D_6(21^4) = 35$.

2)
$$D_N(n) = \binom{N+n-1}{n}$$

[n is given by ▭, and is a totally symmetrical representation i.e. the corresponding tensor $T^{\alpha_1\alpha_2 \dots \alpha_n}$ is symmetrical in all indices]. Examples: $D_3(3) = 10$, $D_6(3) = 56$.

3)
$$D_N(1^k) = \binom{N}{k}\ .$$

Example: $D_6(1^3) = 20$.

The conjugate representation of $(n_1, \dots n_N)$ is $(n_1 - n_N,\ n_1 - n_{N-1},\ \dots\ n_1 - n_2,\ 0)$. If

$$T^{\alpha_1\alpha_2 \dots \alpha_n}_{\beta_1\beta_2 \dots \beta_m}$$

is the tensor corresponding to a given representation, then

$$T^{\beta_1\beta_2 \dots \beta_m}_{\alpha_1 \dots\dots \alpha_n}$$

is associated to the conjugate representation. The representations $(2,1^{N-2})$ are an example of a self-conjugate case.

iii) Reduction of products of two irreducible representations.

We consider firstly the case of the product of an arbitrary representation by the (1). In terms of Young diagrams we have:

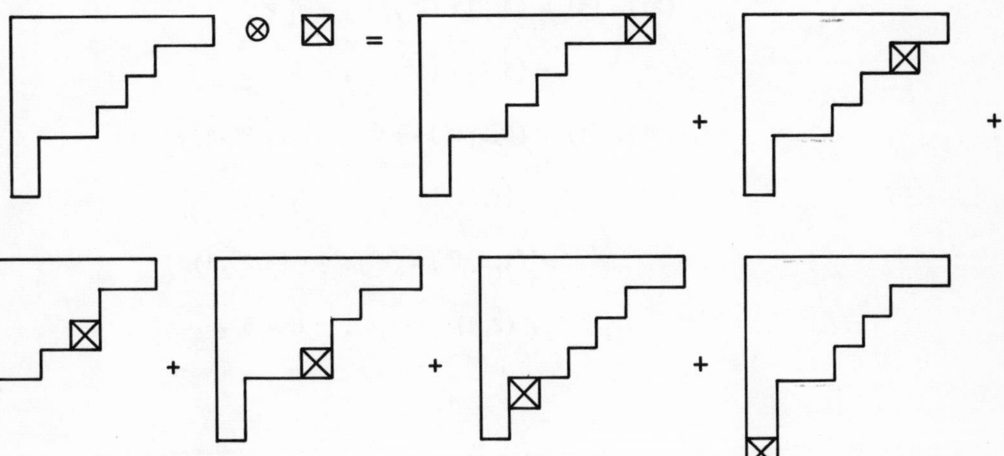

i.e. the box ⊠ has been added to the original diagram in all possible ways which lead to a new allowed partition. For instance,

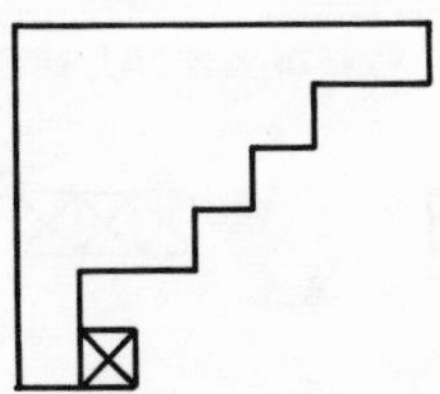

is not a Young tableau. Further, each column with a number of boxes equal to N may be dropped. For instance,

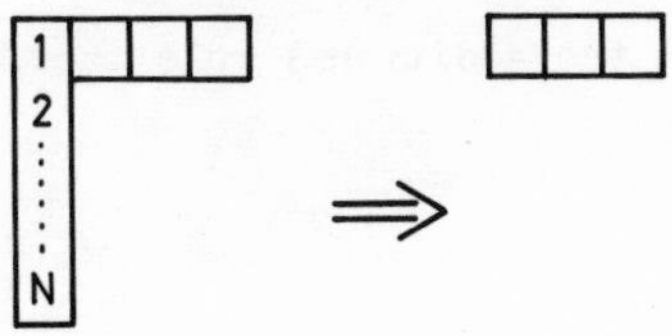

<u>Exercise</u>: Show that

$$(2^2)\times(1) = (3,2)+(2^2,1)\ ,\quad N>3$$

$$= (3,2)+(1^2)\ ,\quad N=3$$

$$(3,2,1^2)\times(1) = (4,2,1^2)+(3^2,1^2)+(3,2^2,1)+$$

$$+ (3,2,1^3)\ ,\quad N>5$$

$$= (4,2,1^2)+(3^2,1^2)+(3,2^2,1)$$

$$+ (2,1)\ ,\quad N=5\ .$$

The second step is to multiply a given representation by a totally symmetrical representation (one row)

$$(n_1,\ \dots\ n_N)\times(n)\ .$$

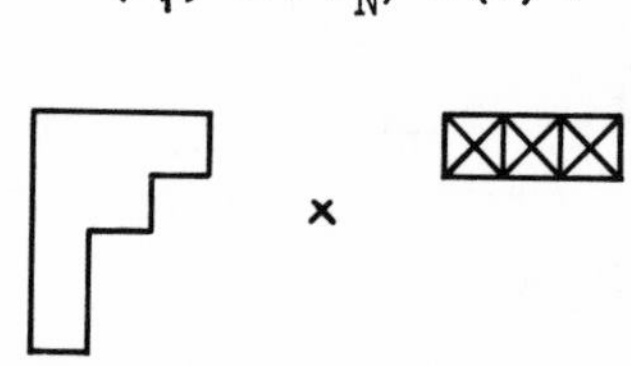

One has to proceed as before, adding each box ☒ to the original diagram in all possible ways, with the additional rule that two boxes ☒ cannot appear on the same column.

<u>Exercise</u> Show that in SU(3) we have

$$(3^2)\times(3) = (6,3)+(4,2)+(2,1)+(0) ,$$

in dimensions:

$$10^{*}\times 10 = 64+27+8+1 .$$

In SU(6)

$$(3^5)\times(3) = 56^{*}\times 56 = 1+35+409+2695$$

$$(2,1^4)\times(3) = 35\times 56 = 56+70+700+1134 .$$

Now, we are in a position to give the general prescription in the case of arbitrary representations. Draw the Young diagrams corresponding to the two representations, put an a on each box of the first row of the Young tableau, an index b on each box of the second row, and so on. Operate with the first row, as in the previous cases, <u>then</u> on each of the diagrams so obtained operate with the second row, <u>then</u> with the third row etc.

Here, we have moreover the additional prescription: reading from right to left at each stage the number of a's must be ≥ the number of b's ≥ number of c's etc. For example (N = 6):

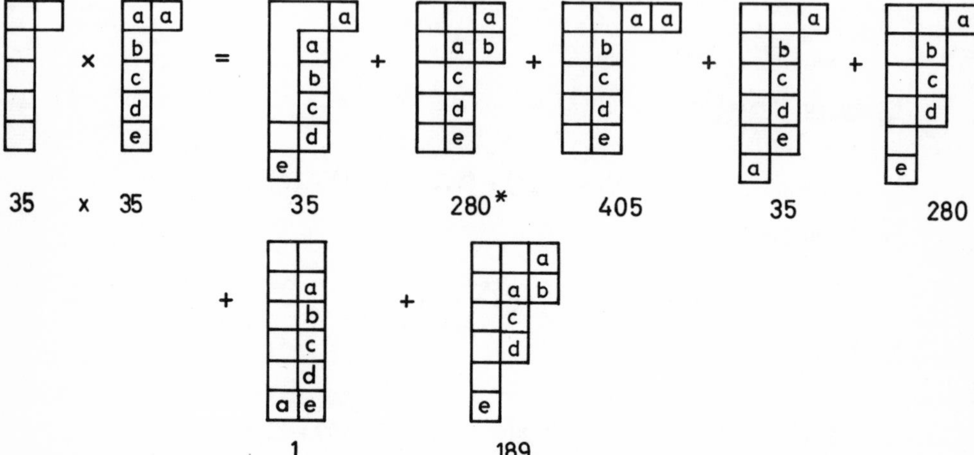

Exercise: For SU(6) check that the tensors $T^{\{\alpha\beta\}}_{\{\gamma\delta\}}$ and $T^{[\alpha\beta]}_{[\gamma\delta]}$ have 405 and 189 components respectively. Here { } and [ ] denote symmetry and antisymmetry respectively. Try also to associate the tensors with the visual-aid Young tableaux. (Hint: antisymmetry in two lower indices corresponds to putting two columns of length 5 on top of each other, modulo 6.)

III. SU(6) WITH APPLICATIONS

We shall now consider the group SU(6). The defining representation has dimension six and the corresponding tensor is S^α, $\alpha = 1, \ldots 6$, the sextet. Instead of one single index α, we shall introduce a pair of indices i,A, i = 1,2, A = 1,2,3. The correspondence between α and the pair i,A is given by the following table

α =	1	2	3	4	5	6
i =	1	1	1	2	2	2
A =	1	2	3	1	2	3 .

For example, to $\alpha = 4$ there corresponds $i = 2$, $A = 1$. The group SU(6) contains $SU(3)\times SU(2)$ as a subgroup. The defining representation of SU(6), denoted by $\underline{6}$, is equivalent to the pair of defining representations of the group $SU(3)\times SU(2)$, denoted by $(\underline{3},\underline{2})$: $(\underline{3},\underline{2})$ means $\underline{3}$ of SU(3) and $\underline{2}$ of SU(2).

If we have a group G and a subgroup $G' \subset G$, each irreducible representation of G is also a representation of G', but, in general, it is not irreducible with respect to G'. In fact, as $G' \subset G$ a vector space may be invariant under G' and not under G. Thus, a representation of SU(6) is, in general, a direct sum of irreducible representations of $SU(3)\times SU(2)$. As an example we shall study the content, in terms of irreducible representations of $SU(3)\times SU(2)$, of the simplest irreducible representations of SU(6):

$$(1) \equiv \underline{6} = (\underline{3},\underline{2}) \; . \tag{1}$$

Now, we have

$$(1)\times(1) = \square \times \square = \square\square + \begin{array}{|c|}\hline \\ \hline \\ \hline\end{array} = (2) + (1^2)$$

$$= (1;1)\times(1;1) = (2+1^2;\; 2+1^2)$$

$$= (2;2) + (1^2;2) + (2;1^2) + (1^2;1^2)$$

i.e.

$$(2) \oplus (1^2) = (2;2) \oplus (1^2;2) \oplus (2;1^2) \oplus (1^2;1^2) \; ,$$

or

$$\underline{21} \oplus \underline{15} = (\underline{6},\underline{3}) + (\underline{3}^*,\underline{3}) + (\underline{6},\underline{1}) + (\underline{3}^*,\underline{1}) \; .$$

By counting the dimensions we have

$$\underline{21} = (\underline{6},\underline{3}) + (\underline{3}^*,\underline{1}) \tag{2}$$

$$\underline{15} = (\underline{3}^*,\underline{3}) + (\underline{6},\underline{1}) \quad . \tag{3}$$

From Eq. (1) we have (conjugate representations have conjugate content)

$$\underline{6}^* = (3^*,2^*) = (3^*,2)$$

as in SU(2), and only in SU(2) each irreducible representation is self-conjugate. We also have

$$\begin{aligned}\underline{6}^* \times \underline{6} &= (\underline{3}^*,\underline{2}) \times (\underline{3},\underline{2}) = \underline{35} + \underline{1} \\ &= (\underline{8}+\underline{1};\ \underline{3}+\underline{1}) = (\underline{8},\underline{3}) + (\underline{8},\underline{1}) + (\underline{1},\underline{3}) + (\underline{1},\underline{1})\ .\end{aligned}$$

Clearly, 1 = $(\underline{1},\underline{1})$, so

$$\underline{35} = (\underline{8},\underline{3}) + (\underline{8},\underline{1}) + (\underline{1},\underline{3})\ . \tag{4}$$

Similarly, we get

$$\begin{aligned}\underline{56} &= (\underline{8},\underline{2}) + (\underline{10},\underline{4}) \\ \underline{70} &= (\underline{8},\underline{4}) + (\underline{8},\underline{2}) + (\underline{1},\underline{2}) + (\underline{10},\underline{2})\ .\end{aligned} \tag{5}$$

Now let us turn to physical applications. The basic vector S^a can be written as

$$S^a = S^{iA} = t^A \chi^i \ ,$$

t^1, t^2, t^3 are the SU(3) quarks or triplets p, n, λ and χ^1, χ^2 are the Pauli spinors $\uparrow$ and $\downarrow$.

Then $S^1 = p\uparrow$, $S^2 = n\uparrow$, $S^3 = \lambda\uparrow$, $S^4 = p\downarrow$, $S^5 = n\downarrow$, $S^6 = \lambda\downarrow$. The quantum numbers of the quarks are listed in the table below:

	q	Y	b	T_3
p	$2/3$	$1/3$	$1/3$	$1/2$
n	$-1/3$	$1/3$	$1/3$	$-1/2$
λ	$-1/3$	$-2/3$	$1/3$	0 .

Note that

$$q = T_3 + \frac{Y}{2} .$$

Now, we have to assign the physical particles to the irreducible representations of SU(6). First, we note that as the baryon number and the parity operators commute with the SU(6) generators, the particles belonging to a given supermultiplet of SU(6) must have the same parity and baryonic number. As far as the bosons are concerned the octet of pseudoscalar mesons and the nonet of vector mesons may be assigned to the 35 representation of SU(6). In fact, from Eq. (4) we see that 35 contains an octet of spin zero, an octet of spin one and a singlet of spin one.

The baryons of parity +1 can be accommodated in the 56 representation, which contains an octet of spin one-half and a decuplet of spin three-half. [See Eq. (5).]

We now study the structure of the tensor B^{abc} corresponding to the 56 representation: B^{abc} is a completely symmetric tensor. Let $S^a(1)\, S^b(2)\, S^c(3)$ denote a three-quark state. Then the tensor

$$\frac{1}{6} \sum_{\substack{\text{perm. of} \\ 1,2,3}} S^a(1)\, S^b(2)\, S^c(3) \qquad (6)$$

has the same transformation properties of B^{abc}. Each baryon may then be regarded as a quark-quark-quark system. This need not have a real physical meaning; it means only that the 56 representation is contained in the tensor product $6\times 6\times 6$. The factor $1/6$ is a normalization factor in order to have the norm conditions

$$\begin{aligned} \| B^{abc} \| &= 1 && \text{if } a = b = c \\ \| B^{abc} \| &= 1/3 && \text{if } a = b \neq c \\ \| B^{abc} \| &= 1/6 && \text{if } a \neq b \neq c \,. \end{aligned} \tag{7}$$

<u>Exercise</u>: Show that

$$\sum_{abc} \| B^{abc} \| = 56 \,.$$

As an example we construct the N^{*+} state with $S_2 = 1/2$. As $q = Y = 1$, it may be built only out of $p\,p\,n$ quarks. The spin-wave function must be of the form $\uparrow\uparrow\downarrow$ in order to have $S_2 = 1/2$. Then, starting from $p^\uparrow p^\uparrow n^\downarrow$ we have to symmetrize and to normalize in both the SU(3) and SU(2) labels (because the SU(3) decuplet and the $3/2$ spinor are both symmetric.) We finally get

$$\begin{aligned} N^{*+\uparrow} &= \frac{1}{3}\Big(p^\uparrow(1)\; p^\uparrow(2)\; n^\downarrow(3) + p^\uparrow(1)\; n^\downarrow(2)\; p^\uparrow(3) + n^\downarrow(1)\; p^\uparrow(2)\; p^\uparrow(3) \;+ \\ &\quad + p^\uparrow(1)\; p^\downarrow(2)\; n^\uparrow(3) + p^\uparrow(1)\; n^\uparrow(2)\; p^\downarrow(3) + n^\uparrow(1)\; p^\uparrow(2)\; p^\downarrow(3) \;+ \\ &\quad + p^\downarrow(1)\; p^\uparrow(2)\; n^\uparrow(3) + p^\downarrow(1)\; n^\uparrow(2)\; p^\uparrow(3) + n^\uparrow(1)\; p^\downarrow(2)\; p^\uparrow(3) \Big) \\ &= \frac{1}{3}\Big(S^1(1)\; S^1(2)\; S^5(3) + S^1(1)\; S^5(2)\; S^1(3) + S^5(1)\; S^1(2)\; S^1(3) \;+ \\ &\quad + S^1(1)\; S^4(2)\; S^2(3) + S^1(1)\; S^2(2)\; S^4(3) + S^2(1)\; S^1(2)\; S^4(3) \;+ \\ &\quad + S^4(1)\; S^1(2)\; S^2(3) + S^4(1)\; S^2(2)\; S^1(3) + S^2(1)\; S^4(2)\; S^1(3) \Big) \\ &= B^{115} + 2B^{124} \,. \end{aligned} \tag{8}$$

As the proton also has $q = Y = 1$, the $S_z = \frac{1}{2}$ state must again be a linear combination of B^{115} and B^{124}. The coefficients are determined by the condition of being orthogonal to N^{*+}. Thus, we may take

$$P = \sqrt{2}(B^{115} - B^{124}). \tag{9}$$

In general, the content of a given component B^{abc} in terms of SU(3) octuplet and decuplet wave functions is given by

$$B^{abc} = \chi^{\alpha\beta\gamma}\, d^{ABC} + \frac{1}{3\sqrt{2}}\Big(\epsilon^{\alpha\beta}\chi^{\gamma}\, X^{ABC} + \epsilon^{\gamma\alpha}\chi^{\beta}\, X^{CAB} \\ \epsilon^{\beta\gamma}\chi^{\alpha}\, X^{BCA}\Big), \tag{10}$$

where d^{ABC} is the SU(3) decuplet wave function (totally symmetric)

$X^{ABC} = \epsilon^{ABD}\, b_D^{\;C}$

ϵ^{ABC} is the Levi-Civita symbol

$b_D^{\;C}$ is the SU(3) baryon-octet wave function

$\epsilon^{\alpha\beta}$ is the Levi-Civita symbol $\begin{pmatrix} 0 & 1 \\ -1 & 0 \end{pmatrix}$

$\chi^{\alpha\beta\gamma}$ = spin $\frac{3}{2}$ wave function (totally symmetric)

(χ^{111} corresponds to $s_z = \frac{3}{2}$, χ^{112} to $s_z = \frac{1}{2}$, χ^{122} to $s_z = \frac{1}{2}$,

χ^{222} to $s_z = -\frac{3}{2}$ and $\|\chi^{111}\| = \|\chi^{222}\| = 1$, $\|\chi^{112}\| = \|\chi^{122}\| = \frac{1}{3}$)

χ^{γ} = spin $\frac{1}{2}$ wave function

$\alpha, \beta, \gamma = 1, 2,$

$A, B, C = 1, 2, 3$

$a = (\alpha A)$, $b = (\beta B)$, $c = (\gamma c)$.

Let us consider the 35 mesons. This may be regarded as the quark-antiquark system. The tensor of the 35 representation is of the form $M_b^{\;a}$, with the trace condition

$$M^a_a = 0 \ .$$

In terms of the pseudoscalar octet and vector nonet, M^a_b may be written as

$$M^a_b = -i\,\delta^\alpha_\beta\, P^A_B - \chi^\alpha_\beta\, V^A_B \ , \tag{11}$$

where P^A_B is the SU(3) pseudoscalar octet $\left(P^A_A = 0\right)$

V^A_B is the SU(3) octet singlet $\left(V^A_A \neq 0\right)$

$\chi^\alpha_\beta = (\vec{\sigma} \cdot \vec{\epsilon})^\alpha_\beta$ is the spin one-wave function, $\vec{\epsilon}$ is the polarization vector $\left(\chi^\alpha_\alpha = 0\right)$.

The i is required by time-reversal invariance as the spin one-wave function changes sign under time reversal, whereas δ^α_β (spin zero) does not.

We may now give one of the most interesting results of SU(6), namely the prediction of proton/neutron magnetic-moment ratio. In the SU(3) case the charge and the magnetic-moment operators are assumed to transform as the same component of an octet. As far as SU(6) is concerned we must also take into account their transformation properties in spin space. The charge is a scalar, whereas the magnetic moment vector behaves like a triplet. The magnetic-moment term in the non-relativistic electromagnetic current is, in fact, proportional to $\vec{\sigma} \cdot \vec{H}$, where $\vec{H}$ is the magnetic field. Then the charge and the magnetic moment must transform like an $(\underline{8},\underline{1})$ and $(\underline{8},\underline{3})$ of SU(3) × SU(2), respectively. As both the $(\underline{8},\underline{1})$ and the $(\underline{8},\underline{3})$ are contained in the 35 representation, we assume that the latter is their transformation property in SU(6).

In the case of baryons of the 56, these assumptions lead to the following expression

$$3e\, B^{\dagger}_{abd}\, B^{abc}\, \delta^\delta_\gamma\, Q^D_C \tag{12}$$

for the charge and

$$3\mu_p \, B^{\dagger}_{abd} \, B^{abc} \, i(\vec{\sigma} \cdot \vec{q} \times \vec{\epsilon})^{\delta}_{\gamma} \, Q^D_C \tag{13}$$

for the magnetic moment. Here, Q^D_C is the charge matrix of SU(3)

$$Q^D_C = \begin{pmatrix} 2/3 & 0 & 0 \\ 0 & -1/3 & 0 \\ 0 & 0 & -1/3 \end{pmatrix}, \tag{14}$$

μ_p is the static magnetic moment of the proton

$\vec{q}$ is the momentum transfer

$\vec{\epsilon}$ is the polarization vector of the vector potential.

An explicit evaluation of expression (13) gives the magnetic moments of the decuplet and octet states in terms of the proton magnetic moment. In particular, we have

$$\frac{\mu_p}{\mu_n} = -\frac{3}{2} , \tag{15}$$

$$\mu(\Delta) = Q(\Delta)\,\mu_p , \tag{16}$$

where Δ is an arbitrary decuplet state and $Q(\Delta)$ is its charge. The result Eq. (15), is in remarkable agreement with the experimental data.

The reason why only one parameter μ_p occurs in the expression of all the baryon magnetic moments in SU(6) is that the product $56^* \times 56$ only once contains 35. This means that one can construct only one scalar out of 56*, 56, 35 tensors. In SU(3) the product 8×8 contains 8 twice; then one has two independent scalars. They are generally taken as follows:

$$\bar{B}^A_B B^B_C M^C_A - \bar{B}^A_B M^B_C B^C_A = \mathrm{Tr}\ (\bar{B}[B,M]) \qquad (17)$$

(this is referred to as the F-type coupling) and

$$\bar{B}^A_B B^B_C M^C_A + \bar{B}^A_B M^B_C B^C_A = \mathrm{Tr}\ (\bar{B}\{B,M\}) \ , \qquad (18)$$

which is referred to as the D-type coupling. So, at the end we have an over-all multiplicative factor and the F/D ratio, which cannot be determined. This shows up when we calculate the magnetic moments of the baryon-octet members in SU(3), where all of them can be expressed in terms of two constants.

Now, if we expand the expression $\bar{B}_{abd}\ B^{abc}$, using formula (10), we find an octet-octet part of pure F type behaving like a spin scalar, and an octet-octet part in the combination 3D + 2F behaving like an SU(2) triplet (and other terms involving the decuplet.) So, the charge term (12) for the octet is pure F type, whereas the magnetic-moment term (13) has an F/D ratio = $^2\!/_3$.

This F/D ratio also occurs in weak interactions. In SU(3) the vector and axial currents form two separate octets. So in SU(6) we can accommodate these currents in the 35 representation: the vector currents in the ($\underline{8}$,$\underline{1}$) and the axial currents in ($\underline{8}$,$\underline{3}$) and ($\underline{1}$,$\underline{3}$). Therefore, for the former ones we have a pure F type and for the latter a D/F ratio still equal to $^3\!/_2$. This is compatible with the experimental value 1.7 ± 0.35.

IV. SU(6,6) AND SUBGROUPS

SU(6) is a static non-relativistic symmetry and therefore cannot be applied to scattering problems in that this involves non-zero momenta and orbital momenta, and off mass shell virtual states. This led to the question whether the approximate dynamical group SU(6) can be embedded in a larger approximate dynamical group in such a way that we can incorporate momentum dependence.

Briefly, the starting point for SU(6,6) is the following. In SU(6), the defining representation is $S^a = S^{\alpha A}$ and we have the SU(6) relation

$$S^*_{\alpha A}\, S^{\alpha A} = \text{invariant}. \tag{1}$$

Consider next the relativistic quark $S^{\alpha A}(p)$, where now $\alpha = 1 \ldots 4$. One asks for a symmetry group that leaves invariant the "relativistic mass term" (mass is what much of supermultiplet theory is about):

$$\bar{S}_{\alpha A}\, S^{\alpha A} = \text{invariant}, \tag{2}$$

which may be looked upon as a covariant generalization of (1). The corresponding group is SU(6,6).

<u>In the presence of interaction</u> [where $(\gamma p)S = imS$ is not valid] this group is intrinsically broken by kinetic energy terms $\bar{S}(\gamma p)S$. Such terms introduce spin-orbit coupling. The motivation for exploring SU(6,6) and related groups was that, where static SU(6) is successful, an <u>effective</u> damping out of these symmetry breaking spin-orbit couplings must be at work. This may have bearing on the three-point functions (vertices). However, it was soon realized that no systematic application of such invariance ideas is possible for all n-point functions. This is because of the unitarity problem, as will be explained later on.

The defining representation of SU(6) is spanned by the vectors $S^a = \chi^\alpha t^A$, where χ^α is a two-component wave function. We introduce the basic states

$$\chi_1^\alpha = \begin{pmatrix} 1 \\ 0 \end{pmatrix}, \quad \chi_2^\alpha = \begin{pmatrix} 0 \\ 1 \end{pmatrix}$$

(thus, for example, $X_1^1 = 1$, $X_2^1 = 0$ etc). These describe states of zero momentum and positive energy, and correspond to the large components of the Dirac spinors

$$u_1^\alpha = \begin{pmatrix} 1 \\ 0 \\ 0 \\ 0 \end{pmatrix} , \quad u_2^\alpha = \begin{pmatrix} 0 \\ 1 \\ 0 \\ 0 \end{pmatrix} .$$

The corresponding negative energy states are:

$$v_1^\alpha = \begin{pmatrix} 0 \\ 0 \\ 1 \\ 0 \end{pmatrix} , \quad v_2^\alpha = \begin{pmatrix} 0 \\ 0 \\ 0 \\ 1 \end{pmatrix} .$$

In these equations α now runs from 1 to 4. In the non-relativistic description, the upper two components of u^α and the lower two of v^α may both be considered as representations of one U(6). Even at $\vec{p} = 0$ this is no longer true in the four-component language. The u^α and v^α now appear as (6,1) and (1,6) representations of the so-called non-chiral group $U(6) \times U(6)$.

We next introduce the following set of functions:

$$D_i^\alpha(0) = (u_1^\alpha, u_2^\alpha, v_1^\alpha, v_2^\alpha) = \delta_i^\alpha .$$

We use the following definitions:

$$p \equiv (\vec{p}, ip_0) ,$$

$$\gamma_\mu = \gamma_\mu^\dagger ,$$

$$\gamma p = \gamma_\mu p_\mu \quad \text{(sum over } \mu)$$

$$\gamma_4 = \rho_3 \otimes I, \quad \vec{\gamma} = \rho_2 \otimes \vec{\sigma}, \quad \gamma_5 = \rho_1 \otimes I ,$$

where $\vec{\rho}$ and $\vec{\sigma}$ are the familiar two sets of Pauli matrices.

For non-vanishing momentum we define the following matrix

$$D_i^\alpha(p,m) \equiv \left[\frac{m - i\gamma p\,\gamma_4}{\sqrt{2m\,(p_0 + m)}}\right]_i^\alpha .$$

This can be written in the following way

$$D_i^\alpha(p,m) = D_\beta^\alpha(p)\,D_i^\beta(0) . \tag{3}$$

D_β^α is called a boost matrix because it transforms $u^\alpha(0)$ into $u^\alpha(p)$ for any i. For instance,

$$u_1^\alpha(0) \Rightarrow \sqrt{\frac{p_0 + m}{2m}} \begin{pmatrix} \chi_1 \\ \dfrac{\vec{\sigma}\cdot\vec{p}}{p_0^{\,o} + m}\,\chi_1 \end{pmatrix} .$$

Similarly, starting from $\bar{u}(p) = u^\dagger(p)\gamma_4$ we may define

$$\bar{\bar{D}}{}_\alpha^i(p) = D_\beta^{*\,i}(p)\,\gamma_{4\alpha}^{\ \beta} . \tag{4}$$

This may be written as

$$\bar{\bar{D}}{}_\alpha^i(p) = \bar{D}_\beta^i(0)\,\bar{D}_\alpha^\beta(p) ,$$

where

$$\bar{D}_\alpha^\beta = \left[\gamma_4\,D^\dagger(p)\,\gamma_4\right]_\alpha^\beta . \tag{5}$$

$\bar{D}$ (single bar!) is the conjugate boost matrix.

We list some useful properties of the boost matrices:

$$
\begin{aligned}
D I \bar{D} &= I , \\
D \gamma_5 \bar{D} &= \gamma_5 , \\
D \gamma_4 \bar{D} &= - \frac{i \gamma p}{m} \\
D \vec{\gamma} \cdot \vec{\epsilon}(0) \bar{D} &= \gamma \epsilon (p) ,
\end{aligned}
\tag{6}
$$

where $\epsilon(p)$ is the four vector defined $\equiv [\vec{\epsilon}(p), i \epsilon_0(p)]$ and

$$
\begin{aligned}
\vec{\epsilon}(p) &= \vec{\epsilon}(0) + \frac{\vec{p}[\vec{p} \cdot \vec{\epsilon}(0)]}{m(p_0 + m)} \\
\epsilon_0(p) &= \frac{\vec{p} \cdot \vec{\epsilon}(0)}{\mu} , \qquad p\epsilon(p) = 0
\end{aligned}
\tag{7}
$$

[$\vec{\epsilon}(0)$ is an arbitrary three vector.] Finally,

$$
D \gamma_4 \vec{\gamma} \cdot \vec{\epsilon}(0) \bar{D} = \frac{\sigma_{\mu\nu} p_\mu \epsilon_\nu(p)}{m} , \tag{8}
$$

where $\sigma_{\mu\nu} \equiv - \frac{i}{2} [\gamma_\mu, \gamma_\nu]$. These matrices are useful because they can transform a static coupling into a dynamical one, for which the restriction $\vec{p} = 0$ is no longer required.

We shall now study the relation between SU(2,2) and SU(4). First of all we remember a theorem which states that finite dimensional representations of SU(m,n) are in a one to one correspondence to the (unitary) representations of SU(m+ n). Two corresponding representations have the same tensor structure and the same dimensions. This theorem is usually referred to as the "unitary trick" (cf. H. Weyl book).

The SU(2) generators obey the following commutation relations:

$$[\sigma_i, \sigma_j] = i\,\epsilon_{ijk}\,\sigma_k\,, \tag{9}$$

and in the defining representation they are the ordinary Pauli matrices. The generators of SU(4) in the defining representation can be taken as

$$\rho_i \otimes I\,,\quad I \otimes \sigma_i\,,\quad \rho_i \otimes \sigma_j\,,\quad j,i = 1,2,3\,.$$

These are 15 linearly-independent Hermitian traceless matrices, and by linear combinations we can get the equivalent Hermitian set

$$T_a \equiv (\gamma_\mu,\ \sigma_{\mu\nu},\ i\gamma_\mu\gamma_5\,,\ \gamma_5)\,. \tag{10}$$

Thus, under infinitesimal transformations of SU(4)

$$\begin{aligned} u^\alpha(p) &\to (1 + i\,\epsilon_a\,\Gamma_a)^\alpha_\beta\, u^\beta(p)\,, \\ u^\dagger_\alpha(p) &\to u^\dagger_\beta(p)\,(1 - i\,\epsilon_a\,\Gamma_a)^\beta_\alpha\,, \end{aligned} \tag{11}$$

the form

$$u^\dagger_\alpha(p)\,u^\alpha(p) = u^*_1\,u^1 + u^*_2\,u^2 + u^*_3\,u^3 + u^*_4\,u^4$$

is left invariant. As was explained earlier, we are interested in the group which leaves $\bar{u}_\alpha(p)\,u^\alpha(p)$ invariant. This is SU(2,2) and its generators may be taken as the following set of matrices

$$\tilde{\Gamma}_a = (i\vec{\gamma},\ \gamma_4\,;\ \sigma_{ij},\ i\sigma_{i4}\,;\ i\vec{\gamma}\gamma_5\,,\ \gamma_4\,\gamma_5\,;\ i\gamma_5)\,. \tag{12}$$

In fact, the transformations

$$u^{\alpha} \rightarrow (1 + i\,\epsilon_a\,\tilde{\Gamma}_a)^{\alpha}_{\beta}\, u^{B} \ , \qquad (13)$$

$$u^{\dagger}_{\alpha} \rightarrow u^{\dagger}_{\beta}(1 - i\,\epsilon_a\,\tilde{\Gamma}^{\dagger}_a)^{\beta}_{\alpha}$$

leave

$$u^{\dagger}\gamma_4\, u = \bar{u}u = u_1^{*}\, u^1 + u_2^{*}\, u^2 - u_3^{*}\, u^3 - u_4^{*}\, u^4 \qquad (14)$$

invariant by definition. The matrix γ_4 plays the same role as the matrix Γ in Section I.

In the same way as we passed from SU(2) to SU(2,2), obtaining a relativistic theory, we may go from SU(6) to SU(6,6). The SU(6) generators are (insofar as they act on the defining representation $\underline{6}$)

$$\sigma_i \otimes I, \ I \otimes F_A, \ \sigma_i \otimes F_A \ ,$$

where F_A are the 8 generators of SU(3). In order to get SU(6,6) we pass first from the basis $S^a = t^A \chi^{\alpha}$ to the basis $u^{A\alpha}(p)$, A = 1,2,3, where now α = 1,2,3,4 along the pattern discussed at the beginning of Section IV. We then consider the group of transformations

$$u^{A\alpha}(p) \rightarrow \left(1 + \epsilon_a \tilde{\Gamma}_a \otimes I + \epsilon_C\, I \otimes F_C + \epsilon_{aC}\, \tilde{\Gamma}_a \otimes F_C\right)^{A\alpha}_{B\beta} u^{B\beta}, \qquad (15)$$

[where $\tilde{\Gamma}_a$ are the matrices defined by Eq. (12)], which leave the form $\bar{u}_{A\alpha}\, u^{A\alpha}$ invariant. This is the group SU(6,6). Other names for this group are $\tilde{SU}(12)$, $SU(12)_{\mathcal{L}}$, $M(12)_{even}$, SV(12). We notice that SU(6,6) does not commute with the free Hamiltonian [$\gamma_{\mu}p_{\mu}$ is Lorentz invariant but does not commute with SU(6,6).] This illustrates the fact that SU(6,6) is an approximate dynamical symmetry.

Now, we study the possible invariant couplings in order to get predictions about form factors. We begin by studying the structure of baryon-antibaryon-meson tensors in SU(6,6). We consider the SU(6) tensor B^{abc}. Replacing the two component Pauli spinors χ^{γ} by the four component Dirac spinors $u^{\alpha}(0)$, and by changing accordingly the other quantities like

$$\epsilon^{\alpha\beta} = \begin{pmatrix} 0 & 1 \\ -1 & 0 \end{pmatrix} \rightarrow \begin{pmatrix} 0 & 1 & 0 & 0 \\ -1 & 0 & 0 & 0 \\ 0 & 0 & 0 & 0 \\ 0 & 0 & 0 & 0 \end{pmatrix} \tag{16}$$

we get $B^{abc}(0) \rightarrow B^{A\alpha B\beta C\gamma}(0)$, where $\alpha,\beta,\gamma = 1,2,3,4$, so that we have 364 components. To obtain $B^{A\alpha B\beta C\gamma}(p)$ we act on each index with the D matrices

$$B^{A\alpha B\beta C\gamma}(p) = D^{\alpha}_{\alpha'}(p)\, D^{\beta}_{\beta'}(p)\, D^{\gamma}_{\gamma'}(p)\, B^{A\alpha B\beta C\gamma}(0)\,. \tag{17}$$

Similarly, by acting with the $\bar{D}$ matrix we get $\bar{B}_{A\alpha B\beta C\gamma}(p)$

$$\bar{B}_{A\alpha B\beta C\gamma}(p) = \bar{B}_{A\alpha' B\beta' C\gamma'}(0)\, \bar{D}^{\alpha'}_{\alpha}(p)\, \bar{D}^{\beta'}_{\beta}(p)\, \bar{D}^{\gamma'}_{\gamma}(p)\,. \tag{18}$$

For the mesons, starting from the formula

$$M^{\alpha A}_{\beta B} = -i\,\delta^{\alpha}_{\beta}\, P^{A}_{B} - \left(\vec{\sigma}\cdot\vec{\epsilon}(0)\right)^{\alpha}_{\beta} V^{A}_{B}$$

we first change

$$\delta^{\alpha}_{\beta} \rightarrow \begin{pmatrix} 1 & 0 & 0 & 0 \\ 0 & 1 & 0 & 0 \\ 0 & 0 & 0 & 0 \\ 0 & 0 & 0 & 0 \end{pmatrix} = \Delta^{\alpha}_{\beta}$$

$$(\vec{\sigma})^{\alpha}_{\beta} \rightarrow \begin{pmatrix} \vec{\sigma} & 0 \\ 0 & 0 \end{pmatrix} = \Sigma^{\alpha}_{\beta} \tag{19}$$

$$M^{\alpha A}_{\beta B}(0) = -i\,\Delta^{\alpha}_{\beta}\, P^{A}_{B} - \left(\vec{\Sigma}\cdot\vec{\epsilon}(0)\right)^{\alpha}_{\beta} V^{A}_{B} \;.$$

In order to have the right parity property we introduce, for $\underline{p} = 0$

$$\mathcal{M}^{\alpha A}_{\beta B}(0) = \gamma_5 \left(\frac{1+\gamma_4}{2}\right)^{\alpha}_{\beta'} M^{\beta' A}_{\beta B}(0) \tag{20}$$

as the tensor describing the mesons. We then apply the boost method (for mass μ) to both indices:

$$\mathcal{M}(p)^{\alpha A}_{\beta B} = D^{\alpha}_{\beta'}\, \mathcal{M}(0)^{\beta' A}_{\alpha' B}\, \bar{D}^{\alpha'}_{\beta}$$

$$= i\left(\left[1-\frac{i\gamma p}{\mu}\right]\left[\gamma\,\epsilon(p)\, V^{A}_{B} - \gamma_5\, P^{A}_{B}\right]\right)^{\alpha}_{\beta} . \tag{21}$$

Now we have the following tensors $S^{a}(p)$, $\vec{S}_{a}(p)$, $\mathcal{M}^{a}_{b}(p)$, $B^{abc}(p)$ and $\bar{B}_{abc}(p)$, and we can construct invariant vertices. The only covariant baryo antibaryon-meson vertex allowed by SU(6,6) invariance is:

$$\bar{B}_{abc}(p_1)\,\mathcal{M}^{c}_{d}(q)\, B^{abc}(p_2) \tag{22}$$

$(q = p_1 - p_2)$. This formula has the following consequences for the $\bar{B}BM$ form factors.

1) $\frac{F}{D} = \frac{2}{3}$ for the coupling of pseudoscalar mesons to baryons belonging to the SU(3) octet. This value is true for <u>all</u> p.

2) $$\frac{G_m(q^2)}{G_m^P(q^2)} \text{ and } \frac{G_e(q^2)}{G_e^P(q^2)} \text{ are independent of } q^2 . \qquad (23) \quad (24)$$

Here, G_m and G_e are the Sachs magnetic and electric form factors respectively, for any member of the baryon octet or for the $(\Sigma^0|\Lambda)$-transition moment. The subscript P refers to the proton case. Within ~ 20% margins, the relations (23) appear to be reasonable for the neutron-proton comparison up to $q^2 \sim (1\ \text{GeV/c})^2$.

3) The ratio between Sachs form factors of the proton turns out to be

$$\frac{G^P_{\text{charge}}(q^2)}{G^P_{\text{magn.}}(q^2)} = 1 + \frac{q^2}{2M\mu} , \qquad (25)$$

where M is the symmetric baryon mass and μ is the symmetric vector meson mass.

Further study has shown, however, that the relations (23) and (24) can be derived under very much weaker conditions than SU(6,6). At the same time, this weakening no longer produces the prediction of Eq. (25). These weaker requirements are related to W spin, to be discussed hereafter.

In a similar way we may study scattering problems, such as $BB\eta\eta$ scattering

$$B + \eta \rightarrow B + \eta .$$

Here SU(6,6) gives several bad predictions such as no Ξ polarization in the reaction $K^- + p \rightarrow \Xi^- + K^+$.

Finally, we remark that SU(6,6) invariance is inconsistent with the unitarity condition. In fact, the sum over intermediate states in the equation

$$\text{Im}\, T_{ab} = T^{\dagger}_{ac}\, T_{cb} \tag{26}$$

brings in projection operators proportional to $i\gamma p - m$, which is not a scalar under SU(6,6), so that the two sides of the above equation have different transformation properties. These unitarity violations are of order $(v/c)^2$, we have "asymptotic unitarity" in elastic regions for $v/c \to 0$. The original motivation for the study of groups like SU(6,6) and related groups such as SL(6,c) (for which there is also the unitarity problem) was to be able to discuss the baryon-meson vertex. To order v/c (P-wave vertex) it is now known that one gets all the "good" results under the weak assumptions that the baryon is boosted in the totally symmetric way described above; and that the form factors satisfy SU(6) relations for $\vec{q}^2 = 0$ only. Thus, to order v/c one is not in bad shape. From the order $(v/c)^2$ on, the unitarity troubles set in, which make illusory, as a matter of principle, any phenomenological comparisons.

IV.1 THE SUBGROUP $SU(6)_W$ OF SU(6,6)

In this lecture I want to give a short survey of what is called $SU(6)_W$. This group (also called "the hybrid collinear group") was first introduced by Lipkin and Meshkov [Phys.Rev. Letters 14, 670 (1965)]. It is a subgroup of SU(6,6). Its essential features can, however, be explained in terms of SU(2,2) so I shall, for the moment, forget about SU(3) indices.

I have given the generators of SU(2,2) above. On the defining representation they are the Dirac matrices. Among these generators there are certain subsets which close upon commutation and, therefore, generate subgroups of SU(2,2). An obvious example is the set σ_{ij} (i,j = 1,2,3), generating a subgroup SU(2) which is the usual spin. Another subset are the operators

$$W_1 = \frac{i}{2}\gamma_1\gamma_5 = \frac{1}{2}\gamma_4\sigma_1$$

$$W_2 = \frac{i}{2}\gamma_2\gamma_5 = \frac{1}{2}\gamma_4\sigma_2$$

$$W_3 = \frac{i}{2}\gamma_1\gamma_2 = \frac{1}{2}\sigma_3 \quad .$$

Because $\gamma_4^2 = 1$ and $[\gamma_4, \vec{\sigma}] = 0$ one sees immediately that the three operators W_1, W_2 and W_3 again have the commutation relations of an angular momentum:

$$[W_1, W_2] = iW_3 \quad \text{(and cyclic permutations)} .$$

Hence, W_1, W_2 and W_3 generate a subgroup of SU(2,2) which is isomorphic to SU(2) but is, of course, a different group. This is the so-called W spin. Exactly as we have done before, we can now consider the operators

$$W_\alpha \otimes \mathbb{1}$$

$$\mathbb{1} \otimes F_A$$

$$W_\alpha \otimes F_A \, .$$

They generate a group which is isomorphic to SU(6). This is $SU(6)_W$. It is, as we have seen, also a subgroup of SU(6,6), but different from the ordinary SU(6).

Why is this group so interesting? To see this, let us look at a problem in which a certain given momentum p is involved, and let us take the z co-ordinate in the direction of that momentum. (In fact, we already defined W spin in a way that singles out the z direction.)

Let us then perform a rotation-free Lorentz transformation along the z direction. This involves the generator M_{34} of the Lorentz group which in our representation is the operator $-i\gamma_3\gamma_4 = \gamma_5\sigma_3$. You can now easily check that this operator commutes with W spin

$$[\vec{W}, \gamma_5\sigma_3] = 0 .$$

This means that unlike the ordinary spin (which does not commute with M_{34}) W spin is not a static spin! If W spin is good in the limit of vanishing momentum, it is also good for all collinear processes, i.e. such processes for which all the momenta involved are in one and the same direction. This is very convenient because it is much easier to handle than the full group SU(6,6), and for certain subsets of states we may be slightly better off with it because it may possibly be a good approximate symmetry even where SU(6,6) is a bad one.

What does the W spin really amount to? Let us see what it does for quarks, and let us again, for convenience only, drop the SU(3) labels. You then see that in the four-dimensional representation the W spin matrices are of the form

$$W_1 = \frac{1}{2}\begin{pmatrix}\sigma_1 & 0 \\ 0 & -\sigma_1\end{pmatrix}, \quad W_2 = \frac{1}{2}\begin{pmatrix}\sigma_2 & 0 \\ 0 & -\sigma_2\end{pmatrix}, \quad W_3 = \frac{1}{2}\begin{pmatrix}\sigma_3 & 0 \\ 0 & \sigma_3\end{pmatrix} .$$

If you look at the defining representation for $\vec{p} = 0$, this representation is reducible and breaks up into two-dimensional spin representations for quarks and antiquarks separately. The point is, however, that unlike ordinary spin, W spin acts differently on quarks and antiquarks. A simplified way to say this is that for quarks W spin is just like ordinary spin, but for antiquarks it is $(-\sigma_1, -\sigma_2, \sigma_3)$. If you get nervous about doing things that way, write down second quantized wave functions, operate carefully with γ_4 and you will get this answer.

For baryons, if you think of them as being built up of three quarks (whether physically or just in a mathematically convenient notation), you see that W spin acts on baryons just like ordinary spin. Mesons, however, are quark-antiquark systems, therefore W spin multiplets are different from ordinary spin multiplets for mesons. To see this explicitly, let us write down the helicity states for vector and pseudoscalar mesons:

$$V_1 = \bar{q}\uparrow q\uparrow$$

$$V_0 = \frac{1}{\sqrt{2}}(\bar{q}\uparrow q\downarrow + \bar{q}\downarrow q\uparrow)$$

$$V_{-1} = \bar{q}\downarrow q\downarrow$$

$$P = \frac{1}{\sqrt{2}}(\bar{q}\uparrow q\downarrow - \bar{q}\downarrow q\uparrow) .$$

The standard way to obtain these states is to start from the one with highest weight (i.e. J_z = max) which in the case of the vector mesons is simply $V_1 = \bar{q}\uparrow q\uparrow$, and to apply the spin-lowering operator $\sigma_- = \sigma_1 - i\sigma_2$. In this way one finds all three states of the J = 1 multiplet. The pseudoscalar meson is a singlet: $\sigma_- P = 0$.

Now, if instead of σ_- you apply $W_- = W_1 - iW_2$ to V_1, you will find to your amusement that $W_- V_1 = P$ and not V_0, and $\sigma_- P = V_-$ and not zero. On the other hand, $\sigma_- V_0 = 0$. In other words, whereas for ordinary spin V_1, V_0, V_{-1} form a triplet and P a singlet, for W spin you find that V_1, P, V_{-1} go together in a triplet and V_0 stays by itself as a singlet.

To which kind of processes can this be applied? I shall only mention a few examples:

i) Decay of a particle at rest into two particles;

ii) vertices. If we go to a Breit frame (this is known as a "brick wall" transformation), we can make the vertex look like this $\rightleftarrows$ ---- . If you look at the results it turns out that all the good results of SU(6,6) on form factors [$D/F = 3/2$ for all q^2, $F_{el}/F_{el}^{(p)} = Q/Q^{(p)}$] are also obtained by using only the subgroup $SU(6)_W$. On the other hand, the relation

$$\frac{G_{el}}{G_{magn.}} = 1 + \frac{q^2}{2M_\mu}$$

does not follow from $SU(6)_W$ alone. You need the whole SU(6,6) to get this relation. We have had some disagreement as to whether this relation is found to hold experimentally or not; to me it seems an advantage of $SU(6)_W$ that this relation must not necessarily be true.

iii) Non-leptonic two-body decays;

iv) forward or backward scattering. Application to forward scattering yields the Johnson-Treimen relations. There is also an interesting prediction [Carter, Coyne, Meshkov, Horn, Kugler and Lipkin, Phys.Rev. Letters 15, 373 (1965)] of the cross-sections for the three processes

$$\pi^- p \rightarrow \pi^- N^{*+}$$

$$\pi^- p \rightarrow \pi^0 N^{*0}$$

$$\pi^- p \rightarrow \pi^+ N^{*-} \ .$$

The ratio of these cross-sections for either forward or backward scattering should be 2 : 9 : 24. Of course, results like this also follow from SU(6,6), but the point is that you do not need the full group to get it. This is the reason why $SU(6)_W$ has been useful, in practical calculations.

DEVIATIONS FROM UNITARY SYMMETRY

D.H. Sharp,
Princeton University.

I. INTRODUCTION

All of the strong interaction symmetries that have been proposed so far have the property that in nature they are broken, either as a result of interactions which do not preserve the underlying symmetry which is assumed to exist or, perhaps, spontaneously.

It is certainly a remarkable fact that a considerable remnant of the symmetry survives the symmetry-breaking process, so that the predicted supermultiplet patterns, mass sum rules and so forth can be identified with reasonable confidence. Nevertheless, it has been clear from the beginning, in the case of SU(3) for example, that the symmetry-breaking effects are generally large and must be included in order to achieve a correct description of any but the crudest features of the hadron spectrum. Aside from this, the study of deviations from SU(3) symmetry is intriguing because the "medium strong", electromagnetic and weak violations of SU(3) themselves appear to follow a characteristic pattern, in that the violations of SU(3) which transform like the components of an octet seem to predominate in nature. We shall refer to this as the pattern of octet enhancement.

In these lectures, I wish to describe an approach to understanding the pattern of SU(3) symmetry breaking within the framework of the bootstrap theory of strong interactions[1-3]. In particular, I want to indicate how the bootstrap theory can provide a mechanism for universal octet enhancement in the strong, electromagnetic and weak violations of SU(3).

Section II is devoted to a review of some of the experimental evidence on violations of SU(3), and how this leads one to infer a pattern of universal octet enhancement. In this section, the general features of the bootstrap theory of the deviations from unitary symmetry[1-3] will be outlined. Here, most of the statements will be independent of specific models or calculational methods.

In Section III, we will discuss an S-matrix method for calculating the change in the position and residue of a bound state pole resulting from a small change in the forces which produce the bound state. These techniques, due to Dashen and Frautschi[4,5], seem well adapted to treat perturbations about an SU(3) symmetric problem.

The techniques discussed in Sections II and III have been applied[3,6,7], in conjunction with a specific reciprocal bootstrap model for the baryon octet (B) and decuplet (Δ), to study the strong and electromagnetic mass shifts in the B and Δ supermultiplets and the strong, electromagnetic and weak shifts in the $BB\pi$ and $B\Delta\pi$ couplings from their SU(3) symmetric values. These results will be surveyed in Section IV.

Before plunging into detail, a few words of caution are in order:

i) Throughout these lectures we shall suppose that SU(3) symmetry is a good starting point for the strong interactions.

ii) We shall frequently talk as if there were some limit in which SU(3) symmetry becomes exact, or in which an SU(3) symmetric reciprocal bootstrap exists. This may or may not be the case. The case in which no SU(3) symmetric reciprocal bootstrap exists can be incorporated into the general formalism of Section II, but the results of the detailed calculations may change.

iii) It should be borne in mind that although the bootstrap idea is very appealing on physical grounds, no really satisfactory way of implementing it mathematically has yet been found. For this reason it is important to remember that the formalism developed in Section II, which does not invoke specific calculational schemes (or strictly speaking even the assumption that all the strongly interacting particles are composite), is on a sounder footing than the detailed quantitative predictions reviewed in Section IV.

iv) As will become apparent in Section II, bootstrap considerations alone (as we presently understand them) will not lead to a complete theory of symmetry-breaking. For example, neither the over-all scale of a violation of SU(3) nor the axis in SU(3) space along which the violation lies are determined by bootstrap requirements.

Also not taken up here is the important question of how the symmetry violation gets started. This is a very problematical point in the case of both strong and weak symmetry breaking.

With these points in mind, let us see to what extent we can develop a coherent picture of the deviations from unitary symmetry.

II. BOOTSTRAP THEORY OF OCTET ENHANCEMENT

1. Basic ideas of octet enhancement

There are a number of experimental facts pertaining to violations of SU(3) which form somewhat of a pattern and demand an explanation. As examples we note that:

i) The electromagnetic mass splittings within the baryon isospin multiplets appear to transform largely like the third component of an octet[8,9].

ii) The non-leptonic weak decays are octet dominated in as much as the $|\Delta I| = \frac{1}{2}$ rule appears to be reasonably well satisfied[10].

These phenomena arise when the electromagnetic and weak currents, respectively, act twice. If we make the customary assumption that these currents transform like components of an octet, then we would expect to find that phenomena which are of second order in the electromagnetic or weak currents contain terms which transform like each of the symmetric representations that occur in $\underline{8} \otimes \underline{8}$, namely, $\underline{1}$, $\underline{8}$ and $\underline{27}$. Instead, we find that these effects are described by octet terms alone.

iii) The medium strong mass splitting between members of a given SU(3) supermultiplet are described (to a remarkable degree of accuracy) by the Gell-Mann-Okubo sum rule[11,12], which is obtained by assuming that the strong violations of SU(3) transform like the eighth component of an octet. The dilemma presented by the success of the Gell-Mann-Okubo sum rule is, of course, compounded by the fact that we are very much in the dark as to the detailed mechanism responsible for strong symmetry-breaking. However, even if the basic mechanism leading to the medium strong mass splitting is pure octet, in contrast to the mechanisms leading to the electromagnetic and weak effects mentioned above, the question of why higher order effects do not introduce terms transforming like a $\underline{27}$ or other representation is still left unresolved.

The facts i) - iii) listed above led Coleman and Glashow[8] to make the important suggestion that octet enhancement is a general feature of SU(3) symmetry breaking, and that there should be a common dynamical mechanism responsible for octet enhancement in the strong, electromagnetic and weak interactions.

Before trying to develop some sort of a theory of octet enhancement, it is instructive to look more closely at the example of the electromagnetic mass splittings of the baryons to see in a little detail what can be inferred about symmetry breaking more or less directly from experiment.

Let us consider[13] the Σ hyperons, which form an isotriplet. If the electromagnetic interactions are turned off, all the members of the isotriplet have a common mass. The electromagnetic interactions cause a shift in the masses from their original common value, so that each member of the isotriplet $(\Sigma^+, \Sigma^-, \Sigma^0)$ acquires a distinct mass. One can break the mass shift operator up into pieces transforming like $I = 0$, $I = 1$ or $I = 2$. The mass shifts transforming like $I = 1$ and $I = 2$ are related to particle masses by

$$\Delta M^{(1)} = (M_{\Sigma^-} - M_{\Sigma^+})/\sqrt{2}\ , \tag{II.1}$$

and

$$\Delta M^{(2)} = (M_{\Sigma^+} + M_{\Sigma^-} - 2M_{\Sigma^0})/\sqrt{6}\ . \tag{II.2}$$

Experimentally, one finds that $\Delta M^{(1)}/\Delta M^{(2)} \approx 7$, a result that one would hardly anticipate on the basis of very crude dynamical estimates of the electromagnetic mass splittings.

In any case, it is clear that the strong interactions must be set up so that the isospin 1 part of the mass splitting is enhanced over the isospin 2 part. Now let us suppose that we can ignore the effect of "medium" strong symmetry-breaking on our analysis. This is done primarily on the basis that the Coleman-Glashow sum rule[14] for the electromagnetic mass splittings of the baryons, derived by neglecting the effects of strong symmetry-breaking, is in fairly good agreement with experiment. Then we can conclude that the enhancement of the isospin 1 part of the Σ mass splittings <u>is a result of the very strong, SU(3) symmetric, interactions</u>.

If, however, the enhancement is due to the SU(3) symmetric part of the strong interactions, we know that if a single isomultiplet is enhanced, the entire unitary supermultiplet to which it belongs will also be enhanced. In the case of the order e^2 electromagnetic mass splittings, we have the choice of supposing that either the $\underline{8}$ or $\underline{27}$ supermultiplet is enhanced. Now the $\underline{27}$ contains both I = 1 and I = 2 isomultiplets, so that if the strong interactions enhanced the $\underline{27}$, both I = 1 and I = 2 mass shifts would be enhanced, in disagreement with experiment. The $\underline{8}$, however, contains I = 1 but not I = 2, so we would infer from this analysis of electromagnetic mass splittings in the Σ isomultiplet that the SU(3) symmetric part of the strong interactions must enhance the octet part of the electromagnetic mass splittings.

One will note that according to the preceding analysis, octet enhancement is a property of the SU(3) symmetric part of the strong interactions, not of electromagnetism. Once one has realized this, it is a natural extrapolation to suppose that the same mechanism that operates within the very strong interactions so as to enhance the octet part of the electromagnetic violations of SU(3), also enhances the medium-strong and weak violations of SU(3). In other words, the octet enhancement mechanism might be independent of the axis in SU(3) space along which the violation lies. In this case one would find a pattern of universal octet dominance.

Cutkosky and Tarjanne[1]) and Dashen and Frautschi[2,3]) have suggested that bootstrap dynamics may provide a mechanism leading to the enhancement of violations of SU(3) which transform like components of an octet. Let us try to see how this might happen in the case of electromagnetic perturbations on an SU(3) symmetric problem[15]).

We shall assume that each hadron may be described as a bound or resonant state of other hadrons, and that our unperturbed strong interaction scattering problem is SU(3) symmetric, so that each supermultiplet appears as a degenerate set of poles in the relevant scattering amplitude. Moreover, we shall suppose that we can work to order e^2 in the electromagnetic interaction, and that the effect of medium strong violations of SU(3) can be ignored.

When the electromagnetic interaction is turned on, the forces producing the bound or resonant states will change in a way that varies from one to another member of a given supermultiplet; as a result, the position and residues of the poles in each supermultiplet will shift from their SU(3) symmetric values.

For the moment we shall disregard the shifts in the residues of the poles, which govern the shifts in the coupling of the supermultiplet represented by the poles to the external particles in the scattering amplitude, and concentrate on the shifts in the positions of the poles.

The mass shifts in a given supermultiplet may be described by a mass shift matrix, as pointed out by Glashow[16]. To understand something of the structure of this matrix, it is illuminating to return for a moment to the example of the electromagnetic mass shifts in the Σ isomultiplet. The mass shifts are generated, to order e^2, when an electric current having the isotopic spin structure $j = e(T_3 + Y/2)$ acts twice, and they will clearly contain terms proportional to $1, T_3$ and T_3^2. It is convenient, however, to classify the mass shifts according to their isospin transformation properties. For this purpose, one expands the mass shift matrix in terms of irreducible tensors in isospin space[17]. In this case one finds a singlet mass correction $\delta M_{1,1}$ transforming like $I = 0$, a triplet correction $\delta M_{3,n}$ transforming like the nth component of an $I = 1$ state, and finally a correction $\delta M_{5,n}$ transforming like the nth component of an $I = 2$ state.

Returning to our SU(3) problem we find, analogously, a mass shift $\delta M_{1,1}$ transforming like a unitary singlet, an octet shift $\delta M_{8,n}$ $(n = 1, \ldots, 8)$ transforming like the nth component of an octet and finally a term $\delta M_{27,n}$ $(n = 1, \ldots, 27)$ transforming like the nth component of a <u>27</u>- plet.

It is useful to separate the contributions to the mass shift of a bound or resonant state into two kinds. The first, called driving terms, includes all intermediate states in which a particle is present which was <u>not</u> present in the unperturbed problem. In the present case, this includes all intermediate states in which a single photon (but, in principle, any number of hadrons) is present. To calculate these driving terms explicitly requires a specific knowledge of the mechanism of symmetry-breaking. An important general feature of the driving terms is that a driving term transforming according to a given irreducible representation effects only the mass shift which transforms in the same way.

In bootstrap theory, a second type of mass shift arises as a result of shifts in the masses of the constituent particles of a bound state, as well as from changes in the binding forces coming from changes in the masses of exchanged particles. These are examples of self-consistent terms[18].

To illustrate these points further, let us specialize in the reaction $B+\pi \to B+\pi$. Here π is the 0^- meson octet and B is the $\frac{1}{2}^+$ baryon octet. The B and Δ ($\frac{3}{2}^+$ decuplet) supermultiplets appear as poles in the appropriate partial waves, and the forces producing the bound state (B) and resonance (Δ) are due to B and Δ exchange.

The combined effect of the self-consistent and driving terms is to give relations for the electromagnetic mass shifts in the B and Δ supermultiplets which, to order e^2, have the form[2]:

$$\begin{aligned}
\delta M^{B^s}_{8,n} &= A_8^{B^s B^s}\,\delta M^{B^s}_{8,n} + A_8^{B^s B^a}\,\delta M^{B^a}_{8,n} + A_8^{B^s \Delta}\,\delta M^{\Delta}_{8,n} + \cdots + D^{B^s}_{8,n} \\
\delta M^{B^a}_{8,n} &= A_8^{B^a B^s}\,\delta M^{B^s}_{8,n} + A_8^{B^a B^a}\,\delta M^{B^a}_{8,n} + A_8^{B^a \Delta}\,\delta M^{\Delta}_{8,n} + \cdots \quad D^{B^a}_{8,n} \qquad \text{(II.3)} \\
\delta M^{\Delta}_{8,n} &= A_8^{\Delta B^s}\,\delta M^{B^s}_{8,n} + A_8^{\Delta B^a}\,\delta M^{B^a}_{8,n} + A_8^{\Delta\Delta}\,\delta M^{\Delta}_{8,n} + \cdots + D^{\Delta}_{8,n}\ ;
\end{aligned}$$

for octet violations. B_s(D) and B_a(F) label the two independent octet terms in the baryon mass.

In writing down these linear equations, we have supposed that it makes sense to work to first order in the symmetry violations. These first order violations are confined to the factors δM and D. Hence, we note the very important fact that the coefficients A_α^{ij} which contain the self-consistent contributions to the mass shifts contain no SU(3) violations. Since the A matrix[2] is SU(3) symmetric, it connects only 8 to 8 and not for example 8 to 27. Moreover n is connected to n, and the A matrix is independent of n. Since the independent octet terms can mix even in unbroken unitary symmetry, the A matrix can connect B_a and B_s.

In the present problem, similar relations must of course be written down for singlet and 27- plet violations of SU(3). Moreover, other terms than those written down explicitly in Eq. (II.3) must, in principle, be included. These would correspond, for example, to pion electromagnetic mass shifts, coupling constant shifts, terms corresponding to the exchange of other particles or resonances and so on. The inclusion of more and more terms would eventually result in an infinite set of linear equations. The solution of these equations requires that they be truncated at some finite stage. What an appropriate approximation for practical computations might be is a matter separate from the points under discussion here, and in particular the specific terms kept in Eq. (II.3) do not play any distinguished role in formulating this theory of octet enhancement.

We may write Eq. (II.3) more compactly as follows:

$$\sum_{\beta} (1 - A_8)^{\alpha\beta}\, \delta M_{8,n}^{\beta} = D_{8,n}^{\alpha} , \qquad \text{(II.4)}$$

where α and β range over the various independent supermultiplets such as B_s, B_a and Δ. A similar equation can be written for the 27- plet violations. These equations can be inverted, providing $D^{\alpha}_{8,n} \neq 0$ and $D^{\gamma}_{27,n} \neq 0$, to find

$$\delta M^{\alpha}_{8,n} = \sum_{\beta} \left(\frac{1}{1-A_8}\right)^{\alpha\beta} D^{\beta}_{8,n}$$

$$\delta M^{\gamma}_{27,n} = \sum_{\delta} \left(\frac{1}{1-A_{27}}\right)^{\gamma\delta} D^{\delta}_{27,n} \quad . \tag{II.5}$$

The problem of octet enhancement can now be stated as follows[2]. Suppose that the driving terms D_8 and D_{27} are of comparable magnitude. How can the octet mass splitting emerge as the dominant ones? In the present formalism, it is clear that this situation can arise if the matrix A_8 has one or more eigenvalues near 1, while A_{27} lacks any such eigenvalues. If this situation obtains, the mass shifts transforming like an octet will contain a large term multiplying the associated eigenvectors, and will be enhanced compared to the <u>27</u> mass shifts.

To actually show that octet enhancement occurs in any particular case, such as the problem of electromagnetic mass shifts discussed above, will thus require the calculation of the A matrix corresponding to violations of the various allowed supermultiplicity (<u>1</u>, <u>8</u> and <u>27</u> in the problem of the order e^2 electromagnetic mass shifts). Before we take up the problem of calculating the A matrix, however, we wish to discuss in the next section a number of important general features of the bootstrap theory of octet enhancement.

2. General features of the bootstrap theory of octet enhancement

Let us first summarize and generalize the conclusions obtained in the preceding section in the case of the electromagnetic mass shifts in the B and Δ supermultiplets.

We study the linear deviations in the masses and coupling constants from their SU(3) symmetric values. These deviations are governed in part by driving terms, whose structure is not in general determined by the strong interactions, and in part by the requirement of self-consistency.

These statements can be expressed by equations of the following schematic form:

$$\delta M_i = \sum_j A_{ij}^{MM} \delta M_j + \sum_j A_{ij}^{Mg} \delta g_j + D_i^M ,$$
$$\delta g_i = \sum_j A_{ij}^{gM} \delta M_j + \sum_j A_{ij}^{gg} \delta g_j + D_i^g ; \qquad \text{(II.6)}$$

where δM_i, δg_i and D_i represent the mass shifts, the coupling shifts and the driving terms respectively. The matrix A_{ij} contains the self-consistent contributions.

As pointed out above, an important feature of the above equation is that A_{ij} contains, to first order, <u>no</u> violations of SU(3). These are contained only in the terms δM_i, δg_i and D_i. For this reason it is useful to expand the various operators which appear in irreducible representations of SU(3), in which case the equations take the general form:

$$\delta M^{\alpha}_{\lambda,n}/M^{\alpha} = \sum_{\alpha'} A^{\alpha\alpha'}_{\lambda}\left(\delta M^{\alpha'}_{\lambda,n}/M^{\alpha'}\right) + \sum_{\beta'} A^{\alpha\beta'}_{\lambda} \delta g^{\beta'}_{\lambda,n} + D^{\alpha}_{\lambda,n}$$
$$\delta g^{\beta}_{\lambda,n} = \sum_{\alpha'} A^{\beta\alpha'}_{\lambda}\left(\delta M^{\alpha'}_{\lambda,n}/M^{\alpha'}\right) + \sum_{\beta'} A^{\beta\beta'}_{\lambda} \delta g^{\beta'}_{\lambda,n} + D^{\beta}_{\lambda,n}, \qquad \text{(II.7)}$$

where λ is the supermultiplicity of the violation and $n = 1, \ldots, \lambda$ is the component of that supermultiplet along which the violation lies. Also α and β run over the various supermultiplets (B, Δ, π) and vertices $(BB\pi, B\Delta\pi)$ and over the independent irreducible representations with given supermultiplicity λ.

A number of important general conclusions about symmetry breaking and octet enhancement can be drawn from these equations.

i) The A matrix is SU(3) symmetric. Hence, it is independent of n, the direction in unitary spin space along which the violation lies, and connects n only to n and λ to λ.

ii) As was illustrated in the previous section, the pattern of symmetry violation is determined by the pattern of eigenvalues and eigenvectors of the A matrix. That is, if we assume that the driving terms are of comparable magnitude, then those mass and coupling shifts associated with an eigenvector whose corresponding eigenvalue is near unity are preferentially enhanced.

iii) The eigenvalues of the A matrix are determined by the SU(3) symmetric strong interactions alone. As a result, they are independent of the detailed structure of the driving terms as well as the axis n in SU(3) space along which the violation lies. These facts have two very important consequences.

a) One sees from this how the octet enhancement pattern can be universal. Suppose the A matrix has an octet eigenvalue near one, and that its eigenvalues corresponding to violations of other supermultiplicity are not near one. Then since A_8 is independent of n, the medium strong violations $(n = 8)$, the electromagnetic violations $(n = 3)$ and the weak violations of SU(3) should all show octet enhancement. Thus, all three violations of SU(3) have a common dynamical origin.

b) Secondly, this means that one can hope to explore the pattern of symmetry breaking without first having to know the detailed mechanism of symmetry violation as given by the driving terms. Although we have a good idea of what the driving terms are in the case of electromagnetic violations, we are very much in doubt as to what the driving terms are in the case of medium strong or weak violations of SU(3).

iv) One should particularly note that in this formalism, octet enhancement is <u>not</u> an automatic consequence of our assumptions. It could well happen that the <u>27</u>, for example, is enhanced or that there is no marked enhancement of any sort. To <u>establish</u> that this theory results in octet enhancement, and in particular to estimate the amount of enhancement, therefore requires detailed calculations of the eigenvalues and eigenvectors of the A matrix.

v) It is the driving terms, not the A matrix, which set the scale of the violation of SU(3). It is to be emphasized that the driving terms, and hence the scale of violation, are not determined by bootstrap dynamics. What is determined by bootstrap dynamics is the <u>pattern</u> of the symmetry breaking; for example, the <u>ratios</u> of various mass shifts are determined but not the absolute magnitude of any particular mass shift.

vi) We have, so far, assumed that all of the driving terms feeding the mass shifts (or coupling shifts) transforming according to different representations are of a comparable order of magnitude. This need not be the case. Let us see how the octet enhancement pattern would be effected if other possibilities are considered.

a) If the medium strong violations of SU(3) are the result of a spontaneous breakdown[1] of SU(3), then $D_{strong} = 0$. In this case, the enhanced eigenvalue must be exactly equal to one in the linear approximation. If it is the octet eigenvalue that is equal to one, then the octet violations will be enhanced as before.

When higher order effects are taken into account, it is no longer necessary for the A matrix to have an eigenvalue exactly equal to one. Moreover, in the case of spontaneous symmetry breakdown, the higher order terms must set the scale of the violation.

b) It may, of course, turn out that the structure of the driving terms for the strong and weak violations of SU(3) favours octet violations. In this case, the observed pattern of octet violations would result partly from the octet enhancement mechanism and partly from the driving terms. This may, in fact, be what is happening in cases where there is apparently almost complete octet dominance, such as the medium strong mass splittings in the $\frac{1}{2}^+$ baryon octet.

vii) A theory of octet enhancement of strong violations of SU(3) would still be required in conjunction with most of the specific mechanisms for strong symmetry-breaking that have been proposed so far.

In Ne'eman's theory[19], for example, strong symmetry breaking arises as a result of a "fifth" interaction carried by a singlet vector meson which is coupled to the hypercharge and baryon currents. In such a theory, the strong mass splittings appear as effects that are of second order in the currents, and the observed octet dominance requires an explanation for the same reasons as in the case of electromagnetic and weak symmetry breaking.

As mentioned above, strong symmetry breaking may occur spontaneously[1]. The dynamics underlying strong violations of SU(3) would then be different in character from that responsible for electromagnetic or weak violations. However, an explanation is still required of why <u>octet</u> violations predominate.

Finally, it may be that the mechanism causing strong symmetry breaking is of a purely octet type. But this still leaves open the question of why this octet symmetry breaking is not affected to a greater extent than it appears to be by higher order terms in the mass and coupling shifts.

viii) The octet enhancement mechanism implicit in Eq. (II.6) is independent of many of the details of the SU(3) symmetric strong interactions, although this fact is not immediately apparent from an inspection of these equations. It can be shown[20], however, that is is not necessary to assume (a) that a complete bootstrap theory which has no CDD parameters exists, (b) that the strong interaction bootstrap equations have a unique SU(3) symmetric solution or (c) that higher order terms in the mass and coupling shifts can be neglected.

A number of other approaches to symmetry breaking are capable of leading to a pattern of universal octet enhancement. These have recently been compared to the bootstrap theory of octet enhancement by Dashen and Frautschi[20]. Their conclusions will be briefly summarized at this point; for details the reader is referred to Ref. 20.

i) The vector mixing theory of Sakurai et al.[21,22], in which one exploits the near degeneracy of the $J = 1^-$ singlet and octet states in the absence of symmetry-breaking effects to infer a large octet-singlet coupling, is again a theory in which octet enhancement follows from properties of the strong interactions. Here, however, a specific set of strong interaction terms are conjectured to be dominant. There is nothing the matter in principle with this assumption. Whether or not it is in fact correct can only be decided by detailed computation. The calculations that have been performed[3] on the B and Δ mass shifts using the bootstrap theory indicate that in this case other terms are at least as important as the vector-mixing terms, and do not support the idea that the whole pattern of SU(3) symmetry-breaking is governed by vector-mixing effects.

ii) In the Coleman-Glashow theory[8], universal octet enhancement is shown to follow from the assumption that an octet of low mass scalar mesons exists and dominates the pattern of symmetry breaking. This theory and the bootstrap theory are more closely related than may be apparent at first sight. For example, it can be shown[20] that if such an octet of low mass scalar mesons exists, then the A matrix as determined by the bootstrap equations will have an octet eigenvalue near one. However, the bootstrap theory appears to be more inclusive in that it does not seem to be the case that the condition that the A matrix have an eigenvalue near unity implies the existence of scalar mesons.

III. S MATRIX PERTURBATION THEORY[23]

1. Introduction

In this section a technique will be described which employs the N/D formalism to study corrections to strong interaction scattering amplitudes which are brought about by the inclusion of an additional, weaker, interaction. These methods, developed by Dashen and Frautschi[4,5], appear to be very useful for the study of symmetry-breaking interactions.

When the detailed mechanism responsible for the perturbations that one wishes to study is known - as is the case with the electromagnetic corrections to the strong interactions - one can use these methods to calculate the actual values of mass and coupling constant shifts. For this reason, these techniques are of interest in their own right, and will be described in some detail here. Their particular application to the calculation of A matrix elements will be illustrated by an example in this section, and more fully in Section IV.

2. First-order equations for mass and coupling constant shifts in potential theory

For simplicity, we shall begin by considering non-relativistic single channel s-wave scattering, for which we define the amplitude

$$A(s) = (e^{i\eta} \sin \eta)/q, \quad s > 0 \tag{III.1}$$

where η is the phase shift, q the momentum and $s = q^2$ the kinetic energy (we have set $\hbar = 1$ and $\hbar^2/2m = 1$). As usual, we assume that A(s) is analytic in the energy plane, with a right-hand cut required by unitarity and a left-hand cut coming from the partial wave projection of the Born amplitude and the double spectral function.

We shall suppose that we have obtained, for some strong potential V, an expression for the amplitude in the N/D form. It is not the purpose of this method to provide a way to compute A(s), N or D. In principle, these are to be regarded as known functions, although in practice one must usually employ (rather drastic) approximations for them. In deriving the perturbation formulae, one must, however, make use of a number of general properties of N and D, which I shall now list:

i) We suppose that N(s) is real for s > 0, and that it contains the left-hand cut of A(s).

ii) D(s) is real for s < 0; it has the right-hand cut of A(s) and the phase $e^{-i\eta}$ along the right cut. We also define D(s) to have no poles or zeros on the physical sheet, except for zeros at the positions of bound states. Thus, in particular, CDD poles are assumed to be absent.

iii) To formulate our assumptions about the asymptotic behaviour of D, we write the D function for a channel containing n bound states in the Omnès form[25]:

$$D(s) = c \prod_{i=1}^{n} \left(s - s_{B_i} \right) \exp \left\{ - \frac{(s - s_0)}{\pi} \int_0^\infty \frac{ds' \, \eta(s')}{(s' - s_0)(s' - s - i\epsilon)} \right\}. \quad \text{(III.2)}$$

We define $\eta(0) = 0$. At large s, one has

$$\begin{aligned} \lim_{s \to \infty} D(s) &\to c\, s^n \exp \left\{ \frac{\eta(\infty)}{\pi} \ln s + \text{constant} \right\} \\ &= c\, s^{(n + \eta(\infty)/\pi)} \, . \end{aligned} \quad \text{(III.3)}$$

In potential theory, Levinson's theorem[26] tells us that

$$[\eta(\infty) - \eta(0)]/\pi = N - n \, , \quad \text{(III.4)}$$

where N is the number of stable particles present in the theory before the strong interaction is turned on and n is the number of stable particles present after the strong interaction is turned on. Putting these statements together we find

$$D(s) \xrightarrow[s \to \infty]{} c\, s^N \, . \quad \text{(III.5)}$$

We shall always suppose $N = 0$. This statement, plus the assumption that there are no CDD poles, expresses rather clearly in potential theory the idea that all the particles in the theory are composite.

Now, let us suppose that the strong potential V is perturbed by a small amount δV. The unperturbed scattering amplitude $A_0(s)$ will correspondingly be changed by an amount $\delta A(s)$ and we introduce

$$A_N(s) = \frac{N_N(s)}{D_N(s)} \equiv \frac{N_0 + \delta N}{D_0 + \delta D} \equiv A_0(s) + \delta A(s) \, . \quad \text{(III.6)}$$

In particular, the residue R and position s_B of a bound state pole in $A_0(s)$ will change as a result of the change in the forces producing the bound state, so that the new amplitude $A_N(s)$ will have a bound state pole at $s_B + \delta s_B$ with residue $R + \delta R$. We can evidently write

$$\delta A(s) = \frac{R + \delta R}{s - (s_B + \delta s_B)} - \frac{R}{s - s_B} \quad \text{near} \quad s = s_B \, . \qquad (III.7)$$

It is the shift in the residue, δR, and the mass shift δs_B which we wish to compute[27]. For this purpose, we shall study the quantity $D_N D_0 \, \delta A(s)$. We observe that $D_N D_0 \, \delta A(s)$ has no poles, since D_0 has a zero at $s = s_B$ and D_N has a zero at $s = s_B + \delta s_B$. Moreover, for $s > 0$, $D_0 = |D_0| e^{-i\eta}$ and $D_N = |D_N| e^{-i(\eta + \delta\eta)}$ so that

$$D_0 D_N \delta A(s) = |D_0| |D_N| (\sin \delta\eta)/q \, . \qquad (III.8)$$

Therefore $D_N D_0 \, \delta A(s)$ has no right-hand cut.

Finally, we shall assume that we can write an <u>unsubtracted</u> dispersion relation[28] for $D_N D_0 \, \delta A(s)$. Application of Cauchy's theorem then gives:

$$D_N(s) D_0(s) \, \delta A(s) = \frac{1}{\pi} \int_{-\infty}^{0} \frac{D_N(s') D_0(s') \, \mathrm{Im} \, \delta A(s')}{s' - s} \, ds' \qquad (III.9)$$

recalling that D has no left-hand cut.

We have not had to use the fact that δV is supposed to be a <u>small</u> perturbation in order to derive Eq. (III.9). However, it is necessary to do so in order to obtain useful expressions for δs_B and δR from this equation.

First, note that to first order in δA and δD the right-hand side of Eq. (III.9) becomes

$$\frac{1}{\pi}\int_{-\infty}^{0} \frac{D_0^2(s')\,\mathrm{Im}\,\delta A(s')}{s'-s}\,ds' \equiv J(s)\,. \tag{III.10}$$

To find δs_B we simply compare Eqs. (III.7) and (III.9) near $s = s_B$. Evidently, $D_o(s) \approx D_o'(s_B)(s - s_B)$ and $D_N(s_B) = \delta D(s_B)$. Also, expanding $D_N(s_B + \delta s_B)$ about s_B to <u>first order</u>, we find

$$\begin{aligned} 0 = D_N(s_B + \delta s_B) &= D_o(s_B + \delta s_B) + \delta D(s_B + \delta s_B) \\ &\approx D_o'(s_B)\delta s + \delta D(s_B)\,, \end{aligned} \tag{III.11}$$

so

$$\delta s_B = \frac{-\delta D(s_B)}{D_o'(s_B)}\,. \tag{III.12}$$

Thus, near $s = s_B$, Eq. (III.9) has the approximate form

$$\delta A(s) = \frac{J(s)}{D_N(s)D_o(s)} \approx \frac{J(s_B)}{\delta D(s_B)D_o'(s_B)(s - s_B)} = -\frac{R}{s - s_B}\,. \tag{III.13}$$

Comparing the residues of the poles at $s = s_B$ on each side of Eq. (III.13), and using Eq. (III.12), we find

$$\delta s_B = \frac{1}{R[D_o'(s_B)]^2}\,\frac{1}{\pi}\int_{-\infty}^{0} \frac{D_o^2(s')\,\mathrm{Im}\,\delta A(s')}{s'-s_B}\,ds'\,, \tag{III.14}$$

which is valid to first order in small quantities.

The shift in the residue is found by comparing Eqs. (III.7) and (III.9) at $s = s_B + \delta s_B$. The algebra is a little more complicated here, however, because to obtain the answer one must first expand all quantities consistently to second order. Clearly, $D_N(s) \approx D_N'(s_B + \delta s_B)[s - (s_B + \delta s_B)]$. Writing $\delta s_B \approx \delta s_B^{(1)} + \delta s_B^{(2)}$ and $\delta D \approx \delta D^{(1)} + \delta D^{(2)}$, where the superscripts indicate the first and second order terms in δs and δD when they are expanded in a small parameter, we have

$$D_o(s_B + \delta s_B) \approx D_o'(s_B)\left(\delta s_B^{(1)} + \delta s_B^{(2)}\right) + D_o''(s_B)\,\frac{(\delta s_B^{(1)})^2}{2}$$

and

$$\begin{aligned} D_N'(s_B + \delta s_B) &= D_o'(s_B + \delta s_B) + \delta D'(s_B + \delta s_B) \\ &\approx D_o'(s_B) + D_o''(s_B)\,\delta s_B^{(1)} + \delta D^{(1)\prime}(s_B)\,. \end{aligned}$$

Expanding Eq. (III.11) to <u>second</u> order gives

$$\delta s^{(2)} = -\frac{\delta D^{(2)}}{D_o'} - \frac{D_o''}{D_o'}\,\frac{(\delta s_B^{(1)})^2}{2} - \frac{\delta D^{(1)\prime}\,\delta s_B^{(1)}}{D_o'}\;.$$

Thus we have

$$D_o D_N \approx D_o'^2\left[\delta s_B^{(1)} + \frac{D_o''}{D_o'}\left(\delta s_B^{(1)}\right)^2 - \frac{\delta D^{(2)}}{D_o'}\right]\,.$$

Next, we observe from Eq. (III.13) that

$$R = \frac{-J(s_B)}{D_o'(s_B)\,\delta D(s_B)} \approx \frac{-J(s_B)}{D_o'(s_B)\,\delta D^{(1)}(s_B)}\left[1 - \frac{\delta D^{(2)}}{\delta D^{(1)}}\right].$$

Finally, we write $J(s) \approx J(s_B + \delta s_B) \approx J(s_B) + J'(s_B)\,\delta s_B$ and combine the above results to find

$$\begin{aligned}\delta R &= -R\,\delta s_B\left(\frac{D_o''(s_B)}{D_o'(s_B)}\right) + \frac{1}{[D_o'(s_B)]^2}\,\frac{1}{\pi}\int_{-\infty}^{0} \frac{D_o^2(s')\,\mathrm{Im}\,\delta A(s')}{(s'-s_B)^2}\,ds' \\ &= \frac{d}{ds}\left[J(s)\left(\frac{s-s_B}{D(s)}\right)^2\right]_{s=s_B}\,,\end{aligned} \qquad \text{(III.15)}$$

in which only terms of <u>first order</u> have been retained.

Note that the reality of δs_B and δR as given by the above equations is ensured by the vanishing of $D_o(s')$ at $s' = s_B$.

The crucial assumption used in deriving these equations was that one could write an unsubtracted dispersion relation[3,29] for $D_N D_o\,\delta A(s)$. The reason is the following. <u>Any value</u> of δR, for example, would be compatible with the assumed <u>analyticity properties</u> of $D_N D_o\,\delta A(s)$. One might then wonder how one can arrive at an equation which purports to <u>determine</u> δR simply by using Cauchy's theorem. The point is that for an arbitrary δR the contribution to Eq. (III.9) from the circular contour at ∞ will <u>not</u> vanish. In fact, this will be true only for δR computed from Eqs. (III.7) and (III.9). In other words, it is the assumed absence of undetermined subtraction constants that permits us to obtain an equation which determines the value of δR.

In deriving the relativistic equations we will have to make the same assumption of rapidly convergent dispersion integrals. It is believed by many that this property is a consequence of the composite nature of the strongly interacting particles. Thus, it may be that the necessary convergence of the integrals ties this method to the bootstrap theory of composite particles. The connection between these two assumptions (rapidly convergent dispersion integrals and composite particles) is, of course, not known in the relativisitc case with anything like the precision that is known in potential theory. Here, it will be recalled, the situation is the following.

We saw that the asymptotic behaviour $D \to$ constant as $s \to \infty$ followed from the assumption that there were no elementary particles (in the sense that there were no stable particles present in the problem before the strong interactions were turned on) provided the D function was defined so as to be free of poles of any sort. Consequently, the possibility of writing an unsubtracted dispersion relation for $D^2\,\delta A(s)$ reduces to a question of the asymptotic behaviour of $\delta A(s)$, which is just the difference of two partial wave amplitudes. Now, starting from the Schroedinger equation, one can prove[30)] that one can write an unsubtracted dispersion relation for the partial wave amplitude if the potential is a superposition of Yukawa potentials or exponential potentials. Hence, for a wide class of perturbing potentials, one can prove that $\delta A(s)$ has the desired asymptotic behaviour.

It is also likely that such a proof can be given using the Mandelstam representation and Regge poles. Thus, for a potential theory containing only composite particles, the dispersion integrals for $D^2\,\delta A(s)$ are rapidly convergent.

On the other hand, if there are elementary particles in the theory, the D function (defined in the way specified above) will have the behaviour $D \to s$ as $s \to \infty$, or worse, and it is most unlikely that one could write an unsubtracted dispersion relation for $D^2\,\delta A(s)$.

Thus, we see that the assumptions that there are only composite particles and that the dispersion integrals are rapidly convergent are tied together very closely in potential theory. On this basis it is conjectured that in relativistic theory the convergence of this method is connected with bootstrap dynamics.

At first sight, Eqs. (III.14) and (III.15) for the mass and residue shift of a bound state pole look very different from the usual expressions for these quantities. It is interesting to note, therefore, that one can show[31] that Eq. (III.14) for the mass shift δs_B, for example, is closely related to the familiar expression obtained from first order Rayleigh-Schroedinger perturbation theory;

$$\delta s_B = \int_0^\infty |\Psi_B(r)|^2 \, \delta V(r) \, dr \; . \qquad \text{(III.16)}$$

Here, Ψ_B is the unperturbed bound state wave function and δV is, as before, the perturbing potential. For details the reader may consult Refs. 24 and 32.

3. First order mass and coupling constant shifts in relativistic scattering theory

The generalization of the results of section 2 to the case of relativistic, elastic scattering is straightforward. Here, we shall consider the s wave elastic scattering of two spinless particles of equal mass, and define the partial wave amplitude

$$A(s) = (e^{i\eta} \sin \eta)/\rho(s) \; , \qquad \text{(III.17)}$$

where s is now the total c.m. energy squared $[s = 4(q_s^2 + m^2)]$ and ρ is a function which removes kinematic singularities. The phase shift η is, in general, complex above the first inelastic threshold, and a convenient measure of the inelasticity is given by

$$I(s) = |e^{2i\eta}|/2\rho(s) .$$

As before we assume that A(s) is an analytic function of s, with a right-hand cut given by physical unitarity and a left-hand cut controlled by unitarity in the t and u channels.

The strong interaction amplitude is supposed to be given in the N/D form, where N and D have analytic properties which are evident generalizations of those stated in section 2. In particular, D may be written as in Eq. (III.2), except that the phase shift $\eta(s)$ is now to be replaced by Re $\eta(s)$. Moreover, we shall again assume that $D(s) \to$ constant as $s \to \infty$; this time without having the understanding of this condition provided by the Levinson theorem in potential scattering. One will note that to apply the N/D method in this case, one must be <u>given</u> Im η along the right-hand cut, as well as Im A along the left-hand cut.

To find the shift in the position and residue of a bound state pole in A(s) when the strong interactions are modified by some weaker interactions, we again write a dispersion relation for $D_N D_0\,\delta A(s)$. In potential scattering $D_N D_0\,\delta A(s)$ had no right-hand cut, but in strong interaction problems a right-hand cut arises (a) because $\delta\eta$ has an imaginary part and (b) because mass shifts of the scattered particles change the kinematic factor by an amount $\delta\rho$.

Taking these changes into account and assuming we can write an unsubtracted dispersion relation for $D_N D_0\,\delta A(s)$, we find

$$D_N D_0\,\delta A(s) = \frac{1}{\pi}\left\{\int_L \frac{D_0^2(s')\,\mathrm{Im}\,\delta A(s')}{s'-s}\,ds' + \int_R \frac{\mathrm{Im}[D_0^2(s')\,\delta A(s')]}{s'-s}\,ds'\right\}, \tag{III.18}$$

where in writing Eq. (III.18) we have used the fact that Im $D(s) = 0$ along the left-hand cut as well as the assumption that we are dealing with a small perturbation. The integral R runs over those cuts given by physical unitarity, L runs over all other cuts.

The equations for δs_B and δR are obtained from Eq. (III.18) in exactly the same way as their potential theory analogues were obtained from Eqs. (III.7) and (III.9). The answers which result are, of course, of exactly the same form, with the right-hand side of Eq. (III.18) replacing the expression for $J(s)$ given in Eq. (III.10). Thus, one finds

$$\delta s_B = \frac{1}{R[D_o'(s_B)]^2} \frac{1}{\pi} \left\{ \int_L \frac{D_o^2(s') \operatorname{Im} \delta A(s')}{s' - s_B} ds' + \int \frac{\operatorname{Im}[D_o^2(s') \delta A(s')]}{s' - s_B} ds' \right\}, \tag{III.19}$$

and

$$\delta R = -R\,\delta s_B \left(\frac{D_o''(s_B)}{D_o'(s_B)} \right) + \frac{1}{[D_o'(s_B)]^2} \frac{1}{\pi} \left\{ \int_L \frac{D_o^2(s') \operatorname{Im} \delta A(s')}{(s' - s_B)^2} ds' + \int_R \frac{\operatorname{Im}[D_o^2(s') \delta A(s')]}{(s' - s_B)^2} ds' \right\}. \tag{III.20}$$

Note that along the right-hand cut we may write

$$\operatorname{Im}[D_o^2 \delta A(s)] = \operatorname{Im}\left[D_o^2 \left(\frac{\delta I\, e^{2i \operatorname{Re} \eta}}{i} + \frac{1}{2i\rho} \frac{\delta\rho}{\rho} \right) \right]. \tag{III.21}$$

Since $D_o^2 = |D_o|^2 e^{-2i \operatorname{Re} \eta}$ along the right-hand cut, this is simply

The contribution of nucleon exchange to this amplitude consists, in the static approximation, of a pseudopole of the form

$$\frac{1}{9}\gamma_{11}\Big/\left(W - 2M_N^{ext} + M_N^{exch}\right) , \qquad \text{(III.28)}$$

where $\gamma_{11} = 3f^2_{\pi NN}/\mu^2$. The label M_N^{ext} refers to the mass of an <u>external</u> nucleon entering in the kinematic factors associated with going from the pole in the crossed channel of the complete amplitude to the corresponding pole in the partial wave amplitude and M_N^{exch} labels the mass of the exchanged nucleon.

For simplicity, we shall only consider the A-matrix element, $A_1^{NN_{exch}}$, corresponding to a mass shift transforming like an isotopic singlet (I = 0). Since the <u>singlet</u> mass shifts preserve SU(2), we may calculate this effect directly by varying the SU(2) symmetric nucleon exchange contribution. From Eq. (III.28) it is clear that to first order an SU(2) symmetric shift in the mass of the exchanged nucleon will change the partial wave amplitude by an amount

$$\delta A(W) = -\frac{1}{9}\gamma_{11}\,\frac{\delta M_N^{exch}}{\left[W - 2M_N^{ext} + M_N^{exch}\right]^2} . \qquad \text{(III.29)}$$

The labels on M_N are now superfluous. One may use this expression for $\delta A(W)$, together with Eq. (III.19), to estimate the effect of nucleon exchange on the nucleon mass shifts transforming like an isosinglet. Carrying out the simple contour integral that arises, one finds

$$\delta M_1 = -\frac{1}{9}\left[\frac{D^2(W)}{D'^2(M_N)(W - M_N)}\right]'\Bigg|_{W = M_N} \delta M_1 , \qquad \text{(III.30)}$$

$$\equiv A_1^{NN_{exch}}\,\delta M_1 \qquad \text{(III.31)}$$

In order to establish some connection between S matrix perturbation theory and the octet enhancement formalism discussed in the preceding section, let us look very briefly at the important problem of the proton-neutron mass difference treated by Dashen[33] and see how contributions to the A matrix and the driving terms arise there.

Let us first consider a typical self-consistent term. As we know from work on the reciprocal bootstrap model[34] for the N and N*, one of the important forces producing a nucleon bound state in πN scattering is generated by nucleon exchange in the crossed channel. The change in mass of the exchanged nucleon which arises when electromagnetic interactions are turned on evidently results in a small change in the forces which produce the nucleon bound state, with a resultant shift in the position of the bound state - that is, in the mass of the nucleon.

We may estimate this effect as follows. In the absence of electromagnetic forces, the two charge states of the nucleon are indistinguishable. One may then consider an SU(2) symmetric model of πN scattering, in which the nucleon is regarded as a bound state pole appearing in the $J = \frac{1}{2}$, $L = 1$, $T = \frac{1}{2}$ channel of the πN partial wave scattering amplitude. This amplitude is defined as

$$A(W) = e^{i\eta} \sin \eta/\rho(W) \qquad \text{(III.27)}$$

with

$$\rho(W) = \frac{W^2}{M_N^2} \left[(W - M_N)^2 - \mu^2 \right]^{-1} q^{-1} .$$

As before, η is the phase shift, W the total c.m. energy and q the c.m. momentum.

$$-\mathrm{Im}[D_0^2\,\delta A(s)] = \delta I|D_0|^2 + \frac{1}{2\rho}\,\mathrm{Re}\left(D_0^2\,\frac{\delta\rho}{\rho}\right) . \qquad \text{(III.22)}$$

We see that in a relativistic problem we must know δI and $\delta\rho$ as well as D and δA on the left.

Finally, if we make the additional assumption that the unperturbed amplitude satisfies elastic unitarity;

$$\mathrm{Im}\,A(s) = \rho(s)|A(s)|^2 , \qquad \text{(III.23)}$$

so that the unperturbed phase shift is taken to be real even above the inelastic threshold, one can simplify Eq. (III.22) further[32]. The perturbed amplitude $A(s)+\delta A(s)$ will satisfy the unitarity equation

$$\mathrm{Im}(A+\delta A) = (\rho+\delta\rho)|A+\delta A|^2 + \sum_i \rho_i|\delta A_i|^2 , \qquad \text{(III.24)}$$

where the second term is the contribution to the absorptive part of the partial wave amplitude coming from new inelastic intermediate states, and ρ_i is the corresponding phase space factor. Combining these equations gives

$$\mathrm{Im}\,\delta A = 2\rho\,\mathrm{Re}(A^*\delta A) + (\delta\rho|A|^2 + \sum_i \rho_i\,|\delta A_i|^2) . \qquad \text{(III.25)}$$

Using the relations $D = |D|e^{-i\eta}$, $A = N/D = (e^{i\eta}\sin\eta)/\rho$ (η real), we may write[32]

$$\mathrm{Im}\,D^2\,\delta A(s) = N^2\,\delta\rho + |D|^2 \sum_i \rho_i|\delta A_i|^2, \quad s > \text{threshold}, \qquad \text{(III.26)}$$

which makes the origin of the contributions to the right-hand cut transparent.

where the subscript 1 has been added to δM to emphasize that we are dealing at this point with I = 0 mass shifts. The factor $[D^2(W)/D'^2(M_N)(W-M_N)]'|_{W=M_N}$ is precisely one in the present case, independent of the shape of the D function for πN scattering.

The above example serves to illustrate how an element of the A matrix may be calculated using the techniques discussed in this section. Actually, the particular term $A_1^{NN_{exch}}$ effects the proton and neutron masses uniformly, since it connects isosinglet mass shifts and therefore does not enter in the proton-neutron mass difference. What is needed here is the element $A_3^{NN_{exch}}$, governing self-consistent mass shifts transforming like I = 1. This does not require a separate calculation because ratios like $A_3^{NN_{exch}} : A_1^{NN_{exch}}$ are fixed by group theory, as explained in Refs. 3 and 5. One finds $A_3^{NN_{exch}} = (-5/3) \cdot A_1^{NN_{exch}} = +5/27$.

It is evident that relations like Eq. (III.31) do not determine the actual value of a mass shift. For this purpose one must introduce driving terms (not determined solely by the strong interactions) to set the scale of the violation.

In the case of the proton-neutron mass difference, the driving terms arise from the difference in the forces which act in the $T_3 = \pm 1/2$ charge states of the $J = 1/2^+$ πN scattering amplitude. Because only the $T_3 = -1/2$ amplitude contains two charged particles, one may anticipate that single photon exchange (that is, the process $N\bar{N} \to \gamma \to 2\pi$) may provide an important driving term, a result that is borne out by detailed calculation[33].

If one evaluates the partial wave amplitude δA_γ for the process $N\bar{N} \to \gamma \to 2\pi$, and uses this result to evaluate Eq. (III.19) for δM (here $\delta M = M_p - M_n$), one finds[33]:

$$\delta M = \frac{\alpha}{f^2}\frac{\mu^2}{M_N} \times \text{[a dimensionless number]} \tag{III.32}$$

$$= D_{\text{electromagnetic}} , \tag{III.33}$$

where $\alpha = 1/_{137}$, $f^2 = 0.08$, μ is the pion mass and M_N the nucleon mass. Here, we see illustrated some characteristic features of the driving terms:

i) they introduce terms into Eq. (III.19) which are <u>not</u> homogeneous in δM so that an equation results from which one can hope to determine the magnitude of δM;

ii) factors are introduced, like the fine structure constant, which are characteristic of the electromagnetic, not the strong, interactions [note that no such factors are present in Eq. (III.30), for example];

iii) the calculation of the driving terms themselves requires one to go <u>outside</u> of the SU(2) symmetric strong interactions.

4. Perturbation formulae for multi-channel scattering problems

For applications to practical problems in particle physics it is necessary to extend the formulae developed above so as to obtain expressions for the mass and coupling constant shifts in a multi-channel problem. In this section, these developments[5)] will be sketched out briefly.

Let us consider the case of n two-body channels. The partial wave amplitude connecting these channels, $\underline{T}(s)$, is a symmetric $n \times n$ matrix. As usual we suppose that this unperturbed amplitude has been obtained in the form $\underline{T} = \underline{N}\,\underline{D}^{-1} = (\underline{D}^T)^{-1}\,\underline{N}^T$, where the matrix functions $\underline{T}$, $\underline{N}$ and $\underline{D}$ are assumed to have analytic properties which are analogous to those in the single channel case.

One may now proceed as before. Along the right-hand cut

$$\mathrm{Im}\,\underline{T}^{-1} = -\,\underline{\rho}\;,$$

where $\underline{\rho}$ is a diagonal $n\times n$ matrix containing kinematic factors. Using the fact that $\underline{N}$ has no right-hand cut, one can show that [5,32,35]

$$\mathrm{Im}(\underline{D}^T\,\delta\underline{T}\,\underline{D}) = \underline{N}^T\,\delta\underline{\rho}\,\underline{N} + \underline{D}^T\!\left(\delta\underline{T}_I^T\,\underline{\rho}_I\cdot\delta\underline{T}_I\right)\underline{D}\;, \qquad \text{(III.34)}$$

where $\delta\underline{T}_I$ is an $m\times n$ matrix if there are m new inelastic states, and $\underline{\rho}_I$ is an $m\times m$ diagonal matrix containing kinematic factors. This is the many channel generalization of Eq. (III.26). With assumptions on $\underline{D}$ and $\delta\underline{T}$ similar to those made in the single channel case, it follows that

$$\underline{J}(s) = \frac{1}{\pi}\int_L \frac{\underline{D}^T(s')\,\mathrm{Im}\,\delta\underline{T}(s')\underline{D}(s')}{s'-s}\,ds' + \frac{1}{\pi}\int_R \frac{\underline{N}^T(s')\,\delta\underline{\rho}(s')\underline{N}(s')}{s'-s}\,ds'\;, \qquad \text{(III.35)}$$

with

$$\delta\underline{T}(s) = \left(\underline{D}^T(s)\right)^{-1}\underline{J}(s)\,\underline{D}^{-1}(s)\;.$$

Suppose that the unperturbed amplitude has a <u>single</u> bound state pole at $s=s_B$. Then, in the neighbourhood of s_B, $\underline{T}\sim\underline{R}/(s-s_B)$, where $\underline{R}$ is a residue matrix which may be written in terms of the couplings f_i of the bound state to the channels $i = 1, \ldots, n$ as $R_{ij} = -f_i\,f_j$.

Comparing the residues of Eq. (III.35) at $s = s_B$ and $s = s_B + \delta s_B$ with those of

$$\delta\underline{T} = \frac{\underline{R}+\delta\underline{R}}{s-(s_B+\delta s_B)} - \frac{\underline{R}}{s-s_B}$$

one finds

$$\underline{R}\,\delta s_B = \left[\lim_{s\to s_B}(s-s_B)(\underline{D}^{-1}(s))^T\right]\underline{J}(s_B)\left[\lim_{s\to s_B}(s-s_B)\underline{D}^{-1}(s)\right] \qquad \text{(III.36)}$$

and

$$\delta\underline{R} = \left(\frac{d}{ds}\right)\left[(s-s_B)(\underline{D}^{-1}(s))^T\,\underline{J}(s)(s-s_B)\underline{D}^{-1}(s)\right]\Bigg|_{s=s_B} \qquad \text{(III.37)}$$

with $\underline{J}(s)$ given by Eq. (III.35).

These formulae can be somewhat simplified if we write

$$\underline{\Delta} = \lim_{s\to s_B}(s-s_B)\underline{D}^{-1}(s) \qquad \text{(III.38)}$$

and

$$\underline{\Delta}' = (d/ds)\left[(s-s_B)\underline{D}^{-1}(s)\right]\Bigg|_{s=s_B} . \qquad \text{(III.39)}$$

Then we may multiply both sides of Eq. (III.36) by $\underline{R}$ and take the trace of both sides of the equation to find

$$\delta s_B = \mathrm{Tr}\left[\underline{R}\underline{\Delta}^T\,\underline{J}(s_B)\underline{\Delta}\right]\Big/\mathrm{Tr}[\underline{R}\underline{R}] . \qquad \text{(III.40)}$$

Using the fact that $\underline{R} = \underline{N}(s_B)\underline{\Delta}$ one can also find an expression for the change in an individual coupling constant. This turns out to be:

$$\delta f_i = -\left(\sum_k f_k^2\right)^{-1}\sum_j f_j\left(\underline{\Delta}'^T\underline{J}(s_B)\underline{\Delta} + \tfrac{1}{2}\underline{\Delta}^T\,\underline{J}'(s_B)\underline{\Delta}\right)_{ij} \qquad \text{(III.41)}$$

We note in passing that one can easily modify this formalism so that it applies to changes in the position and width of a resonance rather than a bound state. In the case of a resonance, $\underline{T}$ and $\delta\underline{T}$ have poles on the second Riemann sheet. The equations for the mass and coupling constant shifts are of the same form as Eq. (III.40) and Eq. (III.41), except that $\underline{D}$ and $\underline{J}$ must be replaced by their values $\underline{D}_{(2)}$ and $\underline{J}_{(2)}$ on the second sheet. However, the integral $\underline{J}_{(2)}$ has a slightly different form

$$\underline{J}_{(2)}(s_r) = \frac{1}{\pi}\int_L \frac{\underline{\underline{D}}^T \operatorname{Im} \delta \underline{\underline{T}}\, \underline{\underline{D}}}{s' - s_r}\, ds' + \frac{1}{\pi}\int_R \frac{\underline{\underline{N}}^T \delta \underline{\rho}\, \underline{\underline{N}}}{s' - s_r}\, ds'$$
$$+ 2i\, \underline{N}^T(s_r)\delta\underline{\rho}\,(s_r)\,\underline{N}(s_r)\ , \qquad \text{(III.42)}$$

where s_r is the position of the resonance. Now, of course, δs_r acquires an imaginary part.

There are two further technical developments that are very important for applications of this method to the study of violations of SU(2) and SU(3) and which should be mentioned at this point.

First, the results have so far all been derived on the assumption that the unperturbed solution contains only a single bound state at a given energy. In cases where the unperturbed set of amplitudes are SU(2) or SU(3) symmetric, however, one will have to deal with unperturbed solutions which have several degenerate poles. In this case, as in the Shroedinger theory, before applying perturbation theory methods one must diagonalize a matrix, the mass-shift matrix, whose eigenvalues are the mass shifts and whose eigenvectors determine the couplings of the physical particles.

Secondly Glashow[16)], and Cutkosky and Tarjanne[1,36)] have shown that in problems involving perturbations of SU(2) or SU(3) symmetric sets of amplitudes the use of group theory results in a substantial simplification of the structure of the mass-shift matrix. Their methods have been adapted to the S matrix perturbation theory formalism. Time does not permit a detailed discussion of these points here; for details, the reader is referred to Ref. 5.

IV. OCTET ENHANCEMENT IN THE B AND Δ SUPERMULTIPLETS[37)]

1. Introduction

It is apparent, even from the very brief examination of the proton-neutron mass difference problem given in the previous section, that to evaluate the A matrix and obtain detailed predictions pertaining to symmetry breaking it is necessary to employ a specific model of whatever system is to be studied. Moreover, in the course of the calculations, one must make a number of approximations and assumptions which go far beyond those made in setting up the general formalism.

In this section we will first outline the model which has been used to study symmetry violations in the B and Δ supermultiplets[3,6,7)]. Then the effect of parity non-conserving weak interactions on the SU(3) symmetric $BB\pi$ couplings will be discussed in some detail. This is one of the "cleanest" problems to which these methods have been applied, and illustrates very well how the general octet enhancement formalism and S matrix perturbation theory can be used together to obtain interesting results on symmetry breaking. Finally, the results of recent calculations[3,6,7)] on the medium strong, electromagnetic and weak violations of SU(3) symmetry in the B and Δ supermultiplets will be surveyed.

In considering the B and Δ supermultiplets, a Chew-Low type of SU(3) symmetric reciprocal bootstrap model is employed. Specifically, we consider pseudoscalar meson-baryon scattering with forces provided by B and Δ exchange; B and Δ also appear as direct channel poles associated with zeros of the relevant denominator functions. The B and Δ

exchange forces are approximated by their long-range parts, which give the well-known "short-cut" in πB scattering. The short-cut is itself approximated by a pole.

The approximation of keeping only the B and Δ intermediate states (neglecting higher resonance and vector meson exchange, for example) and considering only the πB channel serves to truncate the otherwise infinite A matrix. These approximations, while drastic, are typical of those made in dispersion theory calculations. In the case of the reciprocal bootstrap[34] for the N and N*, they do appear to have a certain amount of validity. Unfortunately, a correspondingly reasonable set of approximations has not yet been found for other interesting systems, such as the pseudoscalar meson-vector meson system. Thus, for purposes of practical computation one is, at present, pretty much limited to the B and Δ supermultiplets.

The input parameters of this model are the average masses of the B and Δ supermultiplets, taken from experiment, the D/F ratio and the <u>ratio</u> of the strong $BB\pi$ and $B\Delta\pi$ couplings. The latter two quantities are calculated from the reciprocal bootstrap theory, which gives a range of values for these quantities consistent with experiment.

A final parameter is the curvature of the D function, which gives a measure of the effective range of the forces in the relevant channel. The reciprocal bootstrap also gives a reasonable range of values for the curvature of the denominator function. All of the input parameters appearing in this model are determined solely by SU(3) symmetric strong interaction considerations, and are held fixed in all subsequent calculations of symmetry breaking.

It is recognized that the model just described, which is essentially the static model, is a crude one. However, it is the best model presently available and probably offers the best means to test the further ideas of symmetry breaking which concern us here.

2. Bootstrap theory and the parity non-conserving hyperon decays[37)]

In the present section we shall show how the techniques of Sections II and III may be applied to treat the violations of SU(3) which are brought about by the parity non-conserving weak interactions. These manifest themselves most directly in the non-leptonic hyperon decays.

First, let us see what bootstrap theory has to do with the weak interactions. As usual, we consider the nucleon as a pion-nucleon bound state. Neglecting the parity non-conserving weak interactions, the nucleon would be a purely P wave πN bound state, but when the weak interaction is included the nucleon acquires a small S-wave component. The amount of S wave in the nucleon state is directly related to the parity non-conserving part of the πNN coupling. In a bootstrap theory this S-wave piece of the nucleon state is determined by the parity non-conserving part of the potential which acts between a pion and a nucleon. Since nucleon exchange is an important piece of this potential, the parity non-conserving part of the πNN coupling will react on itself, and the problem will have a self-consistent aspect. Of course, the bootstrap is only part of the story here. There are direct weak interaction effects, possibly associated with intermediate vector-boson exchange, which also contribute to the parity non-conserving part of the πN potential. These parts of the potential form the driving terms which, as we have repeatedly emphasized, set the over-all scale of the parity non-conserving effects and are not determined by self-consistency requirements.

The coupling at a vertex $\bar{B}^i B^j \pi^k$, i, j, k = 1, ..., 8, may be written $(\gamma_5\, g_{ijk} + \delta g_{ijk})$, where g is the strong SU(3) symmetric coupling and δg is a small parity non-conserving coupling shift. We shall assume as before that we can work to <u>first order</u> in the violations of SU(3), hence we will neglect the medium strong and electromagnetic coupling shifts throughout.

In the present problem Eq. (II.6) takes the form

$$\delta g_{ijk} = \sum_{i'j'k'} A_{ijk,i'j'k'} \, \delta g_{i'j'k'} + \cdots + D_{ijk} \ . \qquad \text{(IV.1)}$$

This equation is simpler than the corresponding equation governing the <u>parity-conserving</u> coupling shifts in a number of important respects:

i) the parity non-conserving weak interaction is "off-diagonal" and can produce no mass shifts in lowest order. Hence, there are no mass shift terms in Eq. (IV.1).

ii) It turns out, for dynamical reasons, that the $\bar{B}B\pi$ coupling shifts are nearly decoupled from the parity non-conserving couplings of the other strongly interacting particles.

Next let us see how we can use the fact that the A matrix is invariant under SU(3) to simplify the structure of Eq. (IV.1). Note that in SU(3) δg_{ijk} belongs to the product representation $\underline{8} \otimes \underline{8} \otimes \underline{8}$. Thus, we can label δg by $\delta g^{\beta}_{N,n}$, where N = 1,8,27, ... runs over all the distinct irreducible representations contained in $\underline{8} \otimes \underline{8} \otimes \underline{8}$, n is the component of the representation and β is an index to be specified later which distinguishes between different representations with the same dimension. We can label the driving terms in the same way. Eq. (IV.1) then reads

$$\delta g^{\beta}_{N,n} = \sum_{\beta'} A^{\beta\beta'}_{N} \, \delta g^{\beta'}_{N,n} + D^{\beta}_{N,n} \ , \qquad \text{(IV.2)}$$

or

$$\delta g^{\beta}_{N,n} = \sum_{\beta'} (1 - A_N)^{-1}_{\beta\beta'} \, D^{\beta'}_{N,n} \ . \qquad \text{(IV.3)}$$

In line with the general philosophy discussed in Section II, we look for the eigenvalues of A_N. The component of $\delta g^{\beta}_{N,n}$ lying along the eigenvector associated with an eigenvalue near unity (if there is one) will be enhanced. By studying the enhanced eigenvector, we can find ratios among amplitudes without having to calculate the driving terms explicitly.

Now let us calculate the A matrix. Baryons are, as usual, regarded as bound states in the πB channel. The parity non-conserving $\bar{B}B\pi$ coupling δg will then appear as a factor in the residue of the baryon poles in the (parity non-conserving) partial wave amplitude $\delta T_{ij,kl}$ for the scattering process

$$J = \tfrac{1}{2}^{+}\ (\text{P wave})\ B^{i}\pi^{j} \rightarrow J = \tfrac{1}{2}^{-}\ (\text{S wave})\ B^{k}\pi^{l}\,.$$

Using the partial wave expansion given by Singh[38)], one finds that the direct channel baryon pole in δT has the form

$$\frac{-\sum_{x} g_{ixj}\,\delta g_{xkl}}{(W - M)}\,,$$

where W is the total c.m. energy and M the baryon mass.

Application of Eq. (III.41) for the coupling shift then gives

$$\delta g_{ijk} = -\frac{1}{2\pi i g^2}\sum_{xyzu}\int_L \frac{[X_{i,xy}(W')\,\delta T_{xy,zu}(W')\,Y_{zu,jk}(W')]}{W' - M}\,dW' \tag{IV.4}$$

with

$$g^2 = \sum_{jk} (g_{ijk})^2$$

$$Y_{zu,jk}(W') = [D_-(W')D_-^{-1}(M)]_{zu,jk} \,, \tag{IV.5}$$

$$X_{i,xy}(W') = \lim_{W\to M} \sum_{lm} g_{ilm}(W-M)\ [D_+(W')D_+^{-1}(W)]_{xy,lm} \,, \tag{IV.6}$$

where $D_{\pm}$ is the denominator matrix for πB scattering in the $J = \frac{1}{2}^{\pm}$ state and the contour L runs clockwise around the left-hand cuts in δT. Note that since g_{ijk} is SU(3) symmetric, g^2 must be independent of i.

The next step is the evaluation of the integral in Eq. (IV.4), which will give the A matrix. We shall suppose, on the basis of arguments presented in Section III, that the dispersion integrals are rapidly convergent and that it is therefore an adequate approximation to keep only the nearby singularities in δT.

Supposing that we need only the singularities of δT near $W = M$; we can set $D_-(W') \approx D_-(M)$ and make a linear approximation for $D_+(W')$ in this region to obtain

$$Y_{zu,jk} \approx \delta_{zj}\delta_{uk}, \quad X_{i,xy}(W') \approx (W'-M)g_{ixy} \,. \tag{IV.7}$$

Finally, we need the nearby singularity in δT which arises from baryon exchange. This can be written as

$$\frac{\sum_v \delta g_{xvu} g_{zvy}}{(W-M)}$$

in the static approximation. Inserting this expression for δT into Eq. (IV.4), and using Eq. (IV.7), one finds

$$A_{ijk,i'j'k'} \approx \delta_{kk'} g^{-2} \sum_{x} g_{ii'x} g_{jj'x} . \qquad \text{(IV.8)}$$

This equation for the A matrix contains the heart of the problem. There are two points in connection with its derivation that should be mentioned here.

i) To obtain Eq. (IV.8) it is actually necessary to know only Y(M) and $dx(W)/dW|_{W=M}$. These quantities are given exactly by Eq. (IV.7). Thus, in this particular problem there is no uncertainty present due to our lack of knowledge of the curvature of the D function.

ii) Secondly, recall that we have not included the effect on δg of the parity non-conserving coupling shifts of other strongly interacting particles. The reason for this is the following. In the region where static kinematics are valid, the s channel scattering angle $\cos\vartheta_s$ is equal to the u channel scattering angle $\cos\vartheta_u$. This implies that a P to S wave transition in the u channel crosses only to a P to S wave transition in the s channel. Thus, only u-channel processes which connect P to S waves can contribute to the nearby part of the left-hand cut in this problem. This means that B exchange will contribute but $3/2^+$ decuplet exchange, which appears as a P to D wave transition, will not.

The analysis of Eq. (IV.8) is most easily carried out in the N,n,β representation. We choose β so that for a given SU(3) symmetry violation N,n at a $\bar{B}B\pi$ vertex, β runs over the independent irreducible representations of $\bar{B}B$. Thus, for N = 1, β is $\underline{8}_s$ or $\underline{8}_a$ while for N = 8 β runs over all the $\bar{B}B$ states.

The evaluation of Eq. (IV.8) in this representation makes use of group theory techniques which have not been explained in these lectures. Instead of going into detail on this primarily mathematical problem, we shall just mention some of the more important results[6].

First, one finds that the matrix $A_N^{\beta\beta'}$ is essentially <u>diagonal</u> in β and is <u>independent</u> of N.

The A matrix has <u>two</u> eigenvalues near 1. One corresponds to the diagonal element A^{11}, but this only enhances the coupling $\delta g_{N,n}^{1}$ which cannot contribute to the strangeness changing decays. The only other eigenvalue near one comes from the 2×2 submatrix connecting $\underline{8}_s$ and $\underline{8}_a$. This eigenvalue e, varies from 0.67 to 0.82 as the ratio F/D ranges from ⅓ to ⅔. The associated eigenvector is $\delta g_{N,n}^{8_s} = R(\lambda)\delta g_{N,n}^{8_a}$ where $R(\lambda)$ is a known quantity[6] given in terms of λ. Since this particular eigenvalue remains close to unity, while the others remain far from unity, we expect that for a given N or n, the dominant terms in δg will be $\delta g_{N,n}^{8_s}$ and $\delta g_{N,n}^{8_a}$, with a ratio between these couplings given approximately by the factor $R(\lambda)$. These results, which are <u>independent</u> of any assumptions (including CP invariance) about the transformation properties of the weak interactions, already give three relations among the seven independent S-wave amplitudes describing non-leptonic hyperon decay. These experimental predictions[6] will be briefly surveyed in the next section.

At this point we want to continue with the theory, and note that we have so far shown that $g_{N,n}^{8}$ is enhanced, where N can be any of the SU(3) violations $\underline{1}, \underline{8}_a, \underline{8}_s, \underline{10}, \underline{10}^*$ or $\underline{27}$. The question of which N actually appears in the driving terms can only be answered in a more specific theory of the weak interactions.

To explore one possibility[6] let us consider the SU(3) charge-conjugation[39] properties of the driving terms $D_{N,n}^{8_a}$ and $D_{N,n}^{8_s}$, which drive the enhanced eigenvector. The $\bar{B}B$ octet states $\underline{8}_a$ and $\underline{8}_s$ with $J = 0$ have $\mathcal{C} = +1$. The π octet also has $\mathcal{C} = +1$. The coupling of π to the enhanced $\bar{B}B$ octet then has $\mathcal{C} = +1$ for the symmetric combinations $N = \underline{1}, \underline{8}_s$ and $\underline{27}$, and $\mathcal{C} = -1$ for the antisymmetric combination $\underline{8}_a$, while from $\underline{10}$ and $\underline{10}^*$ one can form two orthogonal states with $\mathcal{C} = +1$ and -1 respectively. Thus, in Cabibbo's theory[10] in which the parity non-conserving driving terms have N = $\underline{1}$, $\underline{8}$ and $\underline{27}$ and $\mathscr{C} = -1$, only the octet violation can drive the enhanced eigenvector, and the resulting hyperon decay amplitudes will obey the $|\Delta I| = \frac{1}{2}$ rule.

3. Survey of results on octet enhancement in the B and Δ supermultiplets[37]

The A matrix coefficients in Eq. (II.6) which couple together the B and Δ mass shifts and the BBπ and BΔπ coupling shifts have been calculated[3,6,7] using the model described in the first part of this section. We shall not present the rather complicated details of these calculations here, but rather survey the results stressing those which are of direct experimental interest.

The A matrix for this problem is not symmetric. In particular, a detailed analysis[7] of Eq. (II.6) shows that the elements of A^{Mg} are very small compared to the elements of the other three sub-matrices A^{MM}, A^{gM} and A^{gg}. As a result, eigenvalues of A^{MM} are to a very good approximation eigenvalues of the entire A matrix. This circumstance provides a justification for a study[3] of the mass splittings in the B and Δ supermultiplets based on an analysis of the submatrix A^{MM}.

Mass splittings in the B and Δ supermultiplets: The principal mathematical result of this calculation is that A^{MM} has no eigenvalue near 1 for a $\underline{27}$ violation of SU(3), while A^{MM} has an eigenvalue near 1 for octet violations. This allows one to predict a number of mass ratios. The results, which depend to a considerable extent on both the F/D ratio and the curvature of the D function, are[2,3]:

$$\delta M_8^{B_s} \Big/ \delta M_8^{B_a} \approx -0.24, \quad \delta M_8^{\Delta} \Big/ \delta M_8^{B_a} \approx 1.30 \qquad \text{(IV.9)}$$

for $\lambda = 0.46$ and a linear D function, while for $\lambda = 0.33$ and a strongly curved D function one finds:

$$\delta M_8^{B_s} \Big/ \delta M_8^{B_a} \approx -0.5, \quad \delta M_8^{\Delta} \Big/ \delta M_8^{B_a} \approx 0.8 \ . \tag{IV.10}$$

For intermediate values of λ and the curvature of D, the values of the mass ratios lie between these extremes. Recall that the octet enhancement pattern is universal, so that the electromagnetic mass shifts are characterized by the same ratios as the strong mass shifts. The experimental data on the strong mass splittings gives for these ratios[3])

$$\delta M_8^{B_s} \Big/ \delta M_8^{B_a} \approx -0.25, \quad \delta M_8^{\Delta} \Big/ \delta M_8^{B_a} \approx 1.25 \ , \tag{IV.11}$$

while from data on the electromagnetic mass splittings one obtains

$$\delta M_8^{B_s} \Big/ \delta M_8^{B_a} \approx -0.4 \pm 0.1 \ . \tag{IV.12}$$

There is reasonable agreement between theory and experiment here, and thus one would say that experiment supports the prediction that the strong and electromagnetic mass shifts lie along an eigenvector of A_8 whose eigenvalue is close to one.

The results on the electromagnetic mass shifts in the B and Δ supermultiplets that have been mentioned so far were obtained on the assumption that one can treat the electromagnetic violations of SU(3) to first order with the effects of strong symmetry breaking appearing as small second order corrections. However, arguments have been given[3,32,40]) which indicate that in this particular case the corrections to the first order calculation corresponding to strong symmetry breaking effects may be rather large. It is important, therefore, to see how the pattern of octet enhancement in the electromagnetic mass splittings resulting from a first order calculation is modified by strong symmetry breaking.

For this purpose one may start from an SU(2) symmetric reciprocal bootstrap for the N-N* system[34], the Σ-Y_1^* system[41,42] and the Ξ-Ξ^* system. Calculation of the relevant A matrix elements, using values for the masses and coupling constants and keeping those channels which appear to be reasonable from the point of view of broken SU(3), yields the following results. In the N-N* system[3,33,43] and the Ξ-Ξ^* system[32,40], the electromagnetic mass shifts transforming like $\Delta I = 1$ are not enhanced over the $\Delta I = 2$ mass shifts. In these cases then, the octet enhancement results are largely washed out by strong symmetry breaking, and an SU(2) symmetric calculation, together with a careful study of the driving terms[32,33,44], is required to give an accurate description of the pattern as well as the magnitude of the mass shifts. In the Σ-Y_1^* system on the other hand, a considerable enhancement of the mass shifts transforming like $\Delta I = 1$ over those transforming like $\Delta I = 2$ is found[32,40], so in this case the result expected on the basis of octet enhancement is left intact. Generally speaking, the results of these calculations improve the agreement between experiment and the theoretical estimates of the electromagnetic mass splittings, and they provide further understanding of the stability of the octet enhancement mechanism under the influence of small perturbations.

Now let us turn to the coupling shifts, the pattern of which is determined by the submatrices A^{gg} and A^{gM}. We shall discuss A^{gg} first. Use of the model and input parameters discussed in Section IV.1 yields the following results[7].

i) The eigenvalue of A^{gg} closest to one is ≈ 0.93 and corresponds to an octet violation of SU(3). This number is quite insensitive to variation of the F/D ratio of the SU(3) symmetric BBπ coupling.

ii) There are a number of other eigenvalues fairly close to one. For octet violations of SU(3), the eigenvalues next closest to one are ≈ 0.66 and 0.49. Furthermore, there is an eigenvalue close to one for 27 violations of SU(3); it is ≈ 0.88.

The effect of these eigenvalues on the coupling shifts is related to the behaviour of A^{gM}. To understand this, we consider the 2×2 schematic problem for octet violations of SU(3);

$$\begin{pmatrix} \delta M/M \\ \delta g \end{pmatrix} = \begin{pmatrix} A^{MM} & 0 \\ A^{gM} & A^{gg} \end{pmatrix} \begin{pmatrix} \delta M/M \\ \delta g \end{pmatrix} + \begin{pmatrix} D^{M} \\ D^{g} \end{pmatrix} . \tag{IV.13}$$

The solution of this problem is

$$\delta M/M = D^{M}/(1-A^{MM}); \quad \delta g = [A^{gM}/(1-A^{gg})](\delta M/M) + D^{g}/(1-A^{gg}) . \tag{IV.14}$$

It has been found[7] that the element of A^{gM} connecting the leadi octet mass eigenvector to the leading coupling shift eigenvector (corresponding to an eigenvalue 0.93) is large (≈ 1.0). Since both A^{MM} and A^{gg} have eigenvalues near one, one sees from the above equations that the coupling shifts lying along the leading octet eigenvector A^{gg} will receive a <u>double enhancement</u>, leading to large coupling shifts. The same methods applied to the other eigenvalues of A^{gg} near one indicate that they will not benefit from double enhancement. In the case of the octet eigenvectors, the corresponding component of $|A^{gM}|$ is down from ≈ 1.0 to ≈ 0.2, while in the <u>27</u> violation case there is no mass shift to feed the coupling shift. So the double enhancement appears to favour the leading octet eigenvector uniquely, and we will study the effect of this eigenvector on the pattern of coupling shifts.

<u>Strong coupling shifts</u>: Our predictions for the medium strong coupling shifts have the form $g + x\delta\hat{g}$, where g is the SU(3) symmetric coupling, $\delta\hat{g}$ is an arbitrarily normalized octet eigenvector of A^{gg} corresponding to the eigenvalue 0.93 and x is an over-all strength parameter.

The parameter x can be fixed in either of two ways.

i) One can fit to the ratio of any two $\frac{3}{2}^+$ resonance decay widths. Of these, the decay $Y_1^* \to \Sigma\pi$ is too poorly known experimentally. The N^* decay width is well known experimentally, but the static model we are using does not reproduce its shape well and thus does not give an accurate estimate for its width. So we choose the ratio $\Gamma(Y_1^* \to \Lambda\pi)/\Gamma(\Xi^* \to \Xi\pi)$.

ii) Alternatively, we can use Eq. (IV.14) to estimate the magnitude of the coupling shifts. We suppose that D^g is small compared to the enhanced δM's, for which quantities we take experimental values. For $(1 - A^{gg})^{-1}$ and A^{gM} we use our calculated values, which gives a value for x in rough agreement with the one obtained above.

Neither estimate is particularly reliable, but estimate ii) is especially uncertain because the quantity $(1 - A^{gg})^{-1}$ is so sensitive to the precise magnitude of the eigenvalue of A^{gg} which is near one. For this reason we have used the estimate for x based on the ratio $\Gamma(Y_1^* \to \Lambda\pi)/\Gamma(\Xi^* \to \Xi\pi)$ in predicting the other coupling shifts.

The results for the resonance decays are given in Table 1, where they are compared with the SU(3) symmetric results and with experiment[45].

Our principal results on the medium strong $BB\pi$ and $B\Delta\pi$ coupling shifts are summarized in Table 2.

Table 1

$B\Delta\pi$ couplings. The experimental g^2 were calculated from the widths Γ, compiled in Ref. 45, and the relation[7]

$$g^2 = \Gamma(\Delta \to B + \pi)\bar{M}_\Delta^2/2q^3 .$$

$B\Delta\pi$ coupling	Experimental g^2	SU(3)-symmetric g^2	$(g + x\delta\hat{g})^2$
$N^* N\pi$	10.7 ± 0.5	12.2	15.9
$Y_1^* \Lambda\pi$	6.1 ± 0.3	6.1 (input)	6.1 (input)
$Y_1^* \Sigma\pi$	1.7 ± 1.7	4.1	1.7
$\Xi^* \Xi\pi$	2.6 ± 0.6	6.1	2.6 (input)

Table 2

$BB\pi$ and $B\Delta\pi$ couplings in broken SU(3). The entries are defined as follows. We write the $\Delta^{\alpha}B_j\pi_k$ coupling as g^{α}_{jk} (α = 1, ..., 10; j,k = 1, ..., 8). The total coupling strength for a given α is defined as $\sum_{jk}(g^{\alpha}_{jk})^2$, while the percentage of the jk state is defined as $(g^{\alpha}_{jk})^2/\sum_{lm}(g^{\alpha}_{lm})^2$. The total couplings and percentages for the $BB\pi$ couplings are defined in an analogous manner. The total couplings are arbitrarily normalized to unity for the (3-3) resonance and the nucleon. The numbers in this table correspond to an F/D ratio of 0.4.

Particle	Component	Threshold (MeV)	SU(3) %	Broken SU(3) %	Normalized total strength
N*(1240)	$N\pi$	1080	50.0	90.9	1.00
	ΣK	1690	50.0	9.1	
$Y_1^*(1380)$	$\Lambda\pi$	1250	25.0	46.9	0.70
	$\Sigma\pi$	1330	16.7	13.1	
	$N\bar{K}$	1430	16.7	30.3	
	$\Sigma\eta$	1740	25.0	5.3	
	ΞK	1820	16.7	4.4	
$\Xi^*(1530)$	$\Xi\pi$	1460	25.0	30.0	0.47
	$\Lambda\bar{K}$	1610	25.0	46.7	
	$\Sigma\bar{K}$	1690	25.0	15.1	
	$\Xi\eta$	1870	25.0	8.2	
$\Omega^-(1680)$	$\Xi\bar{K}$	1820	100.0	100.0	0.28
N(940)	$N\pi$	1080	68.4	85.7	1.00
	$N\eta$	1490	0.1	0.0	
	ΛK	1610	18.7	9.1	
	ΣK	1690	12.8	5.2	

Table 2 (cont.)

Particle	Component	Threshold (MeV)	SU(3) %	Broken SU(3) %	Normalized total strength
Λ(1115)	Σπ	1330	46.8	60.9	0.57
	N$\bar{K}$	1430	37.3	31.9	
	Λη	1660	15.6	6.8	
	ΞK	1810	0.3	0.4	
Σ(1190)	Λπ	1250	15.6	37.9	0.30
	Σπ	1330	14.7	24.9	
	N$\bar{K}$	1430	8.5	11.1	
	Ση	1740	15.6	3.9	
	ΞK	1810	45.6	22.2	
Ξ(1320)	Ξπ	1460	12.8	30.3	0.17
	Λ$\bar{K}$	1610	0.1	0.7	
	Σ$\bar{K}$	1680	68.4	60.4	
	Ξη	1870	18.7	8.6	

These results support the following main conclusions[7]:

i) the "medium strong" coupling shifts are very large, typically ≈ 100% of the SU(3) symmetric values of the couplings. This fact indicates that it is not consistent in an approximate dynamical calculation to include mass shifts while omitting coupling shifts.

ii) Although not apparent from the table, it is found that the BBπ and BΔπ couplings, while shifted by ≈ 100% in magnitude from their SU(3) symmetric values, do not change sign.

iii) A very interesting qualitative rule is apparent. The breaking of SU(3) raises the coupling strengths for low lying channels, and decreases the coupling strengths to high-mass channels. This result tends to support the neglect of high-mass channels in approximate dynamical calculations, even though they might appear to enter in an important way from SU(3) symmetry consideration[46])

iv) The strange-particle components of the nucleon-wave function are reduced, with various experimental consequences.

a) In high-energy nucleon-nucleon collisions[47]), from about 10 BeV to 10^4 BeV, strange-particle production accounts for only $\approx$ 15% of the secondaries. It is clear from Table 2 that in SU(3) strange particles should amount to $\approx$ 30% of the secondaries produced off nucleons, while our broken SU(3) model would predice $\approx$ 15%.

b) From a comparison of the experimental cross-section $d\sigma(\gamma + p \to K^+ + \Lambda^0)/d\Omega$ near threshold with a simple pole model for this cross-section, it appears that the squares of the physical couplings $g^2_{N\Lambda K}$ and $g^2_{N\Sigma K}$ are smaller than their SU(3) values by an order of magnitude[48]). This situation is improved if one includes the symmetry-breaking, which materially reduces the values of the squares of the coupling constants.

v) Note that the η has a very small coupling to all members of the baryon octet and decuplet, and will consequently be very hard to produce.

Electromagnetic coupling shifts: The electromagnetic coupling shifts may also be deduced from the results of this calculation. Since the experimental information on these is so meager, we will not discuss them here.

Non-leptonic weak interactions:

A. Parity conserving coupling shifts

Here, we use the same A matrix as before, but one must observe that to first order the strangeness changing weak interaction produces no physical mass shifts. For this reason A^{MM}, A^{gM} and A^{Mg} drop out of the A matrix, leaving only A^{gg}. A consequence of this is that the leading octet eigenvector no longer receives a double enhancement, and one can expect that the relative influence of the second and third octet eigenvectors and, perhaps, of the 27 eigenvector as well, will be correspondingly greater.

In particular, since we now have an enhanced 27 eigenvector, we do not have a clean prediction of octet dominance in the partiy conserving weak decays. If the 27 eigenvector turns out to be important, it will, of course, lead to violations of the $|\Delta I| = 1/2$ rule for parity-conserving decays. While we cannot make any quantitative predictions concerning the importance of the 27 eigenvector, there are some arguments which suggest that its effect will be somewhat diminished for the observable parity-conserving coupling shifts. These are:

i) The factor $(1 - A^{gg})^{-1}$ favours octet over 27 violations somewhat.

ii) Detailed analysis shows[7)] that the enhanced 27 eigenvector contributes rather less to the observed hyperon decays than the octet eigenvector. That is, the 27 eigenvector happens to have its largest effect on couplings like g_{NNK}, which are not, at present, observable.

Taking these factors together one finds that the $\Delta I = 1/2$ amplitude may be several lines greater than the $\Delta I = 3/2$ amplitude. Thus, an approximate $|\Delta I| = 1/2$ rule can emerge.

B. Parity-violating coupling shifts

This problem, with its various simplifying features, was treated in Ref. 6. Assuming CP = +1, a rather clean dominance of a single octet eigenvector was found, with experimental consequences for the S-wave amplitudes summarized in Table 3. As one can see, the agreement with experiment is quite good. It is to be emphasized that the uncertainties which prevented us from obtaining a clean prediction of octet dominance in the parity-conserving case do not arise in the parity-violating decays.

Table 3

Comparison of theory[6)] and a particular fit to experiment[45,49)] for non-leptonic, parity-violating hyperon decay amplitudes A, defined as in Stevenson et al.[49)].

Decay	Sign of theoretical amplitude	Theoretical $\|A\| \times 10^{-4}$ sec^{-1} $m_{\pi^-}^{-1}$	Experimental $\|A\| \times 10^{-4}$ sec^{-1} $m_{\pi^-}^{-1}$
Λ_0^0	- (input)	2.1 (input)	2.1 ± 0.2
Λ_-^0	+	3.0	3.1 ± 0.1
Σ_+^+	+	0.0	0.1 ± 0.3
Σ_0^+	+	3.5	3.6 ± 0.35 ($\gamma_\Sigma > 0$)
			1.7 ± 0.2 ($\gamma_\Sigma < 0$)
Σ_-^-	+	4.9	3.9 ± 0.15
Ξ_0^0	+	3.1	3.1 ± 0.2
Ξ_-^-	+	4.4	4.1 ± 0.2

This would lead us to suggest that octet dominance and consequently the $|\Delta I| = \frac{1}{2}$ rule may hold to a higher degree of accuracy in the parity violating decays than in the parity conserving ones; <u>a point which should be checked experimentally.</u>

On the basis of the above arguments let us suppose that the enhanced <u>27</u> eigenvector does not make a major contribution to the parity-conserving amplitudes. There remain the three enhanced octet eigenvectors. Keeping these three eigenvectors and the single enhanced octet eigenvector for the parity-violating amplitudes leads to the following results[7]:

i) The squares of the effective couplings which enter into non-leptonic decays are enhanced compared to those entering into leptonic decays by a factor $(1 - A^{gg})^{-2}$, which is of the order of 10 to 100 for the leading octet eigenvalue. This appears to be the case experimentally

ii) As mentioned above, our predictions[6] for the S-wave amplitudes are in good agreement with experiment. In particular, the Lee sum rule[50] for these amplitudes is obtained.

iii) The Σ^+_+ decay is predicted to be pure P wave, while the Σ^-_- decay is mostly S wave.

iv) The detailed predictions which result upon keeping <u>only</u> the leading octet eigenvector for the parity-conserving decays are for the most part in disagreement with experiment. Adding arbitrary amounts of the next two octet eigenvectors, one is left with the sum rule

$$\Sigma^+_+ = 1.8\ \Xi^-_- - 0.8\ \Sigma^0_+ - 0.4\ \Lambda^0_- \quad , \qquad \text{(IV.15)}$$

which still disagrees with experiment. We are investigating the possibility that enhanced eigenvectors, transforming either like <u>27</u> or with peculiar CP properties, may play an important role[7].

REFERENCES

1) R. Cutkosky and P. Tarjanne, Phys.Rev. 132, 1355 (1963).

2) R. Dashen and S. Frautschi, Phys.Rev.Letters 13, 497 (1964).

3) R. Dashen and S. Frautschi, Phys.Rev. 137, B 1331 (1965).

4) R. Dashen and S. Frautschi, Phys.Rev. 135, B 1190 (1964).

5) R. Dashen and S. Frautschi, Phys.Rev. 137, B 1318 (1965).

6) R. Dashen, S. Frautschi and D. Sharp, Phys.Rev. Letters 13, 777 (1964).

7) R. Dashen, Y. Dothan, S. Frautschi and D. Sharp (to be published).

8) S. Coleman and S.L. Glashow, Phys.Rev. 134, B671 (1964).

9) R.A. Burnstein, T.B. Day, B. Kehoe, B. Sechi-Zorn and G.A. Snow, Phys.Rev. Letters 13, 66 (1964).

10) N. Cabibbo, Phys.Rev. Letters 12, 62 (1964).

11) M. Gell-Mann, Phys.Rev. 125, 1067 (1962).

12) S. Okubo, Progr.Theoret.Phys.(Kyoto) 27, 949 (1962).

13) The discussion at this point follows S. Coleman, "Departures from the eightfold way" in the Proceedings of the Third Eastern Conference on Theoretical Physics, University of Maryland, 1964.

14) S. Coleman and S.L. Glashow, Phys.Rev. Letters 6, 423 (1961).

15) The succeeding discussion is based on Ref. 2.

16) S.L. Glashow, Phys.Rev. 130, 2132 (1963).

17) A detailed discussion of the use of these techniques in the present context is given in Ref. 5.

18) To lowest order, a self-consistent mass or coupling shift having definite transformation properties will affect only mass or coupling shifts having the same transformation properties.

19) Y. Ne'eman, Phys.Rev. 134, B 1355 (1964).

20) R. Dashen and S. Frautschi, Phys.Rev. 140, B698 (1965).

21) J.J. Sakurai, Phys.Rev. 132, 434 (1963).

22) L. Picasso, L. Radicati, J. Sakurai and D. Zanello, Nuovo Cimento (to be published).

23) The discussion in Section III is based on Ref. 24.

24) D.H. Sharp, "S-matrix perturbation theory" in the Proceedings of the Seminar on High-Energy Physics and Elementary Particles, International Centre for Theoretical Physics, Trieste, 1965 (to be published).

25) R. Omnes, Nuovo Cimento 8, 316 (1958).

26) N. Levinson, Kgl.Danske Videnskab.Selskab, Mat.-fys.Medd. 25, No. 9 (1949).

27) The particular derivation given below follows C.E. Jones and D.H. Sharp (unpublished). For an alternative derivation see Refs. 4 and 5.

28) This point will be discussed in detail at the end of this section.

29) H.D. Abarbanel, C.G. Callan and D.H. Sharp (to be published).

30) A. Martin, Supplemento Nuovo Cimento 21, 157 (1961).

31) J.M. Cornwall and J.B. Hartle (unpublished).

32) F.J. Gilman, Ph.D. Thesis submitted to Princeton University, June, 1965 (unpublished).

33) R.F. Dashen, Phys.Rev. 135, B 1196 (1964).

34) G.Chew, Phys.Rev. Letters 7, 394 (1961).

35) If T includes all channels, the inelasticity term in Eq. (III.34) is absent.

36) R. Cutkosky and P. Tarjanne, Phys.Rev. 133, B 1292 (1964).

37) The discussion in Section IV is taken, in large measure, directly from Refs. 6 and 7. The author would like to express his appreciation to Drs. Dashen, Dothan and Frautschi for giving their kind permission to make use of this material.

38) V. Singh, Phys.Rev. 129, 1889 (1963).

39) M. Gell-Mann, Phys.Rev. Letters 12, 155 (1964).

40) F.J. Gilman (to be published).

41) B. Kayser, Phys.Rev. 138, B 1244 (1965).

42) B. Kayser and E. Bloom, Bull.Am.Phys.Soc. 10, 18 (1965).

43) S. Biswas, S. Bose and L. Pande, Phys.Rev. 138, B 163 (1965).

44) F.J. Gilman (to be published).

45) A. Rosenfeld, A. Barbaro-Galtieri, W. Barkas, P. Bastien, J. King and M. Roos, Rev.Mod.Phys. 36, 977 (1964).

46) In this connection we note that the pattern of coupling shifts is determined by the eigenvalue of A^{gg} near one. The structure of A^{gg} itself is determined by the exchange forces, and does not contain any direct information about thresholds. Therefore, it is rather remarkable that the coupling shifts, nevertheless, follow the pattern of the thresholds.

47) D. Perkins, Proceedings of the International Conference on Theoretical Aspects of Very High-Energy Phenomena, CERN, 1961, p. 99.

48) C. Peck, private communication.

49) M. Stevenson, J. Berge, J. Hubbard, G. Kalbfleisch, J. Shafer, F. Solmitz, S. Wojcicki and P. Wholmut, Physics Letters 9, 349 (1964);

R. Dalitz, Varenna lectures (to be published).

50) B.W. Lee, Phys.Rev. Letters 12, 83 (1964).

BROKEN SYMMETRIES AND SUM RULES

N. Cabibbo,
CERN, Geneva.

INTRODUCTION

In the last few years there has been an increasing interest in the attribution of approximate symmetries to the world of strongly interacting particles, with respect to both their strong interactions and their weak and electromagnetic ones. The oldest example is that of the SU(2) invariance implied by I-spin conservation. This symmetry is not exact, due to the electromagnetic interactions which break it. Such breaking, however, is, in general, "under control" owing to the smallness of the fine structure constant, and in most cases we can put safe limits on the possible deviations from the exact symmetry.

A different situation occurs in the (broken) SU(3) symmetry, where the origin of the breaking is not yet properly identified, and the breaking itself can be large enough to play an essential role. In fact, predictions based on the SU(3) symmetry seem to be in fair agreement sometimes, while at other times in wild disagreement (especially in processes involving four or more particles) with experimental results.

Still a different situation occurs in dealing with SU(6) and the other proposed symmetries which connect space-time and internal variables. In fact, it has been conclusively shown (McGlinn, O'Raifeartaigh, Coleman) that such symmetries cannot be exact without conflicting with our most fundamental ideal on the structure of the

---

Notes for first three lectures taken by G. Cicogna and E. Remiddi.
Notes for fourth and fifth lectures taken by G. Altarelli.

physical world (four space-time dimensions, relativistic invariance etc.), so that their "exact symmetry" limit is not physically admissible. In his lectures Professor Pais proposed to call them "dynamical" in order to emphasize the difference with respect to the "kinematical" symmetries, for example, SU(2) and, perhaps, SU(3), whose "exact symmetry" limit does not lead, in principle, to logical difficulties.

Apart from the peculiar troubles connected with SU(6)-like symmetries, we have to face the difficult problem of properly defining a broken symmetry. This problem becomes more and more serious in passing from SU(3) to SU(6), which we cannot consider "nearly exact" in any tenable and general sense.

The point of view which I will discuss in these lectures is not entirely new, as it has already been used in atomic and molecular physics. Its possible application to elementary particles theory has been proposed by Gell-Mann in 1962 and by many other authors during the last year. As we will discuss in the following lectures, a broken symmetry is thought to be a manifestation of an algebraic structure of operators which do not (or do not all) commute with the Hamiltonian. Let me begin with a schematic discussion of exact symmetries.

I. EXACT INTERNAL SYMMETRY

Let us consider a set of Hermitian operators $\{F\}$, acting on the space of the physical states such that

a) $\{F\}$ is a linear space, i.e., if $F_1, F_2 \in \{F\}$ then $(a\,F_1 + b\,F_2) \in \{F\}$ (a,b real);

b) $\{F\}$ is closed under commutation, i.e. if $F_1, F_2 \in \{F\}$, then $i[F_1, F_2] \equiv i(F_1 F_2 - F_2 F_1)$ is Hermitian and belongs to $\{F\}$.

We will limit ourselves to the case of finite dimensional $\{F\}$ spaces: $\{F\}$ is then a Lie algebra. We can pick up a basis $F^i (i = 1, \ldots,$ N being the dimension of $\{F\}$), whose components satisfy

$$[F^i, F^j] = i\, c^{ijk}\, F^k \,.$$

The c^{ijk} are the structure constants of the algebra. They are real as the F^i are assumed to be Hermitian.

The operators F^i are the generators of a group G, which also act on the space of the physical states and whose infinitesimal elements are

$$U(\epsilon) = 1 + i \sum_{j=1}^{N} \epsilon^j F^j \,,$$

where $\epsilon \equiv (\epsilon^1, \ldots, \epsilon^N)$ is a set of N real infinitesimal parameters. As the F^i are Hermitian, the $U(\epsilon)$ (and therefore all the elements of the group G) are unitary

$$U^+(\epsilon)\; U(\epsilon) = 1 \,.$$

Let H be the Hamiltonian of a given physical system. If, for any $F \in \{F\}$,

$$[F, H] = 0$$

then also

$$[U, H] = 0$$

for any $U \in G$, and we say that the group G is an "exact internal symmetry" for our physical system.

Given a state $|A>$ such that

$$H|A> = E|A>$$

then $F|A>$ or, alternatively, $U|A>$ also satisfy

$$H(F|A>) = E(F|A>) ,$$

$$H(U|A>) = E(U|A>) .$$

By letting U run over the whole G, the states $\{U|A>\}$ belong to the same eigenvalue of H and provide the basis of a representation of the group. The group must be compact in order to have finite dimensional unitary representations, i.e. only a finite number of particles with the same mass.

II. BROKEN SYMMETRIES

As discussed in the previous section, the ingredients of an exact internal symmetry are a set of operators $\{F\}$ which:

a) form a Lie algebra, i.e. satisfy certain commutation rules;
b) commute with the Hamiltonian.

What is then a broken symmetry?

In these lectures we will discuss the possibility that a broken symmetry is a manifestation of an (exact) algebraic structure of certain operators, not all of which commute with the Hamiltonian. In other words we will keep a) and give up only b).

Two problems arise:

i) How can the existence of an underlying algebraic structure be tested?
ii) How can an exact algebraic structure appear as an approximate symmetry?

In order to show that this is possible, at least in principle, and how things can work in practice, we will illustrate the above points i) and ii) with examples taken from classical quantum mechanics.

For point i) take a system of a spinless particle of mass m in a potential V, not necessarily central (it could be an electron, disregarding the spin, in an atom). A complete set of operators is given by the position $\vec{q} \equiv (q_i)$ (i = 1,2,3) and the momentum $\vec{p} \equiv (p_i)$ (i = 1,2,3) which satisfy

$$[p_i, q_j] = -i\delta_{ij} \mathbb{1} , \tag{II.1}$$

where $\mathbb{1}$ is the identity operator*).

The Hamiltonian H of the system can be written in terms of p_i, q_j as they form a complete set and, in fact, it reads

$$H = \frac{1}{2m} \vec{p}^2 + V(\vec{q}) . \tag{II.2}$$

As is well known, we have

$$-i[H, q_i] \equiv \dot{q}_i = \frac{p_i}{m}$$

$$-i[H, p_i] \equiv \dot{p}_i = -\frac{\partial V(\vec{q})}{\partial q_i} . \tag{II.3}$$

Neither p_i nor q_i commute with H, so that the algebra with basis $q_i, p_i, \mathbb{1}$ cannot generate an exact symmetry.

---

*) It is well known that due to the appearance of a constant operator in r.h.s. of Eq. (II.1), no finite dimensional representation of the algebra $q_i, p_j, \mathbb{1}$ exists. The standard argument for proving this point is to assume that one has a finite dimensional matrix representation of this algebra: by taking the trace of both sides of Eq. (II.1) a contradiction arises.

Let us consider an electromagnetic transition from a state A_i to a state A_j of our system

$$A_i \to A_j + \gamma \ . \tag{II.4}$$

Assume the dipole approximation for this process. Theoretically this is the case if the γ wavelength is large, compared to the size of the system. Experimentally both wavelength and size can be measured, and the validity of the hypothesis can be checked. In fact, it turns out to hold for most of the e.m. transitions in atoms (for instance, in the optical region.) The interaction Hamiltonian H_{em} with the radiative field is then

$$H_{em} = e\,\vec{E}\cdot\vec{D} \ , \tag{II.5}$$

$\vec{E}$ being the electric field and $\vec{D}$ the dipole moment of the system:

$$\vec{D} = \vec{q} \ . \tag{II.6}$$

The transition amplitude T for (II.4) is

$$T = e\,\vec{E}\cdot \langle A_j|\vec{D}|A_i\rangle \ . \tag{II.7}$$

By using Eqs. (II.6) - (II.3), we find

$$[D_i, \dot{D}_j] = \frac{i}{m}\,\delta_{ij}\,\mathbf{1} \ . \tag{II.8}$$

Let us now evaluate the mean value $\mathcal{M}$ on a state $|A\rangle$ of both sides of Eq. (II.8). The r.h.s. gives

$$\mathcal{M} = \frac{i}{m}\,\delta_{ij} \ . \tag{II.9}$$

From the l.h.s. we get

$$\mathcal{M} = \sum_{|B>} \left[<A|D_i|B><B|\dot{D}_j|A> - <A|\dot{D}_j|B><B|D_i|A> \right] ,$$

where $|B>$ runs over a complete set of states.

As $\dot{D}_i = -i[H, D_i]$, substituting we have

$$\mathcal{M} = 2i \sum_{|B>} (E_A - E_B) <A|D_i|B><B|D_j|A> .$$

By comparison with Eq. (II.9)

$$\sum_{|B>} (E_A - E_B) <A|D_i|B><B|D_j|A> = \frac{1}{2m} \delta_{ij} . \qquad \text{(II.10)}$$

If more electrons are involved a sum appears over the electrons, and the r.h.s. of Eq. (II.10) is simply multiplied by the number **Z** of the involved electrons.

Equation (II.10) is the Thomas-Reiche-Kuhn sum rule for electric dipole transition amplitudes [see Eq. (II.7)].

Let us stress that it has been derived by simply assuming an underlying algebraic structure [commutation relations (II.3)], which does not give rise to any symmetry, and dipole approximation without any explicit knowledge of the dynamical structure of the system.

If the dipole approximation is considered to hold, the sum rule provides a test for the commutation relation (II.1). Such a test would not be deemed important in the case of atomic physics, where we have many good reasons for believing the choice of basic variables p and q, and their commutation relations, but the situation is different in particle physics where the "basic variables" are a priori unknown. The Thomas-Reiche-Kuhn sum rule has found useful applications in nuclear physics, as a test of the electric dipole character of the giant resonance in γ-nucleus interactions.

Let us now illustrate point ii) with another classical example. The orbital angular momentum $\vec{L} \equiv (L_i)$ $(i = 1,2,3)$ is defined as

$$\vec{L} = \vec{q} \times \vec{p} .$$

From this definition, and Eq. (II.3), one gets the commutation rules

$$[L_i, L_j] = i\, \epsilon_{ijk}\, L_k . \tag{II.11}$$

Their validity is based upon Eqs. (II.3) only, without any requirement on the Hamiltonian.

Assume that the Hamiltonian can be decomposed in a part H_0 which commutes with $\vec{L}$, plus a part $\eta H'$ which does not, where η is a small parameter, and H' is such that $\eta H'$ can be considered a perturbation with regard to H_0 (no singularities, etc.).

For the sake of simplicity, we will also assume that to any eigenvalue E_i of H_0 corresponds only one eigenvalue $\ell_i(\ell_i + 1)$ of $\vec{L}^2$ (no accidental degeneration), so that the unperturbed states of the system can be written as $|i,r\rangle$, the r index running from 1 to $(2\ell_i + 1)$.

As $[\vec{L}, H_0] = 0$, $\vec{L}$ has no matrix elements between states with different i (box structure):

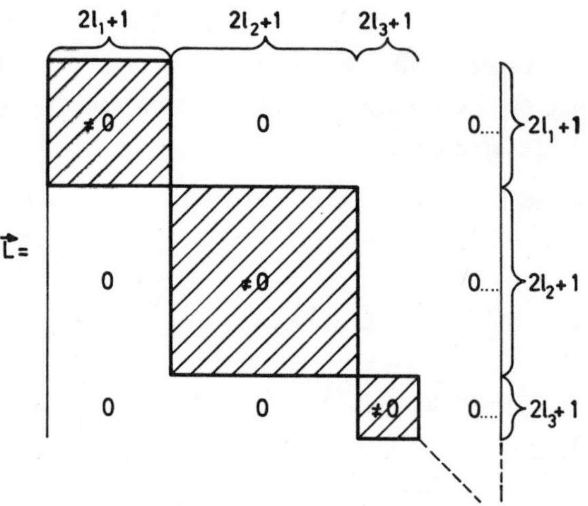

We now introduce the breaking $\eta H'$.

The states $|i,r>$ should be choosen in such a way as to diagonalize H′ in each box. The ith level will, in general, split into $(2\ell_i + 1)$ levels with energies $E_{i,r}$ such that

$$E_{i,r} - E_i = \eta < i,r|H'|i,r > ,$$

and the unperturbed states will get a mixture of other states according to:

$$|i,r> \to |i,r>' = N_{ir} \left(|i,r> + \eta \sum_{j \neq i} \frac{< js|H'|ir >}{E_i - E_j} |js> \right), \qquad \text{(II.12)}$$

N_{ir} being a normalization factor. It is easy to see that

$$N_{ir} = 1 + 0(\eta^2) \ . \qquad \text{(II.13)}$$

It is interesting to evaluate the matrix elements of $\vec{L}$ between the primed states as defined by (II.12). We will indicate them by $(<js|\vec{L}|ir>)'$.

For i = j it is easy to see from Eqs. (II.12) and (II.13) that

$$(<is|\vec{L}|ir>)' = <is|\vec{L}|ir> + O(\eta^2) \ . \tag{II.14}$$

For $i \neq j$

$$(<js|\vec{L}|ir>)' = \eta \left\{ \sum_{\substack{h\neq i \\ u}} \frac{<hu|H'|ir>}{E_i - E_h} <js|\vec{L}|hu> + \sum_{\substack{k\neq j \\ v}} \frac{<js|H'|kv>}{} <kv|\vec{L}|ir> \right\} + O(\eta^2) \ . \tag{II.15}$$

The appearance of such "off box" terms, vanishing in the unperturbed case, is called "leakage".

We emphasize that from Eq. (II.14)-(II.15), the following typical pattern arises

- deviations of the first order in η of the "off box" matrix elements with respect to the unperturbed case;
- deviations (of the second order only) for the matrix elements within the boxes.

In the above example we assumed $H = H_0 + \eta H'$. Of course, a similar decomposition is not always possible, but we can think that the "typical pattern" could, in principle, appear in more general cases.

There are many interesting cases in which a component of $\vec{L}$, say L_3, is conserved (take, for example, $H' \sim \vec{L} \cdot \vec{H}$ if $\vec{H}$, the magnetic field, lies in the direction of the third axis.)

If L_3 is conserved, relabelling the states with its eigenvalues m, Eq. (II.12) reads

$$|i,m\rangle' = N_{im}\left(|im\rangle + \eta \sum_{j\neq i} \frac{\langle jm|H'|im\rangle}{E_i - E_j}|jm\rangle\right) , \qquad (II.16)$$

as $\langle jn|H'|im\rangle = 0$ for $m \neq n$. L_3 keeps its diagonal structure (as in the unperturbed case): looking at Eq. (II.16) we see, in fact, that $L_3|im\rangle' = m|im\rangle'$. No off diagonal matrix elements of L_3 appear.

A similar situation also occurs in elementary particle physics: in dealing with I spin the third component I_3 is conserved even in the presence of electromagnetic interactions; in SU(3), I_3 and Y are conserved by medium strong and electromagnetic interactions, even if these interactions are not SU(3) invariant.

In order to motivate the hope of finding operators having a simple and exact algebraic structure, even if they are not conserved, let us consider I spin in an elementary field theoretical model.

Assume we have two basic particles with the same quantum numbers as those of the proton and neutron, interacting with the electromagnetic field A_μ. These fundamental particles will be described by the field operators $\psi^a_\rho(x)$, where a = 1,2 is an index referring to the internal degree of freedom, ρ = 1, ...4 is the standard spinorial index.

Take a Lagrangian

$$\mathcal{L} = \mathcal{L}_0 + \frac{ie}{2}\,\bar{\psi}(1+\tau_3)\gamma_\mu\psi A_\mu \qquad (II.17)$$

with

$$\mathcal{L}_0 = \bar{\psi}(\gamma\cdot\partial + m)\psi + 1/4\; F_{\mu\nu}F_{\mu\nu} . \qquad (II.18)$$

In quantum mechanics the momentum p is defined as $p = \partial\mathcal{L}/\partial\dot{q}$, and then the canonical commutation relations $[q, p] = i$ are imposed. On the same footing in field theory, the "momentum" conjugate to $\psi_\rho^a(x)$ is $\partial\mathcal{L}/\partial\dot{\psi}_\rho^a(x)$. Such momentum turns out to be $i\psi_\rho^{+a}(x)$ with both the $\mathcal{L}_0$ and the $\mathcal{L}$ Lagrangians. Only in the case of Fermi fields do we get anticommutations instead of commutation rules.

In general, we see that this result still holds regardless of the explicit forms of the interaction, provided only that it does not contain derivative couplings.

The canonical equal time anticommutation relations between fields read:

$$\begin{aligned} &\left\{\psi_\sigma^{b+}(\vec{x}, 0), \psi_\rho^a(\vec{y}, 0)\right\} = \delta_{ab}\,\delta_{\rho\sigma}\,\delta(\vec{x}-\vec{y}) \\ &\left\{\psi_\sigma^b(\vec{x}, 0), \psi_\rho^a(\vec{y}, 0)\right\} = \left\{\psi_\sigma^{+b}(\vec{x}, 0), \psi_\rho^{+a}(\vec{x}, 0)\right\} = 0 \,. \end{aligned} \qquad \text{(II.19)}$$

We define

$$j_\mu^i(x) = \bar{\psi}(x)\,\tau^i\gamma_\mu\,\psi(x) \qquad i = 1,2,3 \qquad \text{(II.20)}$$

and three operators I^i through

$$I^i = \int j^i(x)\, d\vec{x} \,. \qquad \text{(II.21)}$$

Let us note that the Pauli matrices τ^i in Eq. (II.20) are to be considered as sets of c numbers, and that the I^i, Eq. (II.21), are well defined even if they are not conserved. By taking into account Eq. (II.19), the following commutation relations are obtained:

$$[I^i, I^j] = i\,\epsilon^{ijk}\, I^k \,. \qquad \text{(II.22)}$$

Eq. (II.22) enables us to identify the I^i with the generators of the SU(2) group of the I spin. Eq. (II.22) holds independently from the conservati of the involved quantities. Given the explicit form, Eq. (II.17), of the Lagrangian, we see that I^3 turns out to be conserved, while this is not the case for I^1, I^2.

From now on, keeping the main features of the above model, we will assume that the following situation occurs:

i) the Hamiltonian has the form

$$H = H_0 + eH' , \qquad (II.23)$$

where H_0 is full I invariant, eH' is the electromagnetic breaking, to be considered as a perturbation with respect to H_0;

ii) equation (II.22) holds exactly, although only I_3 is conserved and I_1, I_2 are not.

We will show how one can estimate matrix elements of I operators, and how to exploit the information contained in Eqs. (II.22) and (II.23) by means of the Fubini and Furlan method.

As a first step, from Eq. (II.22) we derive the following one

$$[I^+, I^-] = 2I^3 , \qquad (II.24)$$

which is more convenient for further usage. Again, we stress that Eq. (II.24) is assumed to hold <u>exactly</u>, although $I^{\pm} = I^1 \pm iI^2$ are not conserved.

We now take the matrix element $\mathcal{M}$ of both sides of Eq. (II.24) say, between two proton states $|p, \vec{q}\rangle$, $|p, \vec{q}'\rangle$ with momentum $\vec{q}, \vec{q}'$ and $I^3 = \frac{1}{2}$ (since I^3 is conserved this is true independently of electromagnetic interactions.)

The r.h.s. is easily computed:

$$\mathcal{M} = 2 < p\,\vec{q}\,' | I^3 | p\vec{q} > = < p\,\vec{q}\,' | p\vec{q} > = \delta(\vec{q}\,' - \vec{q}) . \qquad (II.25)$$

From the l.h.s. we have

$$\begin{aligned} \mathcal{M} &= < p\,\vec{q}\,' | I^+ I^- | p\,\vec{q} > - < p\,\vec{q}\,' | I^- I^+ | p\,\vec{q} > = \\ &= \sum_{\alpha} < p\,\vec{q}\,' | I^+ | \alpha > < \alpha | I^- | p\,\vec{q} > - \\ &\quad - \sum_{\beta} < p\,\vec{q}\,' | I^- | \beta > < \beta | I^+ | p\,\vec{q} > , \end{aligned} \qquad (II.26)$$

α,β referring to a complete set of states.

The commutation relations (II.22) ensure that $I^+|p,\vec{q}>$, $I^-|p,\vec{q}>$ are eigenstates of I^3 to the eigenvalues $3/2$ and $-1/2$, so that it is enough to keep the states with $I^3 = -1/2$ in the set of the $|\alpha>$ and, in the set of the $|\beta>$, those with $I^3 = 3/2$.

The contribution of neutron states $|n,\vec{k}>$ among the $|\alpha>$ is of particular interest, and we select it from the others:

$$\begin{aligned} &\sum_{\alpha} < p, \vec{q}\,' | I^+ | \alpha > < \alpha | I^- | p\vec{q} > = \\ &= \int d\vec{k} < p\,\vec{q}\,' | I^+ | n,\vec{k} > < n,\vec{k} | I^- | p,\vec{q} > + \\ &\quad + \sum_{\alpha \neq n} < p\vec{q}\,' | I^+ | \alpha > < \alpha | I^- | p\,\vec{q} > . \end{aligned} \qquad (II.27)$$

We put

$$< p, \vec{q}| I^+ | n, \vec{k} > = f \ \delta(\vec{q} - \vec{k}) \ . \tag{II.28}$$

The relative phase of proton and neutron states can be chosen in such a way that f is real. In the exact I-spin limits f = 1 (standar Clebsch-Gordan coefficients technique); the departure of f from unity wi give us a measure of I-spin violation within the proton and neutron state

Combining Eqs. (II.25-II.28), we get

$$\begin{aligned}(1 - f^2) \ \delta(\vec{q} - \vec{q}\,') &= \sum_{\alpha \neq n} <p\vec{q}\,' | I^+ | \alpha > <\alpha | I^- | p\,\vec{q}> - \\ &- \sum_{\beta} < p\vec{q}\,' | I^- | \beta > <\beta | I^+ | p\,\vec{q}> \ .\end{aligned} \tag{II.29}$$

For further manipulations we note that

$$\begin{aligned}\dot{I}^+ &= -i \ [H \ , I^+] \\ &= -ie[H' , I^+] \ ,\end{aligned} \tag{II.30}$$

where use is made of Eq. (II.23). Taking the matrix elements of both sides of Eq. (II.30) between a proton state $|p, \vec{q}>$ and another state $|\alpha>$ not containing a nucleon, we obtain

$$<p\,\vec{q}\,| I^+ | \alpha > = \frac{e}{E_p - E_\alpha} <p\,\vec{q}\,| \ [H' , I^+] \,| \alpha > \ . \tag{II.31}$$

From Eq. (II.31) the off-diagonal matrix elements of I^+ are seen to be of the first order in e [compare with Eq. (II.15)]. Substituting Eq. (II.31) into Eq. (II.29), we find that the deviation of the matrix elements of I^+ within the "nucleon box" [refer to Eq. (II.28)] are of the second order in e. We have thus recovered a situation similar to the one we already discussed in the general case, where no element of the algebra is conserved.

The Fubini-Furlan method has interesting physical applications, such as the one due to Adler and Weisberger, which will be discussed next.

III. THE ADLER-WEISBERGER SUM RULE

The A-W sum rule relates the axial coupling constant in beta decay to an integral over the $\pi^{\pm} + p$ cross-section; it is written as:

$$1 - \frac{1}{g_A^2} = \frac{4M_N^2}{g^2 \pi} \int \frac{W dW}{(W^2 - M_N^2)} \left(\sigma^+(W) - \sigma^-(W)\right), \qquad \text{(III.1)}$$

where $g_A = G_A/G_V$ is the ratio of the axial and vector coupling constant in beta decay, experimentally near to -1.2, $\sigma^{\pm}(W)$ are the total cross-sections of $\pi^{\pm}$ on protons, at total c.m. energy W, but extrapolated to zero pion mass, and g is the renormalized pion-nucleon coupling constant.

In order to derive this formula it is necessary to first give a sketch of the underlying theoretical basis. A more detailed description of the SU(3) approach to weak interactions, and the "partially conserved axial current" (PCAC) can be found, for example, in the Erice lectures (1964) by Feynman, and by Cabibbo (1963).

The weak leptonic decays (including beta decay) are described by a direct interaction among a hadron current, J_μ and a lepton current, ℓ_μ:

$$H_W = \frac{G}{\sqrt{2}} \ell_\mu^{(x)} J_\mu^{(x)} + \text{h.c.} , \qquad \text{(III.2)}$$

G is the Fermi-coupling constant, $G \approx 10^{-5} M_N^{-2}$.

The lepton current is given by

$$\ell_\lambda = \left(\bar{e}\,\gamma_\lambda(1+\gamma_5)\nu_e\right) + (\bar{\mu}\,\ldots\,\nu_\mu)\ , \qquad \text{(III.3)}$$

while in the SU(3) theory the hadron current is written as a combination of two octets, one the vector, the other the axial currents, j_λ^i and g_λ^i; the vector octet is identified with the one which contains the electromagnetic current, $j^3 + (1/\sqrt{8})\,j^8$:

$$J_\lambda = \cos\vartheta\left(j_\lambda^1 + i\,j_\lambda^2\right) + \sin\vartheta\left(j_\lambda^4 + i\,j_\lambda^5\right) \qquad \text{(III.4)}$$

$$+\cos\vartheta\left(g_\lambda^1 + i\,g_\lambda^2\right) + \sin\vartheta\left(g_\lambda^4 + i\,g_\lambda^5\right)\ ,$$

the parameter ϑ measures the ratio of $\Delta S = 1$ over $\Delta S = 0$ amplitudes, and is roughly equal to 0.26.

Furthermore, we assume that

$$\partial_\mu\left(g_\mu^1(x) + i\,g_\mu^2(x)\right) = -i\,\frac{\sqrt{2}\,M_N\,m_\pi^2}{g}\,g_A\,\varphi_{\pi^-}(x)\ . \qquad \text{(III.5)}$$

This is the PCAC hypothesis, introduced in order to explain the validity of the Goldberger-Trieman relations (see Feynman 1964). The meaning of g_A and g has been defined above.

The current J_μ enters beta decay through the matrix element (we assume p and n to have the same momentum)

$$\langle p|J_\mu^{(o)}|n\rangle = \cos\vartheta\,\bar{u}(p)\,(\gamma_\mu + g_A\,\gamma_5\,\gamma_\mu)\,u(n)\ . \qquad \text{(III.6)}$$

The coefficient g_A is what we want to compute [see Eq. (III.1)], while a similar coefficient in front of γ_μ is put equal to one because $j^1 + i\,j^2$ is the I-spin current, and, neglecting electromagnetic corrections is conserved (see previous section for a discussion on deviations of these coefficients from one.)

How can we compute g_A? With the information given up to now, it is impossible to do so. The PCAC does not help, since it contains g_A explicitly, while the fact that g_μ^i transforms as an octet gives only a linear condition on g_A: it duly follows that g_A is linearly related to the similar coefficients in other matrix elements, such as $<p|J_\mu|\Lambda^o>$ etc. In order to fix the value of g_A we need more, at least a quadratic relation for the $g_\mu^i(x)$. Such a relation would follow from the knowledge of the commutation rules of the different $g_\mu^i(x)$. In order to postulate a plausible set of commutation relations, we consider a triplet model, for example, the quark model.

The quarks, p, n, λ, are described by a twelve-component Fermi field

$$\psi(x) = \begin{pmatrix} \psi^p(x) \\ \psi^n(x) \\ \psi^\lambda(x) \end{pmatrix} ,$$

with anticommutation relations:

$$\left\{ \psi_\rho^{i+}(\vec{x},0),\ \psi_\sigma^j(\vec{y},0) \right\} = \delta_{ij}\ \delta_{\rho\sigma}\ \delta(\vec{x}-\vec{y}) \qquad i,j = p,n,\lambda\ . \qquad \text{(III.7)}$$

In this model one can postulate the following explicit forms for the currents: [λ^i being the 3×3 matrices defined by M. Gell-Mann, Phys.Rev. 125, 1067 (1962)]

$$\begin{aligned} j_\mu^i(x) &= \bar{\psi}(x)\,\gamma_\mu\,\lambda^i\,\psi(x) \\ g_\mu^i(x) &= \bar{\psi}(x)\,\gamma_\mu\,\gamma_5\,\lambda^i\,\psi(x)\ . \end{aligned} \qquad \text{(III.8)}$$

This choice has the property of giving J_μ a simple form, very similar to that of ℓ_μ

$$J_\mu(x) = \bar{\psi}^p(x)\gamma_\mu\,(1+\gamma_5)\,\psi^{\tilde{n}}(x) \ ,$$

where

$$\psi^{\tilde{n}} = \cos\vartheta\,\psi^n + \sin\vartheta\,\psi^\lambda \ .$$

From these currents we can define scaler and pseudoscalar generators

$$F^i = \int j_o^i(x)\,d^3x$$
$$D^i = \int g_o^i(x)\,d^3x \qquad \text{(III.9)}$$

and, using Eq. (III.7), we can compute their commutation relations which result:

$$[F^i, F^j] = i\,f_{ijk}\,F^k$$
$$[F^i, D^j] = i\,f_{ijk}\,D^k \qquad \text{(III.10)}$$
$$[D^i, D^j] = i\,f_{ijk}\,F^k \ ,$$

where f_{ijk} are the structure constants of SU(3) (also defined and listed in Gell-Mann, loc.cit.).

The first line in (III.10) affirms that F^i forms an SU(3) algebra, and the second line says that the D^j behave as an octet under SU(3). The third line is the most interesting: it is a quadractic relation for D^i, which is what we need in order to fix the strength of the axial current and compute g_A. According to (III.10) F^i and D^i give the algebra of SU(3) $\otimes$ SU(3), in fact, if we define

$$F^i(\pm) = \frac{1}{\sqrt{2}}\,(F^i \pm D^i)$$

we find that $F^i(+)$ commutes with $F^j(-)$ and they form separately an SU(3) algebra. The algebra of the operators F^i and D^i is usually called the chiral $SU(3) \otimes SU(3)$ to distinguish it from $SU(3) \times SU(3)$ algebras generated by different sets of operators. The chiral $SU(3) \otimes SU(3)$ algebra was proposed by Gell-Mann in the paper quoted above.

We can now give a derivation of the Adler-Weisberger sum rule, Eq. (III.1).

Derivation of the Adler Weisberger sum rule

From Eq. (III.10), if $D^{\pm} = D^1 \pm i\, D^2$, we obtain

$$[D^+, D^-] = 2I^3 . \tag{III.11}$$

As we often stressed, Eq. (III.11) can be exact, in spite of the fact that the $D^{\pm}$ are not conserved. We will exploit it by means of the Fubini Furlan method.

Given two proton states $|p,\vec{q},s>$, $|p,\vec{q}',s'>$, take the matrix element of both sides of Eq. (III.11) between such states, then from the r.h.s. we find

$$\mathcal{M} = \delta_{ss'}\, \delta(\vec{q}' - \vec{q}) . \tag{III.12}$$

Inserting a complete set of states from the l.h.s.

$$\begin{aligned} \mathcal{M} &= \sum_{\alpha} <p,\vec{q}',s'|D^+|\alpha><\alpha|D^-|p,\vec{q},s> - \\ &- \sum_{\beta} <p,\vec{q}',s'|D^-|\beta><\beta|D^+|p,\vec{q},s> = \\ &= I_1 + I_2 - I_3 , \end{aligned} \tag{III.13}$$

where I_1 is the neutron contribution picked up from the states $|\alpha>$, I_2 and I_3 the contribution of the remaining $|\alpha>$ and $|\beta>$ states.

The one-neutron contribution is computed as follows: from Eq. (III.6) we see that

$$<p,\vec{q}\,',s'|[g_0^1(x) + i\; g_0^2(x)]|n,\vec{q},s> =$$

$$= -\frac{i}{(2\pi)^3}\, g_A\, \bar{u}_{ps'}\, \gamma_4\, \gamma_5\, u_{ns}\; e^{i(\vec{q}-\vec{q}\,')\cdot\vec{x}}\; e^{-i(E_n - E_p')t}\;.$$

By integrating over $d\vec{x}$

$$<p,\vec{q}\,',s'|[D^1 + iD^2]|n,\vec{q},s> =$$

$$= \delta(\vec{q}\,' - \vec{q})\, g_A\, \bar{u}_{ps'}\, \gamma_4\, \gamma_5\, u_{ns}\; e^{-i(E_n - E_p')t}\;.$$

Since we are only interested in equal time commutation relations, we put $t = 0$, so that

$$I_1 = \sum_{s''}\int d\vec{q}\,''<p,\vec{q}\,'s'|D^+|n,\vec{q}\,''s''><n,\vec{q}\,''s''|D^-|p,\vec{q}\,s> =$$

$$= \delta(\vec{q}\,' - \vec{q})\, g_A^2\, \bar{u}_{ps'}\, \gamma_4\gamma_5 \left(\sum_{s''} u_{ns''}\, \bar{u}_{ns''}\right) \gamma_4\, \gamma_5\, u_{ps}\;.$$

Substituting

$$\sum_{s''} u_{ns''}\, \bar{u}_{ns''} = \frac{-i\not{q} + M_N}{2E}$$

and using standard techniques, the total neutron contribution is found to be

$$I_1 = \delta_{ss'}\,\delta(\vec{q}' - \vec{q})\, g_A^2 \,\frac{|\vec{q}|^2}{E_p^2} \,. \tag{III.14}$$

For the evaluation of the other terms in Eq. (III.13) we use the analogous of Eq. (II.31), which we write as

$$<p,\vec{q},s|D^+|\alpha> = <p,\vec{q}\,s|\dot{D}^+|\alpha>\,\frac{i}{E_p - E_\alpha}\,.$$

On the other hand,

$$\dot{D}^+ = \int \dot{g}_0^+\, d\vec{x} = \int \partial_\mu g_\mu^+\, d\vec{x}\;,$$

assuming $g_i^+(\vec{x})$ vanishes at infinity.

Using Eq. (III.5) we obtain

$$<p,q,s|D^+|\alpha> = \frac{\sqrt{2}\, M_N m_\pi^2\, g_A}{g(E_p - E_\alpha)} <p,\vec{q},s| \int \varphi_{\pi^-}(\vec{x})\, d\vec{x}|\alpha> =$$

$$= \frac{\sqrt{2}\, M_N m_\pi^2\, g_A}{g(E_p - E_\alpha)}\,(2\pi)^3\,\delta(\vec{q} - \vec{q}_\alpha) <p,\vec{q},s|\varphi_{\pi^-}(0)|\alpha>\,.$$

We can now write

$$I_2 = \frac{2M_N^2 m_\pi^4\, g_A^2}{g^2}\,(2\pi)^6\,\delta(\vec{q} - \vec{q}')\sum_\alpha \frac{\delta(\vec{q} - \vec{q}_\alpha)}{(E_p - E_\alpha)^2}\,\cdot$$

$$\cdot <p,\vec{q},s'|\varphi_{\pi^-}(0)|\alpha> <\alpha|\varphi_{\pi^+}(0)|p,\vec{q}\,s> \tag{III.15}$$

$$= g_A^2\,\delta(\vec{q} - \vec{q}')\,\mathcal{I}_2(\vec{q})\,.$$

Exchanging $\varphi_{\pi^-} \leftrightarrow \varphi_{\pi^+}$, we obtain the corresponding expression for $I_3 \equiv g_A^2\, \delta(\vec{q}-\vec{q}\,')\mathcal{I}_3(\vec{q})$. Collecting the above results, Eqs. (III.12), (III.14), (III.15), we find the following relation for s = s'

$$\frac{|\vec{q}|^2}{E_p^2} - \frac{1}{g_A^2} = \frac{2M_N^2\, m_\pi^4}{g^2}(2\pi)^6 \left\{ \sum_\beta \frac{|\langle p,\vec{q},s|\varphi_{\pi^+}(0)|\beta\rangle|^2}{(E_p-E_\beta)^2}\,\delta(\vec{q}-\vec{q}_\beta) - \right.$$

$$\left. - \sum_\alpha \frac{|\langle p,\vec{q},s|\varphi_{\pi^-}(0)|\alpha\rangle|^2}{(E_p-E_\alpha)^2}\,\delta(\vec{q}-\vec{q}_\alpha) \right\} \qquad \text{(III.16)}$$

$$= \mathcal{I}_3(\vec{q}) - \mathcal{I}_2(\vec{q}) \ ,$$

which can be regarded as a family of sum rules depending on the parameter $|\vec{q}|$.

If we consider the limit of (III.16) for $|\vec{q}| \to \infty$, it turns out that the l.h.s. can be expressed in terms of a total cross-section for "zero mass" $\pi^\pm$ on protons, which gives rise to (III.1). To show this we first divide the sum into contributions coming from states of definite mass, W

$$\sum_\beta \delta(\vec{q}_\beta - \vec{q}) = \int dW^2 \sum_\beta \delta(M_\beta^2 - W^2)\,\delta(\vec{q}_\beta - \vec{q}) \ .$$

Furthermore, given $|\vec{q}| = q >> M_\beta, M_p$ we have

$$E_\beta = \sqrt{q^2 + M_\beta^2} \sim q + \frac{M_\beta^2}{2q}$$

$$E_p \approx q + \frac{M_p^2}{2q}$$

The energy difference

$$E_\beta - E_p \approx \frac{1}{2q}\,(M_\beta^2 - M_p^2) \qquad \text{(III.17)}$$

goes to zero as $q \to \infty$.

If we introduce a fictitious "pion momentum"

$$p_\pi \equiv \left(\frac{W^2 - M_p^2}{2q} \, , \vec{0} \right) \qquad \text{(III.18)}$$

we can rewrite the sum (as $q \to \infty$)

$$\sum_\beta \delta(\vec{q}_\beta - \vec{q}) = \frac{1}{2q} \int dW^2 \sum_\beta \delta^4(q_\beta - q - p_\pi) \; . \qquad \text{(III.19)}$$

Note that these steps (and others to follow) of taking limits an integral are allowed only if the integral converges uniformly in q. In the limit $q \to \infty$ one sees that $(p_\pi)^2$ goes to zero for each W. On the side, let us compute the total cross-section for a π^+ of nonphysical mass $\mu \neq m_\pi$ and four momentum p_π on a proton of momentum q:

$$\sigma^+ = \frac{(2\pi)^2 E_p E_\pi}{[(q p_\pi)^2 - M_N^2 \mu^2]^{1/2}} \sum_\beta |S|^2 \, \delta^4(q_\beta - q - p_\pi) \; .$$

By using the reduction formalism we can write the matrix element $S(p + \pi^+ \to \beta)$ in the form

$$S(p + \pi^+ \to \beta) = \frac{(2\pi)^{5/2}}{\sqrt{2E_\pi}} \; (\Box^2 + m_\pi^2) \; \langle p | \varphi_{\pi+} | \beta \rangle$$

$$= \frac{(2\pi)^{5/2}}{\sqrt{2E_\pi}} \; (m_\pi^2 - \mu^2) \; \langle p | \varphi_{\pi+} | \beta \rangle \; .$$

In the limit of $\mu^2 = 0$ (unphysical zero-mass pions) we then have:

$$\sigma^{\pm} = \frac{m_\pi^4 E_p (2\pi)^7}{|2(q\, p_\pi)|} \sum_\beta |\langle p | \varphi_{\pi^+} | \beta \rangle|^2 \, \delta(q_\beta - q - p_\pi) \ . \qquad \text{(III.20)}$$

Furthermore [see Eq. (III.12)]

$$|2(q p_\pi)| = (W^2 - M_N^2) \frac{E_p}{q} \xrightarrow[q \to \infty]{} (W^2 - M_N^2) \ . \qquad \text{(III.21)}$$

Now it is easy to see that the Adler-Weisberger sum rule follows from Eqs. (III.16) and (III.20) with the help of (III.19), (III.17) and (III.21 If one uses physical cross-sections (not extrapolated to $\mu^2 = 0$), Eq. (III. gives

$$g_A = 1.8 \ .$$

More sophisticated calculations, which include off mass-shell corrections to the cross-sections, give

$$g_A = 1.21 \div 1.23$$

which are also in good agreement with the experimental value.

Let us note that the convergence of the integral appearing in Eq. (III.1) requires the validity of the Pomeranchuk theorem: $[\sigma(\pi^+) - \sigma(\pi^-)] \to 0$ at high energy. Experimentally $\sigma(\pi^+)$ and $\sigma(\pi^-)$ seem to tend, as energy increases, to the same constant value.

IV. SATURATION OF COMMUTATION RELATIONS AND PARTICLE MULTIPLETS

In the previous lectures we have shown how it is possible to test a set of commutation relations among operators by means of the sum rules. It is scarcely necessary to emphasize that in this way it is only possible to test the commutation relations of operators whose matrix elements can be given a physical meaning. Now we can ask: "what is required for an algebra of operators to give rise to an approximate symmetry"? As we will see, we have only a partial answer to this.

Suppose that we have a set of operators which form a Lie algebra defined by the commutation relations

$$[G^i, G^j] = i\, c_k^{ij}\, G^k \; . \qquad \text{(IV.1)}$$

Now, if the set $\{|a>\}$ of states form a representation of this Lie algebra, then we have:

$$< a'|[G^i, G^j]|a'' > = \sum_{A \in \{|A>\}} \{< a'|G^i|a > < a|G^j|a'' > - < a'|G^j|a > < a|G^i|a'' >\} \; , \qquad \text{(IV.2)}$$

and we say that the states $\{|a>\}$ provide the "saturation" of the commutation relations, i.e. the matrix elements on the left-hand side can be evaluated exactly by inserting as intermediate states the vectors of the $\{|a>\}$ set only.

In general it is not possible to find a set of physical states which satisfy this property. But suppose we find a set of physical states $\{|A>\}$ which provide an approximate saturation of the commutation relations, i.e. a set $\{|A>\}$ such that if $|A'>, |A''> \subset \{|A>\}$,

$$\langle A'|[G^i, G^j]|A''\rangle \approx \sum_{A\in\{|A\rangle\}} [\langle A'|G^i|A\rangle \langle A|G^j|A''\rangle - \langle A'|G^j|A\rangle \langle A|G^i|A''\rangle] . \tag{IV.3}$$

This implies that if $|A_\ell\rangle$ ($\ell = 1 \ldots n$> is a basis in $\{|A\rangle\}$, the matrice

$$G^i_{\ell m} = \langle A_\ell|G^i|A_m\rangle \tag{IV.4}$$

give an approximate representation of the algebra of the operators G^i. This need not be an irreducible representation. As an example, the <u>nonet</u> of 1^- mesons could provide saturation for the algebra of SU(3). It is then possible to define multiplets of an algebra which is not conserved as sets of particles which saturate the commutation relations of the algebra.

The requirement of saturation within a multiplet puts some restriction on the amount of leakage: consider a sum rule obtained from (III.1) and (III.4):

$$\langle A_\ell|[G^i, G^j]|A_m\rangle = G^i_{\ell n}\, G^j_{nm} - G^j_{\ell n}\, G^i_{nm} + \sum_{B\notin\{|A\rangle\}} \left[\langle A_\ell|G^i|B\rangle \langle B|G^j|A_m\rangle - \langle A_\ell|G^j|B\rangle \langle B|G^i|A_m\rangle\right] = i\, c^{ij}_k\, G^k_{\ell m} . \tag{IV.5}$$

Given the matrices defined in Eq. (IV.4), this gives some restriction on the leakage, but is not sufficient to exclude it, since only the very special combination

$$\sum_{B \not\subset \{A\}} \langle A_\ell|G^i|B\rangle \langle B|G^j|A_m\rangle - \langle A_\ell|G^j|B\rangle \langle B|G^i|A_m\rangle \qquad \text{(IV.6)}$$

is required to be small. Let us give a specific example, considering again the commutation relation among pseudoscalar generators used to derive the Adler-Weisberger sum rule:

$$[D^i, D^j] = i\,\epsilon^{ijk} I^k \qquad (i,j = 1,2,3)\ . \qquad \text{(IV.7)}$$

Consider the vacuum expectation value of the two members. We note that P and C symmetry imply that

$$\langle\Omega|I^i|\Omega\rangle = \langle\Omega|D^i|\Omega\rangle = 0\ , \qquad \text{(IV.8)}$$

so that the vacuum state can be considered a singlet of the algebra since it gives an exact (assuming P and C) saturation of the same. However, the operators D^i do have matrix elements, in no sense small, among the vacuum and other states. From PCAC [(Eq. (III.5)] it follows that the matrix element of $D^\pm$ among the vacuum, and a state containing a $\pi^\pm$ of momentum $\vec{p}$, is given by

$$\langle\vec{p}, \pi^\pm|D^\pm|\Omega\rangle = (2\pi)^{-3/2}\ \delta^3(\vec{p})\ .$$

However, the presence of this leakage does not disturb the commutation relations:

$$\langle\Omega|[D^+, D^-]|\Omega\rangle = \sum_{\vec{p}} \{|\langle\vec{p}, \pi^-|D^-|\Omega\rangle|^2 - |\langle\vec{p}, \pi^+|D^+|\Omega\rangle^2\}$$

$$= 0 \text{ (by charge conjugation).}$$

This example shows how an arbitrary amount of leakage can be compatible with perfect saturation. Another example is given by the Adler-Weisberger sum rule: the integral over the difference of π^+p and π^-p cross-sections is obtained from a sum equivalent to (IV.6). The integral turns out to give a small correction to g_A (from 1 to 1.18). However, the integral of each of the two cross-sections is a diverging quantity. Then, also in that case the leakage is not small; it is only so arranged that approximate saturation is maintained. However, it is easy to see that a broken symmetry defined by asking saturation in the presence of leakage loses part of its predictive power, and it is not yet clear how it can be applied to multiparticle states (scattering), or if it can be applied at all.

V. A POSSIBLE ALGEBRA OF CURRENTS

In this lecture we illustrate a possible algebra of local currents, the one proposed by B.W. Lee [Phys.Rev. Letters <u>14</u>, 675 (1965)] and by Dashen and Gell-Mann [Physics Letters <u>17</u>, 142 (1965)]. Starting from the quark model these authors introduce 144 currents, subdivided into nonets of vector, axial vector, scalar, pseudoscalar and tensor densities.

Let us consider the quark field $\psi_a = \psi_{\alpha A}$ (A = 1,2,3; α = 1,2,3,4 a = 1,2, ... 12). We assume the following anticommutation relations at t = 0:

$$\left\{\psi_a^*(\vec{x}),\ \psi_b(\vec{x}')\right\}_{t=0} = \delta_{ab}\,\delta^3(\vec{x}-\vec{x}')\,. \qquad \text{(V.1)}$$

We then define the generalized currents:

$$J(A,\vec{x}) = \psi^*(\vec{x})\ A\psi(\vec{x}) \quad (t = 0)\ , \qquad \text{(V.2)}$$

where A is any 12×12 matrix, $J(A,\vec{x})$ is hermitian if, and only if, A is hermitian. By use of relation (V.1), one can compute the commutation relations satisfied by the currents which result

$$[J(A,\vec{x}),\ J(B,\vec{x}')] = J([A,B],\vec{x})\ \delta^3(\vec{x}-\vec{x}')\ . \qquad (V.3)$$

Schwinger has shown that in some cases terms proportional to

$$\frac{\partial}{\partial(\vec{x}-\vec{x}')}\ \delta(\vec{x}-\vec{x}')$$

must be added to the right-hand side of (V.3). But if we introduce the operators

$$G(A) = \int J(A,\vec{x})\, d^3x\ , \qquad (V.4)$$

their commutation relations are well defined as

$$[G(A),\ G(B)] = G([A,B])\ . \qquad (V.5)$$

The group associated to this algebra has the following infinitesimal transformations:

$$U = 1 + i \sum_{G} \epsilon(G)\, G\ .$$

Thus, if we want U to be unitary, G(A) must be hermitian, and this in turn implies $A = A^{+}$, $J(A) = J^{+}(A)$. Thus, we are led to the set of 12×12 hermitian matrices A, and this set generates the compact U(12) algebra. In this way we arrived at the following set of currents:

Vector

$$\left(\vec{j}^{\,i},\ j_0^i\right) \equiv j_\mu^i(\vec{x}) = \qquad\qquad A \equiv i\gamma_4\gamma_\mu \frac{\lambda^i}{2} \quad \text{for } \mu \neq 4;$$

$$= i\,\bar{\psi}(\vec{x})\,\gamma_\mu \frac{\lambda^i}{2}\,\psi(\vec{x}) \qquad\qquad \frac{\lambda^i}{2} \quad \text{for } \mu = 4$$

Axial vector

$$\left(\vec{g}^{\,i},\ g_0^i\right) \equiv g_\mu^i(\vec{x}) = \qquad\qquad A \equiv i\gamma_4\gamma_\mu\gamma_5 \frac{\lambda^i}{2} \quad \text{for } \mu \neq 4;$$

$$= i\,\bar{\psi}(\vec{x})\,\gamma_\mu\gamma_5 \frac{\lambda^i}{2}\,\psi(x) \qquad\qquad \gamma_5 \frac{\lambda^i}{2} \quad \text{for } \mu = 4$$

Scalar

$$s^i(\vec{x}) = \bar{\psi}(\vec{x})\,\frac{\lambda^i}{2}\,\psi(\vec{x}) \qquad\qquad A = \gamma_4 \frac{\lambda^i}{2}$$

Pseudoscalar

$$p^i(\vec{x}) = \bar{\psi}(\vec{x})\,\gamma_5 \frac{\lambda^i}{2}\,\psi(\vec{x}) \qquad\qquad A = \gamma_4\gamma_5 \frac{\lambda^i}{2}$$

Tensor

$$\left(\vec{h}^{\,i},\ \vec{\mathscr{E}}^{\,i}\right) \equiv t_{\mu\nu}^i(\vec{x}) = \qquad\qquad A = \gamma_4\,\sigma_{\mu\nu} \frac{\lambda^i}{2}:$$

$$= \bar{\psi}(\vec{x})\,\sigma_{\mu\nu} \frac{\lambda^i}{2}\,\psi(x) \qquad\qquad \gamma_4\vec{\sigma} \quad \text{if } \mu,\nu = 1,2,3$$

$$\gamma_4\gamma_5\vec{\sigma} \quad \text{if } \mu \text{ or } \nu = 4\,. \tag{V.6}$$

The vector and axial vector densities are subdivided in space and time components, the antisymmetric tensor $t_{\mu\nu}^i$ is divided into two vectors, in analogy with the electromagnetic field $F_{\mu\nu} = (\vec{H}, \vec{E})$.

It is convenient to divide the 144 operators into four groups:

$$\begin{aligned}
&a) \quad \mathbb{1}\left(\lambda^i, \sigma_k \lambda^i\right) && \Longrightarrow \quad j_o^i, g_k^i \\
&b) \quad \gamma_4\left(\lambda^i, \sigma_k \lambda^i\right) && \Longrightarrow \quad s^i, h_k^i \\
&c) \quad i\gamma_4\gamma_5\left(\lambda^i, \sigma_k \lambda^i\right) && \Longrightarrow \quad p^i, \mathcal{G}_k^i \\
&d) \quad \gamma_5\left(\lambda^i, \sigma_k \lambda^i\right) && \Longrightarrow \quad g_o^i, j_k^i \ .
\end{aligned} \qquad (V.7)$$

Terms c) and d) have zero matrix elements between states of the same parity. Then, if we limit ourselves to multiplets which include only particles of the same parity, we see that the largest algebra which we can expect to saturate with particles at rest is the one which has $\gamma_4(\lambda^i,\sigma_k\lambda^i)$ and $\mathbb{1}(\lambda^i,\sigma_k\lambda^i)$ as generators. As the sets of operators $\frac{1}{2}(1+\gamma_4)(\lambda^i,\sigma_k\lambda^i)$ and $\frac{1}{2}(1-\gamma_4)(\lambda^i,\sigma_k\lambda^i)$ separately generate U(6), and mutually commute, we obtain in this case the algebra of $U(6) \otimes U(6)$.

Suppose that a certain set of states at rest saturate approximately the $U(6) \otimes U(6)$ commutation relations of the parity conserving operators. Will the same states at $p \neq 0$ still provide this approximate saturation? To answer this question let us consider the matrix elements of an operator G(A) between states with the same mass. The matrix elements are given by:

$$\begin{aligned}
\langle p'|G(A)|p\rangle &= \langle p'|\int \psi^+ A\psi\, d^3x|p\rangle = \\
&= \frac{1}{(2\pi)^3}\, u^+(p')\,A\,u(p) \int e^{i(\vec{p}\,'-\vec{p})\vec{x}}\, d^3x \\
&= u^+(p')\,A\,u(p)\,\delta^3(\vec{p}\,'-\vec{p}) = u^+(p)\,A\,u(p)\,\delta^3(\vec{p}\,'-\vec{p}) \ .
\end{aligned}$$

The matrix elements $u^{+}(p)\,A\,u(p)$ are obtained from the corresponding ones at $\vec{p} = 0$ by a Lorentz transformation. Then performing a Lorentz transformation along the z axis we obtain the following table: [$\hat{\beta}$ ≡ unit vector in the z direction]

current	$\vec{p} = 0$ limit	$\vec{p} \neq 0$
$\langle j_\mu^i \rangle$	$(0,1)$	$\frac{M}{E}\left(\frac{\vec{p}}{M}, \frac{E}{M}\right)$
$\langle g_\mu^i \rangle$	$(\vec{s},0)$	$\frac{M}{E}\left[\vec{s} - \frac{M-E}{M}\,\hat{\beta}(\hat{\beta}\vec{s})\ ;\ \frac{\lvert p\rvert}{M}\,(\hat{\beta}\vec{s})\right]$
$\langle t_{\mu\nu} \rangle$	$(\vec{h},0)$	$\vec{h}' = \frac{M}{E}\left[\frac{E}{M}\left(\vec{h} - \frac{E-M}{E}\,\hat{\beta}(\hat{\beta}\vec{h})\right)\right]$ $\vec{\mathscr{E}}' = \frac{M}{E}\left[\frac{\lvert p\rvert}{M}\,(\hat{\beta}\times\vec{h})\right]$
$\langle p \rangle$	0	0
$\langle s \rangle$	1	$\frac{M}{E}$

From this table one learns that the matrix elements of the positive parity operators corresponding to the A matrices $\lambda^i, \sigma_3\lambda^i, \gamma_4\sigma_1\lambda^i$, $\gamma_4\sigma_2\lambda^i$ are invariant under Lorentz transformations along the third axis. Other positive parity operators behave like M/E and go to zero when $\vec{p}$ approaches infinity. We must then reject all other terms in order to have an approximate saturation of the commutation relations for $\vec{p} \neq 0$. In fact, if you have, for instance, the commutator:

$$[\gamma_4\lambda^i, \gamma_4\lambda^j] = i f_{ijk}\,\lambda^k$$

and try to saturate it with states at $\vec{p} \neq 0$, you obtain that the left-hand side and the right-hand side of this relation behave in a different way when $\vec{p}$ increases, and while the former decreases, the latter remains constant. Thus, for collinear processes (involving particles all moving along the z axis), we are led to an SU(6) algebra, generated by the operators $\lambda^i, \sigma_3 \lambda^i, \gamma_4 \sigma_{1,2} \lambda^i$ which correspond to the currents $j_0^i, g_3^i, h_1^i, h_2^i$.

In a similar way it can be shown that for states with momentum in the x-y plane we can expect saturation of the commutation relations of $U(3) \otimes U(3)$ only, generated by λ^i and $\lambda^i \sigma_3 \gamma_4$ [coplaner $SU(3) \otimes SU(3)$]. Note that the only condition we required is the saturation of the commutation relations. This is not the same as having a symmetry, because we make no statements about the leakage. This could also be large, even if the commutation relations are saturated. In the case of an approximate kinematical symmetry, i.e. when $H = H_0 + \eta H'$, we were able to prove that leakage is of order η, while the commutation relations are satisfied to order η^2 by "in box" states alone. In this case we cannot say anything about the leakage. The only thing we know is that is must exist. This has been proved by Coleman: his theorem states that if one supposes that the one-particle states are invariant under the group (no leakage), then it follows that the Hamiltonian commutes with the generators of the group, i.e. it is invariant.

We further note that some of the currents we have introduced, namely the p,s and $t_{\mu\nu}$ currents have no physical meaning up to now. This is unfortunate because they are likely to appear in the sum rules, and we do not know what corresponds to their matrix elements in terms of measurable quantities.

DIFFICULTIES OF RELATIVISTIC U(6)

J.S. Bell,

CERN, Geneva.

In these lectures we will speak of U(6) rather than SU(6); the extra symmetry can be identified with baryon-number conservation.

It is well known that it is not easy to make a sensible theory which exactly incorporates both U(6) symmetry and Lorentz invariance. This is an account of these difficulties in so far as they can be expressed in terms of simple models, in terms of quarks, field operators, wave functions and Lagrangians. The abstract group theoretical approach is beyond the competence of this author. The various difficulties were commonly known and discussed among interested people shortly after the first papers on the subject. The author is indebted especially to colleagues in the Theoretical Study Division of CERN for many such discussions. There are also now a number of published papers on the subject; to these I have referred whenever I know about them.

The difficulties occur when you try to make a theory which is Lorentz invariant and has, in some sense, an exact U(6) symmetry built into it. Of course, this is rather academic, because U(6) is not exactly realized in nature; it is manifestly broken - even badly broken. Nevertheless, it is interesting to look for these symmetries. Hitherto, we were familiar with symmetries which could be regarded as becoming exact when some parameter in the Lagrangian was put equal to zero. It

---

Lecture notes by M. Zulauf, Inst. f. theor. Physik, Bern

will probably turn out that the U(6) symmetry is not of that type, and that it is restricted to a very limited range of phenomena. However, in the course of trying to make it exact, you might conceivably learn to what restricted range of phenomena it reasonably might apply. In any case, it is an inevitable question to ask: could God have made U(6) exact if He had so wished?

I. NON-COVARIANT MODELS

1. U(6)

A simple non-relativistic point of departure is the direct generalization of the Wigner supermultiplet theory of nuclei. In that model one says essentially that the forces acting in nuclei do not depend at all on spin and isospin. Analogously for U(6): here we are concerned with a system of many quarks. The total wave function for the system can be regarded as a sum of products of single quark wave functions, and each of these single quark wave functions is a function of position $\vec{r}$ and spin and unitary spin indices

$$\Psi_{\alpha A}(\vec{r}) \quad (\alpha = 1,2; \quad A = 1,2,3) .$$

α refers to spin up and down, A to proton, neutron and λ quark. The symmetry transformations of U(6) are

$$\Psi_{\alpha A}(\vec{r}) \rightarrow U_{\alpha A}^{\alpha' A'} \Psi_{\alpha' A'}(\vec{r}) ,$$

$$\text{shortly } \Psi(\vec{r}) \rightarrow U\Psi(\vec{r}) , \tag{I.1}$$

where U is a 6×6 unitary matrix.

If the symmetry holds, i.e. if the forces between quarks do not depend on spin and unitary spin, then we can replace all quark wave functions in the many particle sum of products by the corresponding primed wave functions, and the new total wave function is just as good as the old one. Especially if the old wave function represents a bound state of specified energy, the new wave function is equally well a bound state with the same energy as before. Likewise, S matrix elements are supposed to be unaltered when initial and final states are transformed.

2. SU(2) - content

Among the symmetry operators U, there are those which are diagonal in unitary spin, viz. the ordinary total spin operators. It is important to note that this total spin is not the total angular momentum of the compound system. If $\vec{\sigma}$ is the spin operator for a single quark, the total angular momentum is

$$\vec{\sigma} + \vec{r} \times \vec{p} \ . \tag{I.2}$$

Thus, when bound states are classified into multiplets under U(6) in this model, the spin quantum numbers involved are not necessarily the total spin of the compound state, but only an intrinsic spin from which orbital contributions have been excluded. It is possible to think that the particles now best known involve bound quarks without orbital motion, so that the distinction is unimportant. But it would be surprising if this would continue indefinitely for higher quark bound-states, especially if there would be excited states where orbital motion occurs. For these the total spin of the particle would not be the object which is referred to in the U(6) classification.

Several authors have considered the possibility that in constructing U(6) one should indeed use as the SU(2) content the total angular momentum (I.2) instead of $\vec{\sigma}$, so that the U(6) classification of bound states would always refer directly to the total spin rather than the intrinsic spin of the many-quark system*). There is a very serious objection to this [H. Lipkin, Phys.Rev. Letters 14, 336 (1965)].

A first, less serious, question is what origin of co-ordinates to use in the construction of the SU(2) generators

$$\vec{\sigma} + \vec{r} \times \frac{1}{i}\frac{\partial}{\partial \vec{r}} \ .$$

This we can overcome with some plausibility by identifying the origin as the centre of mass (in one way or another). But now consider the infinitesimal generators of the full U(6) group, including unitary spin. Concentrate on the one with λ_3, i.e.

$$\left(\vec{\sigma} + \vec{r} \times \frac{1}{i}\frac{\partial}{\partial \vec{r}}\right) \begin{pmatrix} 1 & & \\ & -1 & \\ & & 0 \end{pmatrix} \ .$$

This is an operator which does not act at all on λ quarks, but acts with opposite sign on p's and n's. Thus, the orbital part effects a spatial separation of p's, n's and λ's to an increasing extent with distance from the origin. If one considers, for example, a system composed of two independent parts equally distant from the origin (initial or final state in scattering problem), one sees that this is a symmetry operator which splits them up into separate groups of p's, n's and λ's outside the range of forces between them. The conclusion is that such a symmetry forbids any energy of interaction between different quarks, since no energy would be sacrificed by separating the many-quark system into its component pieces. Therefore, the identification of SU(2) with the full rotation

---

*) The symmetry transformations would then be defined by

$$\Psi(\vec{r}) \rightarrow U\,\Psi(\vec{r}\,') \qquad \vec{r}\,' = R\vec{r}$$

rather than (I.1). $\vec{r}\,'$ is obtained from $\vec{r}$ by a certain rotation operator.

group cannot be regarded as a serious candidate when we look for symmetries that might be exact. Of course, it might conceivably be an approximate symmetry in restricted situations, for example, in compact bound states of many particles with ordinary finite range two-body forces not dominant.

3. Boosting

We will commence the quest for a relativistic theory with the original non-relativistic U(6) model first described. We will restrict ourselves to scattering problems and only think of free particles in initial and final states; since bound states can be described as poles of scattering matrices this is less restrictive than might appear. Then it is useful to work in momentum space. Non-relativistically, each initial or final quark wave function is specified by a momentum $\vec{k}$ and a spinor Ψ_A which depends on spin indices only. Now as far as ordinary spin is concerned we are using Pauli two-spinors rather than Dirac four-spinors. It is important to understand that such a two-spinor description remains perfectly valid even for fast particles. All Dirac states Ψ can be obtained by boosting up particles at rest, i.e. by performing rotation-free Lorentz transformations upon spinors of the form

$$\begin{pmatrix}\Phi\\0\end{pmatrix} \text{ or } \begin{pmatrix}0\\\chi\end{pmatrix}.$$

So we can specify any Dirac state Ψ perfectly by giving the two-spinor Φ appropriate to the parent stationary state. We can then require, even in the relativistic situation of fast particles, that S matrix elements be spin independent in the ordinary sense, i.e. that they be invariant under the U(6) transformations

$$\vec{k} \to \vec{k}' = \vec{k}$$

$$\Phi \to \Phi' = (1 + i\epsilon\sigma\lambda)\Phi .$$

The scheme described above has essentially been proposed by various authors; the objection to it is, that although it is formulated so as to apply to fast particles, it is just not Lorentz invariant, or, alternatively put, that adding the requirement of Lorentz invariance to this U(6) invariance forbids any process [Jordan, Physical Review 137B, 149 (1965)]. This is ultimately because the symmetry operators are defined in terms of two-spinors which are fixed by boosting the given state to rest in some frame of reference, for example the laboratory, and this makes of the laboratory a preferred system.

Let us pursue this in more detail in a simple example, the scattering of a spin ½ by a spin 0 particle. We will leave out U(3) indices here, for the point is concerned only with the hypothetical ordinary-spin independence of the scattering amplitude. The initial and final states of the spin ½ particle are specified by Dirac spinors u and u′ with four momenta k and k′. The spin 0 states are specified simply by four momenta K and K′. The general Lorentz invariant parity conserving matrix element is

$$T = f(s,t)\ \bar{u}'u + g(s,t)\ \bar{u}' i\gamma(k+K)u\ . \qquad (I.3)$$

Other invariant forms reduce to the above in virtue of the Dirac equation $(i\gamma k + M)u = \bar{u}'(i\gamma k' + M) = 0$ and four-momentum conservation $k + K = k' + K'$. The solution of the Dirac equation has the form

$$u = \sqrt{E+M}\begin{pmatrix} \Phi \\ \dfrac{\vec{\sigma}\cdot\vec{k}}{E+M}\,\Phi \end{pmatrix} \qquad (p_0 > 0)$$

$$v = \sqrt{E+M}\begin{pmatrix} -\dfrac{\vec{\sigma}\cdot\vec{k}}{E+M}\,\chi \\ \chi \end{pmatrix} \qquad (p_0 < 0)$$

where Φ is a two-spinor. These states are obtained by boosting from rest the states

$$\begin{pmatrix}\Phi\\0\end{pmatrix} \text{ or } \begin{pmatrix}0\\X\end{pmatrix}$$

so that Φ is precisely the two-spinor referred to in the formulation of the U(6) symmetry. (For the appropriate boost matrix, see Professor Pais' lectures.)

In terms of Φ, the general Lorentz invariant amplitude (I.3) reads

$$\frac{T}{\sqrt{E+M}\,\sqrt{E'+M}} = f(s,t)\,\Phi'^{*}\left(1-\frac{\vec{\sigma}\cdot\vec{k}'}{E'+M}\,\frac{\vec{\sigma}\cdot\vec{k}}{E+M}\right)\Phi$$

$$- \; g(s,t)\;\;(k_0+K_0)\Phi'^{*}\left(1+\frac{\vec{\sigma}\cdot\vec{k}'}{E'+M}\,\frac{\vec{\sigma}\cdot\vec{k}}{E+M}\right)\Phi \qquad \text{(I.4)}$$

$$+ \; g(s,t)\Phi'^{*}\left(\vec{\sigma}\cdot(\vec{k}+\vec{K})\,\frac{\vec{\sigma}\cdot\vec{k}}{E+M} + \frac{\vec{\sigma}\cdot\vec{k}'}{E'+M}\,\vec{\sigma}\cdot(\vec{k}+\vec{K})\right)\Phi.$$

For this to meet the U(2) spin-independence hypothesis laid down, there should be no σ's between Φ'^{*} and Φ. However, this requires

$$f = g = 0\ .$$

Thus, the only way to make the spin dependence cancel out is to make the scattering amplitude completely zero.

4. Spin independence in the centre-of-mass system

One can save the theory discussed in Section I.3 as regards Lorentz invariance by specifying that the spin independence is required only in some particular co-ordinate system determined by the momenta of the particles. For this, one can take the centre-of-mass system (c.m.), for example, which is a Lorentz covariant notion. In that frame

$$\vec{k} + \vec{K} = 0 \ .$$

The last term in (I.4) vanishes, and for the others to be spin independent we get

$$f(s,t) + g(s,t)(k_0 + K_0) = 0 \ . \tag{I.5}$$

With (I.5) and the total energy variable $\sqrt{s} = (k_0 + K_0)_{c.m.}$, the invariant amplitude (I.3) reduces to

$$\begin{aligned} \frac{1}{2} \frac{T}{\sqrt{E+M}\,\sqrt{E'+M}} &= f(s,t)\,\bar{u}'\left(1 - \frac{i\gamma(k+K)}{\sqrt{s}}\right)u \\ &= f(s,t)\,\Phi'^{*}\,\Phi \ . \end{aligned} \tag{I.6}$$

This result is readily generalized to the case of quarks by including unitary spin indices. The invariant amplitude with c.m. U(6) invariance is

$$\frac{1}{2} \frac{T}{\sqrt{E+M}\,\sqrt{E'+M}} = f(s,t)\,\bar{u}'^{\alpha}\left(1 - \frac{i\gamma(k+K)}{\sqrt{s}}\right)u_{\alpha} \ .$$

However, the explicit use of the c.m. frame to form symmetries is very dangerous. There are two objections:

i) Crossing symmetry The c.m. system is not a crossing invariant concept. It is defined by zero total initial momentum or zero total final momentum. However, crossing interchanges some momenta between initial and final states. If the c.m. spin independence is required in both crossed and uncrossed channels, difficulties can be expected. For example, with the reaction (quark + antiquark) → (scalar + antiscalar) one would require in (I.3) $g = 0$, as compared with $g = -f/\sqrt{s}$ from the other channel.

ii) Locality Suppose we again take a system which consists of two separated parts at such a distance that there is no interaction between them. Let one of them move with arbitrary high velocity: then each moves arbitrarily fast in the c.m. frame, and although the two pieces do not interact, the velocity of one of them affects the internal dynamics of the other. Thus, this symmetry also cannot be regarded as one which might have been exact in a world with local causality.

It will be useful later to note one difficulty which this scheme does not have - namely unitarity. There is no trouble in making the scattering amplitude unitary in this theory. For example, (I.6) is unitary if f(s,t) has the standard form of a phase-shift expansion

$$f = C \sum (2\ell + 1)\, e^{i\delta_\ell} \sin \delta_\ell \, P_\ell(\cos \vartheta) .$$

In summary we have seen that it is indeed possible to formulate a Lorentz invariant scattering theory which has a certain U(6) symmetry in the c.m. frame. However, there are difficulties with crossing symmetry and locality (in the sense that two mutually distant systems may be supposed not to interact, so that their common c.m. should have no significance for the internal dynamics of either). We are led, therefore, to study attempts to set up theories which are relativistic and local from the beginning.

II. COVARIANT MODELS

1. U(12) and $\tilde{U}(12)$

The problem is to construct a relativistic theory which in some sense has a relativistic spin independence. Let us start by trying to make a field theory of such relativistic symmetries. In the following we will largely disregard the unitary symmetry which in discussing spin independence is an irrelevancy. A Lagrangian field theory would start from

$$\mathcal{L} = \bar{\Psi}(M + \gamma_\mu \partial^\mu)\Psi + \mathcal{L}_{\text{interaction}}, \tag{II.1}$$

and the equal time anticommutation relations

$$\{\Psi^+(\vec{x}), \Psi(\vec{y})\} = \delta(\vec{x} - \vec{y}). \tag{II.2}$$

The problem is to find symmetry transformations which preserve the Lagrangian as well as the commutation relations. We will see that only groups satisfying certain parts of the problem can be found.

A) The commutation relations (II.2) are invariant under transformations with unitary matrices

$$\Psi'(\vec{x}) = U\Psi(\vec{x}), \quad \Psi^{+\prime}(\vec{x}) = \Psi^+(\vec{x})\,U^+ \tag{II.3}$$

which in some sense mix states of different spin. The unitary 4×4 matrices which act on the Dirac spinors $\Psi(\vec{x})$ form a group, the unitary group in four dimensions U(4), which, if we add the U(3) symmetry, is enlarged to U(12). In order that such a group should be a symmetry of the theory, the Lagrangian also should be invariant; this gives the following additional restrictions on the matrices U

$$U^+\beta U = \beta, \quad U^+\alpha U = \alpha$$

or with $U^+ = U^{-1}$

$$[U, \beta] = 0 = [U, \alpha]. \tag{II.4}$$

There is only the trivial non-vanishing solution $U = 1$. Thus, the group U(12) - or any subgroup of it - does not preserve the quadratic part of the Lagrangian. There is also no cancellation from the interaction part of the Lagrangian, since the latter involves more than two fields.

B) Another group of matrices $\tilde{U}(4)$ or $\tilde{U}(12)$, respectively, is defined as the group that leaves invariant the mass term $M\bar{\Psi}\Psi$ in the Lagrangian (II.1 The condition for that is

$$U^{+}\beta U = \beta . \qquad \text{(II.5)}$$

This equation defines the so-called pseudo-unitary group, which in mathematical literature is often denoted by $U(2,2)$. But under this group neither the remaining kinetic term in the Lagrangian nor the commutation relations are invariant. So $\tilde{U}(12)$ also is not a candidate for a symmetry of a quark field theory.

C) One could still try to demand at least that the interaction term in the Lagrangian (II.1) has $U(12)$ or $\tilde{U}(12)$ symmetry (for example, the four-fermion interaction $\bar{\Psi}\Psi\bar{\Psi}\Psi$ would be $\tilde{U}(12)$ invariant). This would be useful in the Born approximation, where the scattering amplitude is given immediately by the matrix elements of the interaction term. However, in strong interaction physics the Born approximation would not be expected to be very relevant. Something called renormalized Born approximation does have an honourable place in strong interaction physics, giving poles in scattering amplitudes. However, there is no reason to think that the renormalized process, which certainly involves the zero-order Lagrangian, would respect $U(12)$ or $\tilde{U}(12)$ symmetries of the bare vertices. There is, nevertheless, a considerable body of literature in which especially $\tilde{U}(12)$ is employed in this way for effective vertex functions; it seems difficult to find any rational basis for that.

2. $U(6)\times U(6)$

Having failed to make a detailed field theoretic model in which some kind of exact spin independence is incorporated, let us see if this spin independence can be developed directly for the S matrix. The initia and final states of the S matrix are specified by momenta and wave functions which we again take to be ordinary Dirac spinors, forgetting for the moment about the unitary spin. If one starts with a solution of the Dirac equation

$$(M + i\gamma k)\Psi = 0\ , \tag{II.6}$$

and performs on Ψ a transformation (II.3) with a matrix U of U(12) or $\widetilde{U}(12)$, one sees that, in general, Ψ' is simply not an admissible state. This is the same phenomenon as the failure of the kinetic part of the Lagrangian to be symmetric. However, for certain restricted momenta, certain unitary subgroups of equally U(12) as well as $\widetilde{U}(12)$ again provide through the transformation good solutions of (II.6).

For particles at rest, for example, the Dirac equation reduces to

$$(\beta M - E)\Psi = 0$$

$$\begin{pmatrix} M-E & 0 \\ 0 & -M-E \end{pmatrix}\begin{pmatrix} \Phi \\ \chi \end{pmatrix} = 0\ . \tag{II.7}$$

The solutions are either $\chi = 0$ and $E = +M$ or $\Phi = 0$ and $E = -M$, corresponding to quarks and antiquarks. The requirement that $U\Psi$ should still be a solution of (II.7) gives

$$U^{-1}\beta U = \beta \quad \text{or} \quad [U,\beta] = 0\ , \tag{II.8}$$

i.e. U must be of the form

$$U = \left(\begin{array}{c|c} U_a & 0 \\ \hline 0 & U_b \end{array}\right), \tag{II.9}$$

where U_a and U_b are arbitrary. This means that the transformation should not mix the little and big components. The requirement that the norm be still conserved makes U_a and U_b unitary matrices: the corresponding group is referred to as $U(2)\times U(2)$ or $U(6)\times U(6)$, respectively.

Thus, as far as the Dirac equation (or the Wigner-Bargmann generalization of it to higher spin) is concerned, we can require invariance under the group $U(6) \times U(6)$, which are two $U(6)$ symmetries, one for quarks and one for antiquarks [for $U_a = U_b$, $U(6) \times U(6)$ coincides with $U(6)$]. Very loose speaking, we can say that for particles which are compounds of quarks and antiquarks this leads to higher multiplicity than the static $U(6)$ symmetr Thus, the mesons are assigned to a 36 rather than a 35. The 56 baryons, compounds of three quarks, continue to belong to a 56.

3. $U_W(6)$

Consider those special situations in which all particles move in line, as in forward or backward scattering. The Dirac equation for this case reads

$$(M\beta + \alpha_z k_z - E)\Psi = 0 \ , \qquad \text{(II.10)}$$

where we have taken all momenta in the z direction. Then requiring that $U\Psi$ be still a solution of this equation we get

$$[U,\beta] = 0 = [U,\alpha_z] \ . \qquad \text{(II.11)}$$

Again we take U unitary to conserve the norm. By this we have defined the $U_W(2)$ or $U_W(6)$ groups. Here, we still ask what are the infinitesimal generators of these groups. Writing $U = 1 + i\,\delta U$, we can expand the Hermitian δU in the 16 matrices

$$1, \beta, \gamma_5, i\beta\gamma_5, \vec{\sigma}, \vec{\alpha}, \beta\vec{\sigma}, ; \beta\vec{\alpha} \ .$$

Of these, those which commute with β and α_z are

$$1, \sigma_z, \beta\sigma_x, \beta\sigma_y \ . \qquad \text{(II.12)}$$

They have the same commutation relations as the Pauli matrices which generate the U(2) group of rotations of a two-spinor at rest: $1, \sigma_z, \sigma_x, \sigma_y$. The W in $U_W(2)$ reminds us that this group is not exactly the rotation group for particles at rest, but a certain group for collinearly moving particles.

4. $U(3) \times U(3)$

A more general scattering situation, in fact, the general scattering situation for two-body scattering in the c.m., is that all momenta lie in a plane. The Dirac equation for this case reads, if we take the plane of the momenta to be perpendicular to the x direction:

$$(M\beta + \alpha_z k_z + \alpha_y k_y - E)\Psi = 0 . \tag{II.13}$$

The invariance condition for unitary transformations requires

$$[U,\beta] = [U,\alpha_z] = [U,\alpha_y] = 0 . \tag{II.14}$$

This defines the group $U(1) \times U(1)$ or, including the unitary spin, $U(3) \times U(3)$. The infinitesimal generators of this group contain only the two commuting matrices

$$1 \text{ and } \beta\sigma_x . \tag{II.15}$$

Summary: Up to now the quest for possible symmetries of the S matrix has brought three unitary groups

$$U(6) \times U(6),\ U_W(6),\ U(3) \times U(3)$$

which satisfy the minimal kinematic conditions for considering them at all: they transform free Dirac quarks again into free Dirac quarks. All three groups can be regarded equally as subgroups of U(12) or $\widetilde{U}(12)$. (Check the generators!) Before discussing the difficulties of using these groups we will present another viewpoint which reduces to the same thing in the end.

5. Inhomogeneous $\tilde{U}(12)$ or $\tilde{U}(12) \times T_{143}$

Literature: Bell and Ruegg, CERN preprint TH 559;
Matthews et al. (Imperial College preprint ICTP/65/18).

We recall the Lorentz invariance of the Dirac equation (II.6). Under a Lorentz transformation Ψ goes into

$$\Psi' = S\Psi ,$$

where S is a certain 4×4 matrix and, in addition, the momentum k is transformed into

$$k_\mu' = L_{\mu\nu} k_\nu .$$

For Eq. (II.6) to be invariant we must have sufficiently

$$S^{-1} i\gamma_\mu k_\mu' S = i\gamma_\mu k_\mu$$

which defines the relation between the transformation S of Ψ, and the trans- formation L of k. Remember that S cannot be chosen unitary, but that one can always require it to satisfy

$$S^+ \beta S = \beta .$$

The Lorentz transformations S form, therefore, a subgroup of $\tilde{U}(4)$. The distinguishing feature is that for Lorentz transformations $S^{-1}\gamma_\mu S$ is a linear combination of only four γ's, whereas for a general matrix of $\tilde{U}(4)$ each γ_μ is transformed into a combination of all 15 matrices Γ_μ ($\mu = 1, \ldots, 15$). To generalize the idea of Lorentz invariance to $\tilde{U}(4)$, let us formally associate with each of the 15 Γ's a momentum k. Then we can define a 'super-Lorentz' transformation of those momenta for which we demand

$$\tilde{U}^{-1} \sum_{\mu=1}^{15} i\Gamma_\mu k_\mu' \tilde{U} = \sum_{\mu=1}^{15} i\Gamma_\mu k_\mu . \qquad \text{(II.16)}$$

The corresponding super-Dirac equation which is left formally invariant under the $\widetilde{U}(4)$ group is

$$\left(M + i \sum_{\mu=1}^{15} \Gamma_\mu k_\mu\right)\Psi = 0 .$$

We can still further generalize by restoring the unitary spin indices such that there are 143 matrices which transform into one another: the direct products of 16 4×4 matrices with the 9 λ's, less the unit matrix. Thus, we then have 143 'momenta'. What is to be done with all these extra momenta? We can simply suppose that they are zero for physical particles. Thus, we could specify that the S matrix elements are to conserve the total 143 component momenta (i.e. to have translational symmetry in the space of 143 dimensions). This would mean full $\widetilde{U}(12)$ symmetry with the momenta as well as the spinors. However, the only part of this S matrix that we use in physics are the elements between states with physical momenta $k_5 = k_6 = \ldots = k_{143} = 0$. On the face of it, this scheme seems totally out of touch with reality. But let us see what it means in practice. Suppose we start with an S-matrix element with specified physical momenta. In general, a transformation of $\widetilde{U}(12)$ will take this physical configuration of momenta into an unphysical one. Therefore, $\widetilde{U}(12)$ invariance of the scattering matrix brings physical elements to unphysical ones. Only in certain special cases will certain subgroups of $\widetilde{U}(12)$ provide transformations between physical configurations only. It is no accident that these situations are those already just discussed, from the point of view of invariance of the physical Dirac equation:

- all particles at rest $U(6)\times U(6)$
- all momenta collinear $U_W(6)$
- all momenta coplanar $U(3)\times U(3)$

Thus, as regards actual restrictions on physical scattering amplitudes, this scheme amounts to the one discussed above without introducing extra momenta. However, the scheme in 143 dimensions may have some use because it gives a way of constructing manifestly Lorentz invariant and crossing symmetric matrix elements with the required symmetry.

6. Inhomogeneous GL(6,c)

The schemes so far described are more properly called relativistic $U(6)\times U(6)$ rather than relativistic $U(6)$. To eliminate the extra degree of freedom whereby quarks and antiquarks are transformed independently, one can, additionally to (II.8), require commutation with γ_5

$$[U,\gamma_5] = 0 \ . \tag{II.17}$$

In a γ representation where $\gamma_5 = \begin{pmatrix} & -1 \\ -1 & \end{pmatrix}$ this requires that for the matrix (II.9) $U_a = U_b$. Then this group is precisely $U(6)$.

It is useful to study the conditions (II.8) and (II.17) in a representation where

$$\gamma_5 = \begin{pmatrix} 1 & \\ & -1 \end{pmatrix} \quad \beta = \begin{pmatrix} & 1 \\ 1 & \end{pmatrix} .$$

Commutation with γ_5 requires U to be of the form $U = \begin{pmatrix} A & \\ & B \end{pmatrix}$, where A and B are any 6×6 matrices. The remaining condition then gives

$$A^+B = B^+A = 1 \quad \text{or} \quad B = (A^+)^{-1} \ .$$

Thus, the transformation depends on a single arbitrary 6×6 matrix A, and for this reason the group is called GL(6,c): the group of general (G) linear (L) transformations in 6 dimensions with complex (c) coefficients. This group has been used especially by Rühl (CERN TH 505, 1964 and TH 515, 1965), with the trivial restriction to unit determinant in A, giving SL(6,c).

We still mention the appropriate subgroups of GL(6,c):

- particles at rest U(6)
- particles in line U(3) × U(3)
- particles in plane U(3) .

III. COLEMAN'S THEOREMS

1. The statement

We saw that it was rather difficult to make a field theoretic version of U(12) or $\tilde{U}(12)$ symmetries in which those symmetries were explicitly built into the Lagrangian. Therefore, we were reduced to trying to impose symmetries upon the S matrix, and we found that this was sensible for only certain subgroups. The Cal-Tech group*) has attempted, nevertheless, to retain some connection with field theory. Consider the observation that the set of operators

$$A(V) = \int d^3x \; \Psi^+(\vec{x}) \, V \, \Psi(\vec{x}) \, , \qquad \text{(III.1)}$$

where the Ψ's are quark field operators, has the commutation relations

$$[A(V_i), A(V_j)] = A([V_i, V_j]) \; . \qquad \text{(III.2)}$$

That is to say that the algebra of the A's is identical with the algebra of the infinitesimal generators V of the group. They therefore proposed that perhaps these were indeed the representations of the symmetries we are concerned with in the field theoretic Hilbert space.

Let $|1\rangle$ be a one-particle state in Hilbert space corresponding to a wave function Ψ, and $|1'\rangle$ another corresponding to Ψ'; then a transformation formerly specified in terms of wave functions by

---

*) R.F. Dasken and M. Gell-Mann, Physics Letters 17, 142 and 145 (1965).

$$\Psi' = (1 + i\epsilon_j V_j)\Psi \qquad \text{(III.3)}$$

can now be written equally

$$|1'\rangle = \Big(1 + i\epsilon_j A(V_j)\Big)|1\rangle . \qquad \text{(III.4)}$$

Now, as we stated already, only the subgroups $U(6) \times U(6)$, $U_W(6)$ and $U(3) \times U(3)$ in restricted physical situations are admissible. There is no reason to expect these difficulties to disappear when the discussion is transferred to Hilbert space, and it is not surprising that Gell-Mann et al. find by a study of Lorentz invariance of the commutatio relations that it is not possible to act with (III.1) on any one-particle state and have a symmetry transformation. It is only reasonable to attribute symmetry character to the mentioned subgroups: only restricted sets of operators (III.1) acting on restricted states might conceivably be symmetries. For example, if $|1, \vec{p} = 0\rangle$ is specifically a state of one particle at rest, they assume that

$$A(V_i)|1, \vec{p} = 0\rangle = |1', \vec{p} = 0\rangle \qquad \text{(III.5)}$$

[A belonging to the algebra of $U(6) \times U(6)$] is again a one-particle state and a member of the same multiplet as the original ("no leakage"). The same was assumed to hold for the $U_W(6)$ and $U(3) \times U(3)$ groups for one-particle states with appropriately restricted motion.

However, it was shown by Coleman (CERN preprints TH 591 and TH 595) that the general assumptions just formulated were totally inadmissible, not because they would be only approximately valid, but because they were conceptually incorrect. Coleman's arguments go in the following three stages:

i) It is shown that if the one-particle states, appropriate to the multiplets of the specified subgroups, are converted into other one-particle states of the same multiplet by certain A's of the subgroup, then the vacuum is invariant under these A's:

$$A|0\rangle = 0 \quad \text{or} \quad e^{iA}|0\rangle = |0\rangle .$$

ii) It is shown that if the vacuum is invariant, then these A's actually commute with the Hamiltonian and are symmetry operators in the conventional unrestricted sense and, therefore, apply to any state whatever, not only those of the subgroups. Thus, "invariance of the vacuum is invariance of the world".

iii) This is a contradiction because we already knew that these operators were not fully acceptable as symmetries, but only for certain restricted states.

Thus, the initial hypothesis is untenable!

In what follows it will be tacitly supposed that the existence of particles with zèro mass can be ignored.

2. First theorem

To prove: invariance of the one-particle states is invariance of the vacuum. We give here a simplified version of the corresponding theorem of Coleman, using somewhat weaker assumptions. The theorem depends on a certain locality hypothesis which we formulate as follows. Consider the matrix element

$$\langle A'B'| \int d^3x \Psi^+ V\Psi |AB\rangle \qquad \text{(III.6)}$$

with A, B, A′ and B′ arbitrary particle systems. Now, specify the initial and final states to consist of two distinct localized parts such that the groups A and A′ are widely separated, by some distance λ, from the groups B and B′. Then, since we are concerned with an integral of a local operator, the following hypothesis is very plausible

$$< A'B' | \int \Psi^+ V \Psi \, d^3x \, | AB > \quad \xrightarrow[(\lambda \to \infty)]{}$$

$$< B' | B > < A' | \int \Psi^+ V \Psi \, d^3x | A > + < A' | A > < B' | \int \Psi^+ V \Psi \, d^3x | B > . \qquad \text{(III.7)}$$

This should be true for operators which have no vacuum expectation value, which is enough for us. If it is not implied by the axioms of axiomatic field theory, axiomatic field theory needs more axioms. Roughly what is being said here is that the integral can be thought of as the sum of two pieces, one of which acts in the region A, A′, and the other in the region B, B′.

Let us now take for A and A′ one-particle states, for B just vacuum, and for B′ any state other than vacuum. From this

$$< 1'B' | \int \ldots | 1 > \quad \xrightarrow[(\lambda \to \infty)]{} \quad < B' | 0 > < 1' | \int \ldots | 1 >$$

$$+ < 1' | 1 > < B' | \int \ldots | 0 > .$$

Now, apply this when V is any generator of $U(6) \times U(6)$ [remember (III.6)]. The left-hand side is zero by hypothesis, for we have initially a one-particle at rest and finally a many-particle state at rest (actually here localization comes in!). The first term on the right is zero because of $< B' | 0 > = 0$. Thus, we have

$$< B' | \int \ldots | 0 > = 0 \quad \text{or} \quad \int \Psi^+ V \Psi \, d^3x | 0 > = 0$$

i.e. the operator $\int d^3x \, \Psi^+ V \Psi$ annihilates the vacuum.

A delicate point is that the localization of the state $|1>$ is not entirely compatible (uncertainty principle) with having a particle precisely at rest. However, we can take the wavepacket to be as extended as we please, and so the particle as nearly at rest as we please, and subsequently make λ very much larger than the extent of the wavepacket. We leave to those who like such considerations a careful reformulation of the argument, or of the axioms, to supply any desired degree of rigour.

Intuitively the origin of this result is very clear: in an infinite space a one-particle state is nearly everywhere a vacuum. A transformation which takes it into another one-particle state must leave the vacuum unaltered.

3. Second theorem

To prove: the invariance of the vacuum is the invariance of the world. Consider the particular $U(6) \times U(6)$ generators associated with the space components of the axial-vector current (we omit again the unitary-spin generators)

$$V = \beta i \gamma_\mu \gamma_5 \quad (\mu = 1,2,3) .$$

We now have by hypothesis and theorem 1 that if

$$J_\mu(x) = \Psi^+ \beta i \gamma_\mu \gamma_5 \Psi , \quad \int d^3x J_\mu(\vec{x}) \, |0> = 0 \quad (\mu = 1,2,3) \qquad \text{(III.8)}$$

or for any state B with zero momentum [in virtue of (III.8) being translationally invariant]

$$< B, \vec{p} = 0 | J_\mu(\vec{x}) | 0 > = 0 \quad (\mu = 1,2,3) .$$

Then it follows that

$$< B, \vec{p} = 0 | \partial_\mu J_\nu(\vec{x}) - \partial_\nu J_\mu(\vec{x}) | 0 > = 0 \quad (\mu = 1,2,3,4) . \qquad \text{(III.9)}$$

To prove this consider, for example, the first term $< B, \vec{p} = 0 | \partial_\mu J_\nu | 0 >$. For $\mu = 0$, $\nu = 1,2,3$ we get zero in virtue of (III.8), for the time derivative introduces only a factor which is the energy of state B. For $\mu = 1,2,3$ the space derivative gives as a factor the momentum of B, which is zero: thus (III.9). We can go from states with $\vec{p} = 0$ to arbitrary states by Lorentz transformations; since the various components of the tensor $\partial_\mu J_\nu - \partial_\nu J_\mu$ transforms into one another, the matrix elements remain zero. Thus, it must be zero without restriction on the final state:

$$(\partial_\mu J_\nu - \partial_\nu J_\mu) | 0 > = 0 . \tag{III.10}$$

Now the following theorem [Federbush and Johnson, Phys.Rev. <u>120</u>, 1926 (1960)] exists: any local operator which annihilates the vacuum is zero. The essential idea is that of crossing symmetry, which relates matrix elements from the vacuum

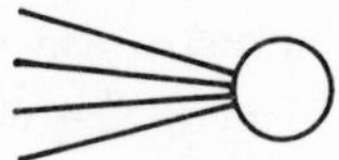

to other matrix elements

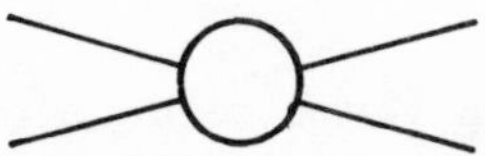

by analytic continuation. Then if all elements from the vacuum are zero, <u>all</u> matrix elements are zero. Then from (III.10)

$$\partial_\mu J_\nu - \partial_\nu J_\mu = 0$$

whence, in particular,

$$\frac{\partial}{\partial t} \vec{J} + \frac{\partial}{\partial \vec{x}} J_0 = 0$$

so that

$$\frac{\partial}{\partial t} \int \vec{J}\, d\vec{x} = 0$$

or

$$\left[\int \vec{J}\, d\vec{x},\, H \right] = 0 \ ,$$

where H is the Hamiltonian.

Thus, $\int \vec{J}\, d\vec{x}$ would be a conventional symmetry operator applicable on arbitrary states, which we know to be incompatible with Lorentz invariance.

One can equally prove that if for the fourth component of a vector current

$$\int d\vec{x}\; J_0(\vec{x})\, |0> = 0$$

then the integral again commutes with the Hamiltonian. As applied to quantities like charge, hypercharge, etc., this is not, of course, a difficulty.

Example: no interaction

It is instructive to illustrate the Coleman difficulties in a trivial case - a field theory of quarks without interaction. As the U(3) symmetry is not relevant we will speak of a single quark field.

First, let us compute the action on a single particle at rest of the space integral of a component of the axial current

$$\int d\vec{x}\; \Psi^{+}(\vec{x})\, \sigma_z\, \Psi(\vec{x})\; |\vec{k} = 0> . \qquad \text{(III.11)}$$

The resultant state is a superposition of states of one particle at rest, and of one particle at rest plus a pair with equal and opposite momenta. The two parts may be represented by diagrams:

and

respectively. From the familiar expansion

$$\psi = \frac{1}{\sqrt{V}} \sum_{k,\sigma} \left\{ a(\vec{k},\sigma) e^{ikx} u(\vec{k},\sigma) + b^{+}(\vec{k},\sigma) e^{-ikx} v(-\vec{k},\sigma) \right\} ,$$

where V is the quantization volume and u and v are positive and negative energy solutions of the Dirac equation, one obtains for the probability amplitude of the state with an extra pair with momenta $\vec{k}$ and $-\vec{k}$, spins σ and $-\sigma$,

$$u^{*}(\vec{k},\sigma)\, \sigma_z \; v(\vec{k},\sigma) \, . \tag{III.12}$$

The total contribution of such states to the norm of (III.11) is

$$\sum_{\vec{k},\sigma} |u^{*} \sigma v|^{2} \to V \int \frac{d\vec{k}}{(2\pi)^{3}} \, |u^{*} \sigma v|^{2} .$$

This goes to infinity as V goes to infinity, whereas the norm of the one-particle component of (III.11) remains finite. Thus, (III.11) is not at all a one-particle state, and the importance of many-particle states is proportional to the volume of the vacuum.

In contrast, if we consider the fourth component of the vector current, σ_2 in (III.12) is replaced by the unit operator. Because u and v are orthogonal all contributions of this kind are zero. That is to say, the non-interacting vacuum is indeed an invariant of the operator.

We can also show in this simple model that although the integrals of currents are not, in general, symmetry operators, in certain circumstances they behave as if they were. Consider the commutator relationship

$$\left[\Psi^{+} A \Psi, \int d\vec{x}\,\Psi^{+} B \Psi\right] = \Psi^{+}[A,B]\Psi ,$$

where A and B are any Dirac matrix. Take the matrix elements of this between one-particle states of momenta $\vec{k}$ and $\vec{k}'$

$$< \vec{k}' | \left[\Psi^{+} A, \Psi, \int d\vec{x}\,\Psi^{+} B \Psi\right] |\vec{k}> = < \vec{k}' | \Psi^{+}[A,B]\,\Psi | k> . \qquad \text{(III.13)}$$

Introduce a complete set of states $|n>$, so that the left-hand side becomes

$$\sum_{n} <\vec{k}' | \Psi^{+} A \Psi | n > < n | \int d\vec{x}\,\Psi^{+} B \Psi | \vec{k}>$$

$$- \sum_{n} <\vec{k}' | \int d\vec{x}\,\Psi^{+} B \Psi | n > < n | \; \Psi^{+} A \Psi | k> . \qquad \text{(III.14)}$$

Now, if the integral were indeed a symmetry operator it would be permissible to restrict the states $|n>$ to single-particle states belonging to the same multiplet as the initial or final state. We can show that this remains the case in the non-interacting model, provided B is restricted to be a generator of the appropriate subgroup

$$U(2) \times U(2) \quad \text{or} \quad U_W(2) \quad \text{or} \quad U(1) \times U(1)$$

or with unitary spin included

$$U(6) \times U(6) \quad \text{or} \quad U_W(6) \quad \text{or} \quad U(3) \times U(3)$$

according to whether $\vec{k}$ and $\vec{k}'$ are specified to be zero, to lie in a given line, or in a given plane.

Consider, for example, the first term in (III.14). The intermediate states that occur will either be one-particle or one-particle plus a pair. We wish to show that in the conditions mentioned only the one-particle contributions, diagrammatically

need be retained. The other contributions may be represented diagrammatically as either

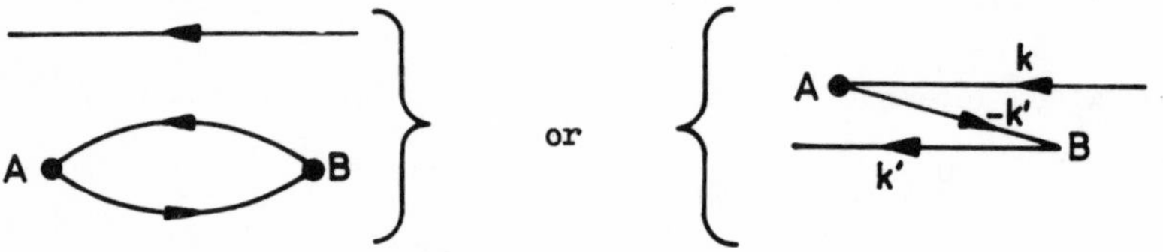

The contribution of the first type, which arises only when initial and final states are identical, differs only in a way to be allowed for below from the vacuum expectation value

$$< 0 | \Psi^{+} A \Psi \int dx \; \Psi^{+} B \Psi | 0 > \; .$$

When the commutator is formed this leads to the vacuum expectation

$$< 0 | \Psi^{+} [A,B] \Psi | 0 > \; ,$$

which, in general, is zero by Lorentz invariance, and can, in any case, be subtracted from both sides of (III.13).

A residual term from the first set of diagrams arises because of the exclusion principle, which forbids that the particle of the pair should be in the same state as the initial and final particle, so that this contribution has to be subtracted from the vacuum expectation value. The remaining pair diagrams also involve the creation from the vacuum, by $\int \Psi^{+} B\Psi$, of a pair in which the particle has the final momentum $\vec{k}'$, and the antiparticle momentum $-\vec{k}'$. In both cases matrix elements of B

$$u^*(\vec{k}')\ B\ v(\vec{k}')$$

are involved. These vanish when B is restricted to the appropriate subgroup. These subgroups are defined, precisely, by the condition that $B\,v(\vec{k})$ is again a solution of the Dirac equation with the same energy as v, in this case negative, so that it is orthogonal on the positive energy solution u.

Resumé on Coleman's theorems

Coleman has destroyed the original idea that certain integrals of currents acting on certain states should be symmetry operators. However, the bulk of the Cal-Tech programme requires only that they should behave as if they were symmetry operators in certain matrix elements of certain commutators. As illustrated above by a trivial example, Coleman's theorems are then irrelevant, and the plausibility or implausibility of the procedure for interacting fields is not altered by these theorems.

IV. UNITARITY

1. General remarks

Conservation of probability requires the unitarity of the S matrix:

$$S^+S = 1$$

or

$$(1 - S^+)(1 - S) = (1 - S) + (1 - S^+) . \qquad \text{(IV.1)}$$

The transition operator T is defined to have matrix elements

$$(b|1 - S|a) = i(2\pi)^4 \delta^4 (P_b - P_a)(b|T|a)$$

$$(c|1 - S^+|b) = -i(2\pi)^4 \delta^4 (P_c - P_b)(c|T^+|b) .$$

Taking matrix elements of (IV.1) between a and c, and summing over a complete set of states b

$$\sum_b (c|T^+|b)(2\pi)^4 \delta^4 (P_c - P_b)(b|T|a) = i(c|T - T^+|a) . \qquad \text{(IV.2)}$$

It is assumed here that $P_c = P_a$, otherwise we would have to include explicitly another energy momentum conserving δ function.

The difficulty of combining (IV.2) with our higher symmetries may be stated in two equivalent ways. The first is to note that we have different symmetry groups in different situations [$U_W(6)$, $U(3) \times U(3)$, etc.] while on the left-hand side we have a summation over all kinematical situations. Thus, if T and T^+ on the left have the required form, there is no guarantee that $T - T^+$ calculated from (IV.2) will have the required form. Thus, the unitarity condition threatens to be much more restrictive in conjunction with the new symmetries than with old ones. From another point of view, the new symmetries can be given unrestricted validity by introducing extra unphysical momenta. But the summation over b in (IV.2) is restricted to physical momenta only and is, therefore, not invariant.

Despite the fact that unitarity threatens to be very restrictive, it is not obvious a priori that the conditions posed by it cannot be met. In fact, it will appear that only in combination with crossing symmetry does it make unmeetable demands. That it does so has been shown for a special example which we will discuss. Unfortunately, no general treatment of this problem has so far appeared, but only some special examples discussed by means of special tricks.

2. Quark-scalar scattering

Consider the scattering of a particle which has neither spin nor unitary spin - a scalar - on a quark. As regards quark U(3) indices, there is only one way to couple them so let us forget about them, i.e. consider a spin-½ singlet. We then apply the symmetry $I\tilde{U}(4)$ [or $\tilde{U}(4) \times T_{15}$]. The general covariant amplitude is, in terms of initial and final Dirac spinors $\bar{u}'$ and u,

$$\bar{u}' \{ f + g\, i\Gamma_\mu (k_\mu - k_\mu') \} u \, , \qquad \text{(IV.3)}$$

where the μ summation is from 1 to 15; k and k′ are initial and final momenta of the scalar particle. Specifying that k and k′ are <u>physical</u> momenta this reduces to

$$\bar{u}' \{ f + g\, i\gamma_\mu (k - k')_\mu \} u \, . \qquad \text{(IV.4)}$$

But this is, in fact, the general form permitted by Lorentz invariance alone; other covariant forms reduce to it by means of the Dirac equations. Thus, in this case the imposition of the higher symmetry imposes no restriction; nor then does it make any trouble with unitarity.

Actually in the original underdeveloped $\tilde{U}(12)$ [or $\tilde{U}(4)$ when we discard U(3) indices] scheme the second term in (IV.3) would not be allowed, because momenta were not supposed to transform. There is, then, indeed a unitarity difficulty*). The permitted amplitude is

$$(b|T|a) = \bar{u}_b f u_a , \tag{IV.5}$$

where f is any function of the kinematical variables. Disregarding normalization factors we have in terms of two spinors Φ_a, Φ_b

$$u_a = \begin{pmatrix} \Phi_a \\ \dfrac{\sigma\cdot k_a}{E_a + M} \Phi_a \end{pmatrix} \qquad \bar{u}_b = \begin{pmatrix} \Phi_b^*, & \Phi_b^* \dfrac{-\sigma\cdot k_b'}{E_b + M} \end{pmatrix}$$

We work in c.m., and adopt units so that $E_a + M = E_b + M = 1$. Then

$$\begin{aligned} (b|T|a) &= f_{ba}\Phi_b^* \{1 - \sigma\cdot k_b\ \sigma\cdot k_a\}\Phi_a \\ (b|T^+|a) &= (a|T|b)^* \\ &= f_{ba}^*\Phi_b^* \{1 - \sigma\cdot k_b\ \sigma\cdot k_a\}\Phi_a , \end{aligned}$$

where f_{ba} is a function of energy and of the cosine of the angle between k_a and k_b:

---

*) M.A. Beg and A. Pais, Phys.Rev. Letters 14, 509 (1965);
R. Blankenbeckler et al., Phys.Rev. Letters 14, 518 (1965).

$$\sum_b (c|T^+|b)(2\pi)^4\,\delta^4\,(P_c - P_b)(b|T|a)$$

$$\propto \int d\Omega_b\, \Phi_c^* \left\{ (1-\sigma\cdot k_c\,\sigma\cdot k_b)(1-\sigma\cdot k_b\,\sigma\cdot k_a) f_{cb}^*\, f_{ba} \right\} \Phi_a\,.$$

Working only to lowest non-trivial order in k, and using

$$\begin{aligned} 1-\sigma\cdot k_b\,\sigma\cdot k_a &= 1-k_b\cdot k_a - i\sigma\cdot k_b\times k_a \\ &= 1-i\sigma\cdot k_b\times k_a + \cdots\quad, \end{aligned}$$

the unitarity condition requires

$$\begin{aligned} &(\mathrm{Im}\ f_{ca})\,(1-i\sigma\cdot k_c\times k_a) \\ &\propto \int d\Omega_b\, f_{cb}^* f_{ba}\Big(1-i\sigma\cdot (k_c-k_a)\times k_b\Big) \end{aligned}$$

with a proportionality constant independent of angle and spin. In particular, with a proportionality constant independent of spin

$$\begin{aligned} &1-\tfrac{1}{2}\,i\,\sigma\cdot (k_c-k_a)\times (k_c+k_a) \\ &\propto \int d\Omega_b\, f_{cb}^*\, f_{ba}\Big(1-i\sigma\cdot (k_c-k_a)\times k_b\Big)\,. \end{aligned}$$

Write

$$k_b = \frac{k_b\cdot (k_a+k_c)}{|k_a + k_c|^2}\,(k_a+k_c) + \cdots\quad,$$

where the extra terms are proportional to $k_a - k_c$, which does not contribute to the cross product, and $k_a\times k_c$, which cannot contribute to the integral by inversion symmetry. Then we need

$$\frac{\int d\Omega_b\, f^*_{cb}\, f_{ba}\, \frac{2k_b\cdot(k_a+k_c)}{|k_a+k_c|^2}}{\int d\Omega_b\, f^*_{cb}\, f_{ba}} = 1 .$$

This is not possible. For example, with $k_a = k_c$ (forward scattering) it reads

$$\frac{\int d\Omega\, |f|^2 \cos\vartheta}{\int d\Omega\, |f|^2} = 1 ,$$

which is not possible because $|\cos\vartheta| \leq 1$.

Thus, for underdeveloped $\tilde{U}(12)$ unitarity is impossible even in this trivial case. On the other hand, with $I\tilde{U}(12)$ there is no problem, but neither is there any restriction on the scattering amplitude above those dictated separately by U(3) and Lorentz invariance.

3. Quark-quark and quark-antiquark elastic scattering in $\tilde{U}(12)\times T_{143}$ or $U(3)\times U(3)$

Alles and Amati, Nuovo Cimento **39**, 758 (1965) have shown that when crossing symmetry is also required the unitarity condition cannot be met at the same time as $\tilde{U}(12)\times T_{143}$ symmetry in this particular example. In their argument the unitary spin indices, as well as the ordinary spin indices, play an essential role and cannot be ignored.

Consider first quark-quark scattering. We will work directly in the centre-of-mass system, and use at once the relevant subgroup $U(3)\times U(3)$. The argument, of course, does not depend on whether we regard this as a subgroup of $\tilde{U}(12)\times T_{143}$ or of U(12), or of anything whatsoever.

Remember that the generators are

$$\lambda, \ \sigma_z \beta \lambda,$$

where the z direction is chosen to be the normal to the scattering plane. We need work explicitly with only the big components of the spinors. For these, β has the eigenvalue +1 and our effective generators are just

$$\lambda, \ \sigma_z \lambda \ \text{ or } \ (1 \pm \sigma_z)\lambda \ . \tag{IV.6}$$

This means that we have two separate U(3) symmetries, one referring to particles with spin up, and the other to particles with spin down. First, for the scattering there are the following spin transitions, if one takes into account that the total number of up spins and the total number of down spins must be conserved:

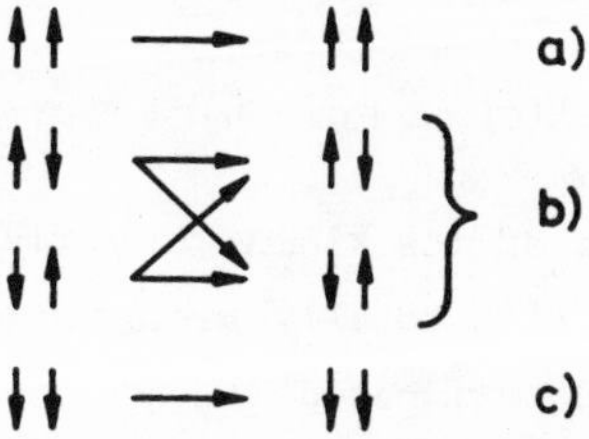

From the point of view of spin combinations there are six amplitudes. Now we still have to add the unitary symmetry U(3). For the first case a) we have two triplets which can be combined to a singlet or an octet which makes two amplitudes. For the four spin transitions of b) there is only one U(3) amplitude at a time, because in this case the unitary transformations on up and down quarks being independent, I can contract indices in only one way. For c we have again two unitary amplitudes, so in sum we get eight independent amplitudes for the scattering [without $U(3) \times U(3)$ symmetry there would be more!]

A particular choice for the operators that can enter the transition matrix is the following (the index 1 or 2 refers to the first or second quark with whose indices σ will be contracted):

$$\begin{array}{llll} 1 & \vec{\sigma}_1 \cdot \vec{n} & \vec{\sigma}_2 \cdot \vec{n} & \vec{\sigma}_1 \cdot \vec{n}\vec{\sigma}_2 \cdot \vec{n} \\ P & \vec{\sigma}_1 \cdot \vec{n}P & \vec{\sigma}_2 \cdot \vec{n}P & \vec{\sigma}_1 \cdot \vec{n}\vec{\sigma}_2 \cdot \vec{n}P \; . \end{array} \qquad \text{(IV.7)}$$

Here $\vec{n}$ is the direction normal to the scattering plane ($\vec{n} \sim \vec{k}_1 \times \vec{k}_2$) and P is the complete exchange operator: it exchanges the momenta as well as the spin and unitary spin indices of the two quarks so that, in the matrix elements, contractions between 1 and 2 can occur. Conservation of statistics requires the transition amplitude to be symmetric in 1 and 2, so the number of coefficients reduces to 6

$$T \sim A + B(\vec{\sigma}_1 \cdot \vec{n} + \vec{\sigma}_2 \cdot \vec{n}) + C\, \vec{\sigma}_1 \cdot \vec{n}\, \vec{\sigma}_2 \cdot \vec{n}$$

$$+ DP + E(\vec{\sigma}_1 \cdot \vec{n} + \vec{\sigma}_2 \cdot \vec{n})P + F\, \vec{\sigma}_1 \cdot \vec{n}\, \sigma_2 \cdot \vec{n}\, P .$$

A, B, ..., F are functions of the kinematic variables s and t. However, as we are concerned only with totally antisymmetric initial and final states, the condition can be imposed

$$PT = -T$$

so there are effectively only three independent coefficients, A, B and C:

$$T \sim A + B(\vec{\sigma}_1 \cdot \vec{n} + \vec{\sigma}_2 \cdot \vec{n}) + C\, \vec{\sigma}_1 \cdot \vec{n}\, \vec{\sigma}_2 \cdot \vec{n}$$

$$- \Big(A + B(\vec{\sigma}_1 \cdot \vec{n} + \vec{\sigma}_2 \cdot \vec{n}) + C\, \vec{\sigma}_1 \cdot \vec{n}\, \vec{\sigma}_2 \cdot \vec{n} \Big) P \; . \qquad \text{(IV.8)}$$

Finally, let us write this in the form

$$T = t(1 - P) = (1 - P)t$$

and proceed to study unitarity.

We have

$$T^+ = t^+(1 - P)$$

and

$$T^+T = t^+(1 - P)(1 - P)t$$
$$= 2t^+ t(1 - P) \ .$$

Thus, the unitarity equation needs essentially

$$2\int d\Omega_b \ t^+ t(1 - P) = i(t - t^+)(1 - P) \,,$$

where the integration is over intermediate momenta. Since t is the unit operator in unitary spin space, and P is proportional to the exchange operator in that space, the terms with and without P must be separately equal. Thus

$$\int d\Omega_b \ 2t^+t = i(t - t^+) \,,$$

which is much simpler.

When the integration is performed, the left-hand side can, of course, be expanded as a sum of terms at most bilinear in σ_1 and σ_2. In general, this will include terms of the type

$$\sigma_1 \cdot k_a \sigma_2 \cdot k_c + \sigma_1 \cdot k_c \sigma_2 \cdot k_a$$
$$\sigma_1 \cdot k_a \sigma_2 \cdot k_a + \sigma_1 \cdot k_c \sigma_2 \cdot k_c$$

which are not present on the right. The requirement that these terms have zero coefficients imposes constraints on A,B and C. In particular, Alles and Amati find in the limit $k_a \to k_c$ restrictions of the form

$$\int d\Omega |B - C \cos \vartheta|^2 = 0$$

whence

$$B = C \cos \vartheta \qquad (IV.9)$$

and

$$0 = \mathrm{Re} \int d\Omega [CA^* + CC^* \cos^2\vartheta] \sin^2 \vartheta \qquad (IV.10)$$

It should be noted that there is no difficulty in meeting these requirements, for example, by setting $B = C = 0$. This particular amplitude, only $A \neq 0$, can, in fact, be made to satisfy all the unitary conditions, and it is a general rule (discussed by Bell, CERN preprint TH. 573) that unitarity is consistent with the symmetry so long as crossing is disregarded.

However, according to crossing symmetry, quark-antiquark scattering is described by an analytic continuation of the amplitude for quark-quark scattering. Unitarity in the quark-antiquark channel then imposes further restrictions. To find them, one makes a covariant amplitude which reduces to (IV.8), with (IV.9), in the quark channel. This involves two amplitudes F_1 and F_2, related to A and B, in terms of which (IV.10) becomes

$$\mathrm{Re} \int d\Omega \, F_1^* F_2 \sin^2 \vartheta = 0 \ . \qquad (IV.11)$$

This same covariant amplitude is used in the crossed channel, reduced to a two-component form in the centre-of-mass system for that channel, and unitarity applied. Amati and Alles find then the additional necessary condition:

$$[(\sqrt{S} + 2m)^2 - u] F_2 + [(\sqrt{S} - 2m)^2 - u] F_1 = 0 \ ,$$

where s, t, u are the usual kinematical variables. Then (IV.11) becomes

$$\int d\Omega \, |F_1|^2 \, \frac{(\sqrt{S} - 2m)^2 - u}{(\sqrt{S} + 2m)^2 - u} \sin^2 \vartheta = 0 \; .$$

The kinematical factor is positive in the quark-quark channel, so we have finally

$$F_1 = F_2 = 0 \; .$$

There is no scattering.

As observed by the authors mentioned there are a number of conceivable ways around this difficulty. Quarks might not exist. Even if they do exist, they are presumably heavy, so that quark and antiquark can annihilate into a number of ordinary particles; then there would not be any elastic channel for quark-antiquark scattering, however low the initial centre-of-mass kinetic energy. Thus, it would be more satisfactory to demonstrate such a contradiction with multiplets of ordinary particles; such work is proceeding. Finally, we might have to fall back on the more modest group, inhomogeneous GL(6,c) [or the equivalent chain of subgroups U(6), $U(3) \times U(3)$, U(3)] for which no unitarity crossing difficulty has yet been demonstrated. Since this last group imposes new restrictions on two-body scattering in the forward direction only, there may not, indeed, be any such difficulty.

General references

Everyone will agree, for varying reasons, that it is less work to give such lectures than to document them. We therefore refer here only to the extensive bibliographies in the lectures of Professor Pais, and in the following recent works: R. Delbourgo et al., "The $\widetilde{U}(12)$ symmetry", Trieste preprint IC/65/57; W. Rühl, "The SL(6,c) model", Trieste seminar 1965, paper SMR 2/20; R.Oehme, "Current algebras and approximate symmetries", Chicago preprint EFINS 65-90.

CP VIOLATION*)

J. Prentki,
CERN

I. INTRODUCTION

Last year at the Dubna conference the very important result of the experiment of Christenson et al.[1]) was announced. They found that the long lived component K_L of the K_0, $\bar{K}_0$ system decays into $\pi^+\pi^-$ with a branching ratio of about $2 \cdot 10^{-3}$ with respect to $K_L \to$ all charged. As is well known, the usual theories of this system incorporating CP invariance forbid this decay absolutely; in fact, the absence of this decay was considered as the best evidence for CP conservation in all types of known interactions. The most direct interpretation of the result of Christenson et al.[1]) was to admit CP violation in weak interactions. However, it was very soon realized that the occurrence of $K_L \to 2\pi$ does not necessarily imply that CP is violated. Many physicists were reluctant to abandon CP or T invariance, and it is therefore not astonishing that many models were proposed allowing $K_L \to 2\pi$, but with intrinsic conservation of CP. It seems to me worth while to discuss some of them briefly, and to show how during this year the situation has evolved in the direction of a general belief in the breakdown of CP invariance.

The most beautiful hypothesis along this line was introduced by Bell and Perring[2]) and also by Bernstein et al.[3]) who supposed the existence of a long-range force that, analogously to electromagnetic interactions, couples differently to particles and antiparticles, but for instance couples to hypercharge rather than charge. The galaxy

*) Reproduced from the Proceedings of the Oxford International Conference on Elementary Particles (1965) by kind permission of Dr. Stafford.

would induce a potential at the earth, and due to the preponderance of matter over antimatter in our galaxy this would give rise to an apparent CP violation. As the potential energy of K_0 and $\bar{K}_0$ would be different, the eigenstates of the Hamiltonian would be no longer K_{02} (CP = -1) and K_{01} (+1) but some states K_L and K_S differing by a small amount from K_{02} and K_{01}

$$K_L = K_{02} + \epsilon\, K_{01}$$

$$K_S = K_{01} - \epsilon\, K_{02} \;.$$

Clearly then K_L may decay into 2π, with a rate determined by the parameter ϵ. This parameter can be calculated in terms of the galactic potential field, and it turns out that if this galactic field is a vector field, the quantity ϵ is proportional to the particle energy. Now, the rate $K_L \to 2\pi$ is proportional to ϵ^2, and we have the peculiar result that the branching ratio varies with the square of the energy of the particle in the lab. system, or more precisely, in the system at rest with respect to the galaxy. This is not astonishing because the presence of an external field can, of course, aside from an apparent PC violation, also lead to an apparent violation of Lorentz invariance. Two experiments[4,5]) have been performed to test this prediction, at Nimrod with a beam containing neutral kaons with an energy of up to 5 GeV, and at CERN with a beam containing kaons of about 10 GeV. The branching ratios found were $2.08 \pm 0.35 \cdot 10^{-3}$ and $2.23 \pm 0.25 \cdot 10^{-3}$, respectively, whereas an E^2 dependence would have aimpled $13.4 \pm 3.0 \cdot 10^{-3}$ and $1.6 \pm 0.3 \cdot 10^{-1}$ using the experimental value of Christenson et al.[1]) as input. These measurements are compatible with constant ϵ, and exclude the galactic theory with E^2 dependence.

Another model was proposed by Lévy and Nauenberg[6]). They postulated the existence of a vector particle s with a mass smaller than the K_L-K_S mass difference, and with PC = -1. The K_{02} decay can then proceed according to the following graph:

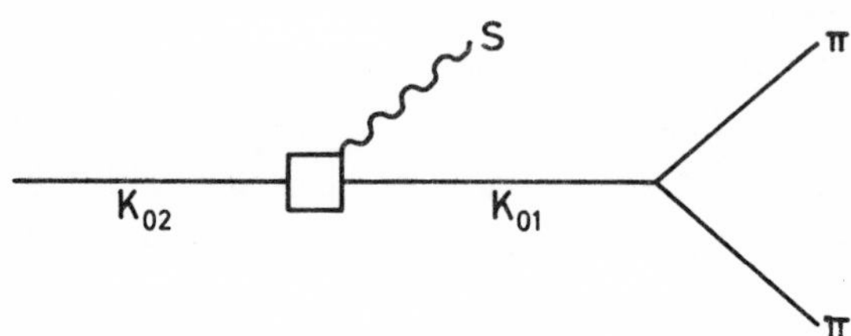

(The Lévy-Nauenberg theory suffers severely, and the Bell-Perring-Bernstein-Cabibbo-Lee theory to some extent, from a difficulty pointed out by Weinberg[7]. This difficulty occurs in all cases of very light vector bosons coupled to imperfectly conserved currents. We do not develop this objection, since these theories are now dead empirically.)

As the final state of the K_{o2} decay is $\pi^-\pi^+s$ this theory leads to an obvious prediction, namely that there can be no interference effects between the final states of K_{o2} ($= \pi^+\pi^-s$) and those of the K_{o1} ($= \pi^+\pi^-$) final state obtained by regeneration of the K_{o2}.

Recently, Fitch[8] has demonstrated interference effects, thus excluding the Lévy-Nauenberg hypothesis. I would like to add that the above-mentioned experiments also exclude all models where new particles K_o' degenerate with K_o or π' degenerate with π are introduced on the basis of some higher symmetry. In the first case the decay rate should be energy dependent, as the K_o' production cross-section relative to the $K_o, \bar{K}_o$ production should be a function of the energy: in the second case no interference effects are expected.

In conclusion, we may say that almost all theories or models introducing new particles or new fields are already excluded. An exception is the model of Gürsey and Pais (unpublished) which - leaving aside any cosmological interpretation - is a special case of the super-weak theory of Wolfenstein, to be discussed later on.

There exists still another class of models that conserve CP and lead to $K_L \to 2\pi$. The price which one has to pay in order to save CP becomes extremely high. I will mention only one of these possibilities[9] leading to very definite experimental predictions. In proving that $K_{02} \to \pi^+\pi^-$ is forbidden by CP one must use Bose statistics for pions, because CP on the $\pi^+\pi^-$ system exchanges the pions. If Bose statistics are not valid, the state $\pi^+\pi^-$ is no more an eigenstate of CP. However, the $\pi^0\pi^0$ state continues to have definite CP, = +1. Thus in such a theory $K_L = 2\pi^0/(\pi^+\pi^-) = 0$. This possibility is excluded by the interference experiment because no interference effect is expected between the symmetric $\pi^+\pi^-$ from K_{01} and the antisymmetric $\pi^+\pi^-$ from K_{02}.

The remaining schemes are even more unpleasant. For instance, in one of these models the fundamental principle of superposition is violated[10], in another one, a shadow universe[11] is introduced.

It seems therefore reasonable to admit that CP or T conservation is not an absolute law of nature. In what follows I will adopt the point of view that CP is violated, CPT being conserved. The question is then to examine where and how CP can be violated, and what can be the experimental implications of the various models proposed in this context.

II. GENERAL REMARKS

The various models which were introduced in order to explain CP violation can be divided in four classes covering all possible cases.

1) CP is violated by an interaction H_V which conserves P and S, but violates C, with a coupling constant somewhat smaller than the coupling constant of strong interactions.

2) CP violation is intimately connected with the electromagnetic interactions of hadrons, P and S being conserved.

3) The CP violation appears at the level of weak interactions with a coupling constant smaller than the CP conserving coupling, and with $\Delta S < 2$.

4) There exists some superweak interaction, with $\Delta S = 2$

In case 1) the CP violation in weak interactions is due to the interplay of the CP conserving interaction H_W and the C violating interaction, H_V. It occurs as a second order effect $H_V - H_W$. Due to the weakness of the coupling constant occurring in H_V the CP violating part should be smaller than the CP conserving part.

Similarly, in case 2), the difference mainly being that the CP violation is connected with the occurrence of photons, virtual or real. The expected order of magnitude of violation in $K_0 \rightarrow 2\pi$ decays is about $\alpha = 1/137$, which is of the correct order of magnitude.

In case 4), CP violation in weak interactions is due to the contribution in first order of H superweak to the mass matrix describing the decaying states K_L and K_S of the $K_0, \bar{K}_0$ system.

Theories of type 1)

Let us begin with a discussion of case 1 and some of its experimental implications[12)]. As a first remark, we note that according to a theorem by Soloviev, Pais and others[13)], one may find it plausible that no C violation occurs in SU_3 conserving interactions. To be more precise, this would be true if the fundamental Lagrangian would be a Yukawa type interaction between baryons and pseudoscalar mesons or vector mesons. Cabibbo[14)] has recently reanalysed the problem and has deduced a number of theorems of this nature for matrix elements rather than for interaction Lagrangians; it appears that in many cases of practical interest C violation even in very strong interactions only shows up at the level of SU_3 breaking interactions.

We will then suppose that the CP violation enters on the level of SU_3 breaking interactions, and for definiteness assume also that I spin is conserved. This is not absolutely necessary: generally speaking we can say that I spin is conserved up to about 1%, and if the C violating interactions break isospin also the associated coupling constant must be very small.

The best systems for experimental tests are obviously systems that are eigenstates of C or of G. We discuss therefore the mesonic resonances and the very important $\bar{p}p$ system.

π^0 decay

$\pi^0 \to \gamma \to e^+e^-$ is forbidden in lowest order of α because of partiy or gauge invariance.

$\pi^0 \to 3\gamma$ is now allowed. The rate for this transition was estimated by F. Berends[14] and found to be very small, about 10^{-6} times $\pi^0 \to 2\gamma$ at most. This reaction was recently remeasured at CERN[15] and at Dubna and found to be smaller than $10^{-5}(\pi^0 \to 2\gamma)$. Note that this decay can also occur in the case of class 2) C violation.

η decay

The η decay offers interesting possibilities for observing C violating interactions in semi-strong and also electromagnetic interactions. The main decay mode $\eta \to 3\pi$ is most probably of electromagnetic nature, because the final state has predominantly I spin 1 (and the same C as the η) as is seen from the uniformity of the Dalitz plot. Any C violating mode $\eta \to 3\pi$ could interfere with this main mode and give rise to striking effects. From the fact that G parity of the final 3π system is -1, and from the relation $G = (-1)^I . C$, one sees that C violating interactions lead to states with I = 0 and I = 2. Leaving out the uninteresting I = 3 final state, the following table may be constructed:

ΔI	Strength	C behaviour
0	gk^3 or e^2k^3	violated
1	e^2	conserved
2	ge^2k or e^2k	violated

Here g is the coupling constant of the C violating I spin conserving interaction; we added also the case that the C violation is in electromagnetic forces. k represents angular momentum barrier effects:

$$k = \frac{m_\eta Q}{M^2}$$

where m_η is the η mass (550 MeV), Q is the average kinetic energy of the pions, about 50 MeV, and M is some unknown reference mass, certainly larger than the pion mass. In certain models discussed later, where some basic C violating interaction involving the ρ meson is assumed, one finds that M is the ρ mass.

From the above table it is seen that angular momentum barriers play a very important role, in particular in the case of a theory with $\Delta I = 0$. It is difficult to estimate the magnitude of the interference effects without specific models, to be mentioned in a moment. Moreover, final state interactions play an important role; without final state interactions the relative phase of C conserving and C violating amplitudes is $\pi/2$, implying no interference. Nevertheless, some qualitative statement can be made. C conservation implies identical π^+ and π^- energy spectra, and a measure of C violation may be the ratio

$$R = \frac{N_b(E_{\pi^+} > E_{\pi^-})}{N_b(E_{\pi^-} > E_{\pi^+})}$$

$R \neq 1$ implies C violation. This ratio was experimentally studied, but the situation is not very clear. The world data show a possible 8% effect, two standard deviations away from no effect. If such an effect would be established, a careful study of the Dalitz plot could reveal whether mode 0 or 2 is the interfering one. From the table we deduce the following:

i) if $g \leq 10^{-2}$ no effect is expected;

ii) if $g \sim 10^{-1}$ both $\Delta I = 0$ and $\Delta I = 2$ modes may become interesting;

iii) a C violating, I spin breaking ($\Delta I = 2$) interaction with $g \sim 10^{-2}$-10^{-3} (which may or may not be connected with electromagnetic interactions) could give rise to an important $\Delta I = 2$ amplitude.

Most probably g must be smaller than 10^{-2} (as will be discussed later) so only if $\Delta I = 2$ is present, may effects be expected.

Various models were proposed[16] in order to get a better insight into this question. These models led to various predictions not always in mutual agreement. They give, however, already now some limitations on the coupling constants and some relations that can be useful. I will now discuss very briefly the model of Barrett et al.[17]

Let us suppose that C violation is due to an interaction $\Delta I = 0$ of the form

$$g_\eta \rho_\mu^i \left(\eta \partial_\mu \pi^i - \partial_\mu \eta \pi^i \right) ,$$

where i refers to I spin. The η decay could then proceed according to the graph

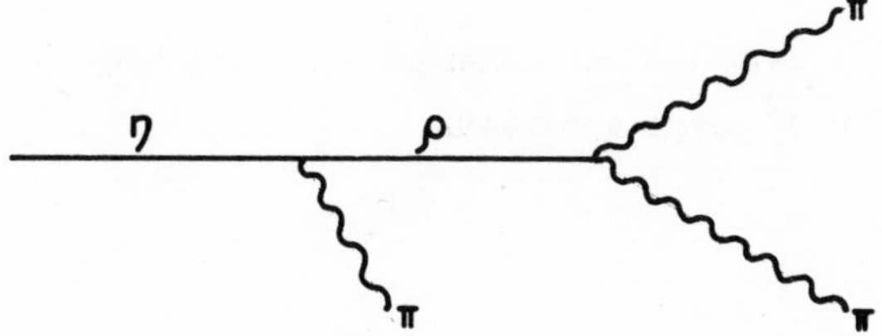

The $\Delta I = 0$ C violating amplitude can be explicitly calculated in terms of g_η and $g_{\rho\pi\pi}$ known from $\rho \to 2\pi$. The $\Delta I = 1$ C conserving amplitude is known: it is not energy dependent, and its magnitude may be obtained by comparing $\eta \to 3\pi$ with $\eta \to 2\gamma$ which through SU_3 may be related to the decay $\pi \to 2\gamma$. The interference term is then a function of g_η and of the difference of the phases of the final state interactions for the $I = 0$ and $I = 1$ states, respectively.

On the other hand, the C violating decay $\eta \to \pi^0 e^+ e^-$ may occur:

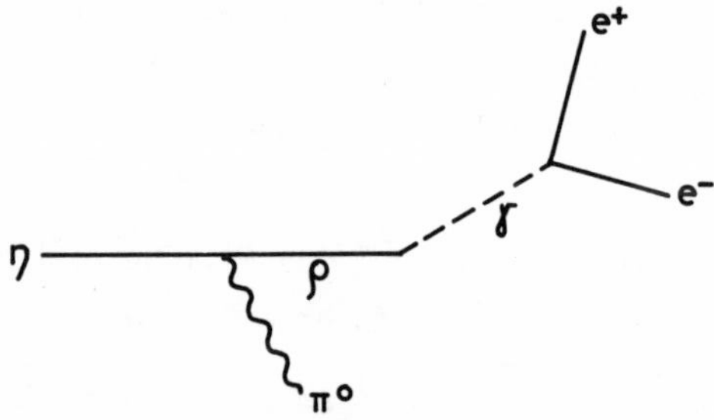

which can be calculated, and depends on g_η and the known $g_{\rho\gamma}$ coupling constants. From the experimental data[18,19]

$$\frac{\eta \to \pi^0 e^+ e^-}{\eta \to 3\pi} < 1\%$$

and the estimate $\Gamma_\eta \sim 300$ eV one can deduce

$$\frac{g_\eta^2}{4\pi} < 4 \cdot 10^{-2}$$

and from this an asymmetry in $\eta \to 3\pi$ of maximally a few percent is found.

Obviously a detailed experimental investigation of the $\eta \to 3\pi$ decay would be of great interest.

$\underline{X^{o}}$

X^{o} has the same quantum numbers as the η, the only difference being that η is supposedly member of an SU_3 octet, while the X^{o} is an SU_3 singlet. Everything said about η may be repeated here. The interesting mode $X^{o} \to 3\pi$ is experimentally known to be less than 5%[20]. The X^{o} width is known to be below 4 MeV. This gives already a limit on a coupling $g_{X^{o}} X^{o} \rho^{i} \pi^{i}$, namely:

$$\frac{g_{X_o}^2}{4\pi} < 10^{-2} .$$

If we use Dalitz and Sutherland's[21] estimate

$$\Gamma_{X^o} \sim 0.1 \text{ MeV} - 0.5 \text{ MeV}$$

(which is reasonable in view of the 20% rate $X^{o} \to \pi\pi\gamma$) we get the much more severe limit:

$$\frac{g_{X_o}^2}{4\pi} < 4 \cdot 10^{-4} - 2 \cdot 10^{-3} .$$

Barrett et al.[17] applying their model to the decay $X^{o} \to \pi e e$ find

$$\frac{X^{o} \to \pi e e}{X^{o} \to \eta \pi \pi} < 5 \cdot 10^{-6}$$

if C violating interactions conserve isospin. This is an important result, because detection of this effect will mean that C violation occurs in electromagnetic interactions.

There are several other decay modes of mesons and vector mesons that could be interesting. They have been discussed in many of the above-mentioned papers.

Let us now discuss $p\bar{p}$ annihilation at rest. The initial states are $^1S_0$ and $^3S_1$ with C = 1, -1, respectively. If the antiprotons are not polarized there is no interference between these two channels. It follows that the spectra of a particle and its antiparticle must be rigorously identical if C is conserved. If C is violated, one can expect an interference term, as in the η case, between the C conserving and C violating amplitudes which may give, for example, different spectra for π^+ and π^- or K^+ and K^-. Last year the possibility of such an effect in the channel $p\bar{p} \to K^{\pm}\pi^{\mp}K_0$ was discussed[22)], the spectra of π^- and π^+ having a slightly different structure. Further studies[23)] of this channel with a better statistics resulted in spectra that are now compatible with CP conservation. A very significant result in this respect was obtained recently by the Columbia group[24)] who have analysed 40,000 events of $p\bar{p}$ annihilation into pions, in all possible channels. No significant difference was observed in π^+ and π^- spectra within a few percent. The results of these experiments indicate that if there is a violation of C in strong interactions, it is probably not larger than 1-2% in amplitude for reactions involving nucleons and pions, and a somewhat higher limit, but still very low, for interactions involving strange particles. This means that $g^2/4\pi$ is of the order of less than 10^{-4}. Personally I consider this result as much more convincing than estimates made on the basis of nuclear physics, where time reversal was seen to hold within a few percent. In the latter case it was always possible to construct models (for example, based on SU_3) such that C violation was connected intimately with the interaction of strange particles. Due to the fact that strange particles are heavy and that the involved energies are small, a suppression of C violation in nucleon-nucleon interactions could be made plausible. (Also arguments as given by Cabibbo[13] apply here to some extent.)

Such arguments cannot be used in the case of the $p\bar{p}$ system where richness of different channels and energies available absolutely exclude such considerations. In the preceding we have seen that all results concerning η and especially X^0 are pointing in the same direction. During this year, we have then learnt that if a "semi-strong" C violating interaction with $\Delta I = 0$ (and also $\Delta I = 1,2$) is responsible for CP violation, its coupling constant is most probably very small, in fact $\sim 10^{-2}$. Such an interaction is not excluded but obviously an experimental proof of its existence by a study of strong, electromagnetic and weak processes becomes more and more difficult, because effects will, generally speaking, (barring some special cases) be at most of the order of a few percent or less.

Theories of type 2)

In this class of theories the electromagnetic interaction of baryons and mesons with photons is responsible for the observed CP violation. In several papers[16,17,25,26] this possibility was studied. Here, I will discuss only the phenomenological side of the problem[*]. The electromagnetic current J_μ can in general be split up into two parts

$$J_\mu = I_\mu + K_\mu$$

where

$$C I_\mu C^{-1} = -I_\mu$$

$$C K_\mu C^{-1} = +K_\mu$$

K_μ represents the C violating part, and of course

*) In a series of papers[27] T.D. Lee has shown that from a purely theoretical point of view such a possibility is not in contradiction with any fundamental principle, that these C violating electromagnetic interactions can be deduced by a principle of minimal electromagnetic interaction from a C conserving Lagrangian for strong interactions, and that quite interesting speculations can be made on this subject.

$$\frac{\partial J_\mu}{\partial x_\mu} = 0 \ .$$

The electromagnetic interaction is as usual $A_\mu J_\mu$, and in some sense one can have a "maximal violating" theory. The first important result of B.F.L.[25] was that the evidence on C or T conservation in nuclear physics cannot be considered as compelling. Consider the matrix element of J_μ between two nucleon states N and N'

$$< N' | J_\mu | N > = i e \bar{u}_{N'} [\gamma_\mu F_1 (Q^2) + i(p_\mu^1 + p_\mu) F_2 (Q^2) +$$

$$(p_\mu^1 - p_\mu) F_3 \ (Q^2)] u_N$$

with obvious notation. The term with $F_3(Q^2)$ is C violating. From conservation of current, one finds $F_3(Q^2) = 0$ for nucleons on the mass shell. This term will not be seen in a study of the nucleon form factor. In nuclear physics, one believes that it is a good approximation to consider the nucleus as a collection of strong interacting physical nucleons, at least as far as electromagnetic transitions are concerned, and further to neglect other terms that may occur in the nucleon photon vertex if the nucleons are off the mass shell. An estimate of possible C violating effects under these conditions comes out to be of the order of a few percent, which is compatible with all known results. A more precise measurement of, for instance, an E_2, M_1 interference term in appropriate B-γ-γ transitions would be quite interesting.

The discussion of implications of such a theory on decays of resonances, etc., can be done on lines similar to those of the preceding section. Of course, one must select now cases where electromagnetic interactions are involved.

Nothing new has to be said concerning the π^o. The η decay is very interesting. If indeed K_μ is as strong as I_μ the expected interference effects in $\eta \to \pi^+\pi^-\pi^o$ could be very large. Indeed, in second order an electromagnetic interaction can induce in general both a $\Delta I = 0$ and $\Delta I = 2$ C violating transition. According to the table, the potential barrier effects are rather small for the $\Delta I = 2$ channel. However, the I spin structure of K_μ is not known, it is, in general, a combination of an isoscalar and an isovector, but not necessarily the same as for I_μ. For instance, K_μ could be dominantly an isoscalar, in which case the transition $\eta \to 3\pi$ will mainly go to the $\Delta I = 0$ state, where the potential barrier plays an essential role. In the latter case the expected effects will be very small. Thus, up to now, no definite prediction on the magnitude of possible interference effects can be made, and if, for example, only a very small or zero C violation is observed in this process, one can perhaps interpret this as an indication of an isoscalar structure of K_μ.

Another interesting process is $\eta \to \pi^o e^+ e^-$. Again, the expected order of magnitude of the branching ratio as estimated by B.F.L.[25] is large:

$$R_1 = \frac{\eta^o \to \pi^o e^+ e^-}{\eta^o \to 2\gamma} \sim 0.04 |< r^2 > m_\eta^2|^2 \sim 1$$

if $< r^2 >$ is set to be the mean square radius of the proton. Experimentally it has meanwhile been established[18,19] that

$$R_1 < 1\% .$$

For at least two reasons this cannot be considered as evidence against the theory:

i) if K_μ is a pure isoscalar, this transition is forbidden in lowest electromagnetic order;

ii) it can be shown[13] that in the limit of exact SU_3 the matrix element $<\eta_0|K_\mu|\pi^0> = 0$, provided that K_μ transforms as a member of an octet. Thus one expects a strong depression.

With regard to the second possibility, Feinberg[28] has done an amusing calculation showing that this can indeed be the case, and moreover obtaining relations between $\eta \to \pi e e$ and $X \to \pi$ (or η) $e e$. The idea of this calculation is as follows: X^0 and η^0 can mix, as they have the same quantum numbers. Insisting on an exactly satisfied Gell-Mann/Okubo mass formula for the unperturbed octet, one obtains X and η in terms of X_1 (singlet) and η_8 (octet):

$$X = X_1 \cos\vartheta + \eta_8 \sin\vartheta$$
$$\eta = - X_1 \sin\vartheta + \eta_8 \cos\vartheta ,$$

where $\mathrm{tg}\,\vartheta = 0.18^{21}$. But $<\eta_8|K|\pi_0> = 0$ if K_μ transforms as an octet. If, moreover, we assume that K_μ transforms under SU_3 as the regular current I_μ one obtains:

$$< X_1|K_\mu|\pi^0> = \sqrt{3} < X_1|K_\mu|\eta_8> .$$

All three decays can then be described in terms of the mixing angle ϑ and a single form factor $f_1(Q^2)$:

$$< X_1|K_\mu|\pi^0> = f_1(Q^2)\left[(X_\mu + \pi_\mu) - \frac{m_X^2 - m_\pi^2}{Q^2}(\pi_\mu - X_\mu)\right] ,$$

where X_μ and π_μ and X and π four momenta, and $f_1(0) = 0$. Assuming that $f_1(Q^2)/Q^2$ is approximately constant, one writes

$$f_1 = -\frac{e}{6} Q^2 <r^2>_{pr} \cdot \lambda ,$$

where $<r^2>_{pr}$ is the proton mean square charge radius, and λ a parameter of the order of 1 if K_μ is of the same order as I_μ. A numerical calculation gives

$$\eta \to \pi e e \sim 2.5 \, \lambda^2 \ \text{eV}$$

$$X \to \pi e e \sim 2.5 \, \lambda^2 \ \text{keV}$$

$$X \to \eta e e \sim 80 \, \lambda^2 \ \text{eV} .$$

Remembering $\Gamma_\eta \sim 300$ eV and $\Gamma_X \sim 100$ keV, we find branching ratios of the order of a few percent for pionic modes. Experimentally[19], it has been established:

$$X \to \pi e e < 1.3\% \qquad (\text{theory} \sim 2.5\%)$$

$$X \to \eta e e < 1.1\% \qquad (\text{theory} \sim 0.1\%) .$$

We may remark here that an isoscalar K_μ would not give rise to $X \to \pi e^+ e^-$.

Let us finally mention that a study of possible $\pi^+\pi^-$ asymmetry in $\eta \to \pi^+\pi^-\gamma$ and especially $X \to \pi^+\pi^-\gamma$ may be interesting. Any interference between the P wave (for the pion system) C conserving and D wave C violating modes will give different $\pi^+\pi^-$ spectra. Preliminary experimental results on about 200 events $X \to \pi\pi\gamma$ are compatible with no asymmetry. This is a test for the presence of an isoscalar part in K_μ.

Similar reasoning can be made for radiative decays of other resonances, like for instance $\varphi \to \rho\gamma$ or ω, $\varphi \to \pi^+\pi^-\gamma$. These transitions provide also more information concerning the possible I spin structure of K_μ. Unfortunately, the branching ratios are not very favourable.

In conclusion, I would say that many effects which were in the beginning expected to be large, are found to be small. This does not mean that the theory is wrong, as I have tried to indicate above.

New, but presumably difficult to obtain, experimental results are needed in this domain.

Before leaving the domain of electromagnetic interactions, I would like to mention a hypothesis forwarded by Salzman and Salzman[29] and further studied by Wu[30]. They suppose that the intermediate vector boson of weak interactions possesses a non-zero electric dipole moment, and a P violating but C conserving interaction:

$$ie\lambda\, \epsilon_{\mu\nu\xi\sigma}\, J_{\mu\nu}\, \varphi^{x}_{\xi}\, \varphi_{\sigma}$$

(with $\lambda \sim 1$) is specified. The influence of this interaction on strong or electromagnetic processes can be shown to be negligible, but it is claimed that in weak interactions CP or T violating effects of the order α can be expected. These predictions are, of course, quite different from those from the B.F.L. model.

The consequences of various theories on the electric dipole moment of the neutron has been studied by several authors[28,29,31,32]. If P and T are violated, the neutron may have an electric dipole moment. From dimensional analysis, we find:

$$d = \kappa\, e\, G_F m_p \sim 4 \cdot 10^{-19}\ \text{cm} \times e\ ,$$

where κ is a parameter of the order 1 to 10^{-1} for theories with CP violation in electromagnetic interactions (including the Salzman et al. model), or for theories admitting a strong CP violation in weak interactions (to be discussed later on). For a class 1 theory with $g \sim 10^{-2}$ and for theories giving a small CP violation in all weak interactions one has $\kappa \sim 10^{-2} - 10^{-3}$. For a superweak theory κ is negligible. These rough guesses are in agreement with the more refined calculations given in Refs. 29 and 31. The present experimental upper limit is $2.9 \cdot 10^{-20}$, and as quoted in Ref. 31 Ramsey proposes an experiment that will detect an electric dipole moment of as low as $5 \cdot 10^{-24}$. Clearly this is of great theoretical interest.

III. WEAK INTERACTIONS

There is only a very limited amount of experimental information on CP properties of weak interactions. The only firmly established result is the CP violating $K_{02} \to 2\pi$ decay, and no significant effects in other reactions have been observed. It is possible that the smallness of the observed CP violation in $K_{02} \to 2\pi$ is an accident, and that elsewhere important effects occur. It is also possible that the effects are always small. This leaves us with almost no limitation on theoretical possibilities, and accordingly many models have been proposed. I will now briefly discuss some of these models, mainly to demonstrate the spirit in which the problem has been approached.

Let me, however, first review the experimental situation and indicate where within reasonable time an essential improvement can be expected.

(a) The phase between V and A currents in neutron decay should be zero if CP is conserved. An experiment done by Burgy et al.[33]) gives an upper limit of 10%.

(b) Cronin and Overseth[34]) have measured the β parameter of $\Lambda \to p + \pi$ and find a value consistent with CP conservation. The experimental error is considerable, namely about 30° in the phase between P and S wave amplitudes. New experiments on this subject are being done.

(c) In V-A theory the $K_{\mu 3}$ matrix element can be written in terms of two form factors f and ξ:

$$f(Q^2)[K_\lambda + \pi_\lambda) + \xi(Q^2)(K_\lambda - \pi_\lambda)]\, \ell_\lambda ,$$

where ℓ_λ is the lepton current:

$$\ell_\lambda = \bar{u}\gamma_\lambda (1+\gamma_5) u .$$

If T holds Im $\xi = 0$. A non-zero Im ξ gives rise to a polarization of the muon perpendicular to the decay plane. The experimental situation[35] is not yet very significant.

(d) If CP is conserved the Dalitz plots for $K^+ \to 3\pi$ and $K^- \to 3\pi$ must be identical. An old measurement by Ferro-Luzzi et al.[36] indicates that the slopes in the plots of τ^+ and τ^- decays are equal within 10%. Apparently CP is not strongly violated. The accuracy of this measurement will probably be much improved (up to 1%) in the near future, and thereby will be an extremely valuable piece of information.

(e) The $\Delta S = \Delta Q$ rule in K^0_{e3} decay was established assuming CP conservation. Recently Aubert et al.[37] and Baldo-Ceolin et al.[38] reanalysed their data without this assumption, and they arrived at the conclusion

$$\frac{A_{\Delta S = -\Delta Q}}{A_{\Delta S = \Delta Q}} = 0.22^{+0.16}_{-0.11} \quad \text{and } 0.4 \pm 0.2 \text{ respectively,}$$

whereby the $\Delta S = -\Delta Q$ amplitude is about 90° out of phase with the $\Delta S = \Delta Q$ amplitude. The whole effect is, however, only two standard deviations away from no CP violation and no $\Delta S = \Delta Q$ currents.

A model in this direction was proposed by Sachs[39] who supposed equally strong amplitudes for $\Delta S = \Delta Q$ and $\Delta S = -\Delta Q$, but a maximal violation of CP in the $\Delta S = -\Delta Q$ current. Although the above-mentioned experiments exclude the case of equal strength quite strongly, it is nevertheless still possible that a 20%, maximally CP violating $\Delta S = -\Delta Q$ current is responsible (via the mass matrix) for the observed $K_L \to 2\pi$ decay.

(f) The phenomenological analysis of the K_0, $\bar{K}_0$ system was given by Wu and Yang[40], and more recently, by Wolfenstein[41]. This, and also the experimental situation will be discussed by Bell and Steinberger. Here I will mention only briefly those results of the analysis which will be needed for the discussion below.

The long- and short-lived components K_L and K_S may be given in terms of one (assuming CPT) complex parameter $\epsilon = |\epsilon| e^{i\delta}$

$$|K_{S \atop L}> = \frac{1}{\sqrt{1 + |1+\epsilon|^L}} \left(|K_0> \pm (1+\epsilon)|\bar{K}_0> \right),$$

ϵ arises from the so-called mass matrix, and we have chosen phase conventions such that the amplitude for $K_0 \to 2\pi$(I spin = 0) is real. One of the important results of Wu and Yang concerns the decay rates of K_L into $\pi^+\pi^-$ and $\pi^0\pi^0$. They find:

$$\eta_{+-} = |\eta_{+-}| e^{i\chi} \equiv \frac{A(K_L \to \pi^+\pi^-)}{A(K_S \to \pi^+\pi^-)} = \frac{1}{2}(\epsilon + \epsilon')$$

$$\eta_{00} = |\eta_{00}| e^{i\psi} \equiv \frac{A(K_L \to 2\pi^0)}{A(K_S \to 2\pi^0)} = \frac{1}{2}(\epsilon - 2\epsilon'),$$

where ϵ' is directly related to the CP violation in the $\Delta I = 3/2$ amplitude with respect to the $\Delta I = 1/2$ amplitude:

$$\epsilon' = i\sqrt{2} \exp\left[i(\delta_2 - \delta_0)\right] \cdot \mathrm{Im} \frac{A_2}{A_0},$$

where A_0 and A_2 are the amplitudes for $K_0 \to 2\pi$, I = 0 or 2, respectively, and where δ_0 and δ_2 being the I = 0, 2 scattering phase shifts. ϵ and ϵ' are small, at most of the order of 10^{-2}. The quantity $|\eta_{+-}|$ is essentially given by the experiment of Christenson et al.[1]. The experiments being performed measure the phase χ of η_{+-}, i.e., the phase of $\epsilon + \epsilon'$, and the magnitude of η_{00}, i.e., $|\epsilon - 2\epsilon'|$.

The implications of theories of class 1) and 2) on the weak interactions may now be stated. In a class 1) theory with $g = 10^{-2}$ we expect everywhere 1% effects, except in the K_L case. Aside from the $\Delta I = \frac{1}{2}$ transition, we have a $\Delta I = \frac{3}{2}$ and a $\Delta I = \frac{5}{2}$ $K \to 2\pi$ amplitude. Assuming that there are no accidental cancellations, we may believe that the magnitude of the $\Delta I = \frac{3}{2}$ and/or $\frac{5}{2}$ amplitudes is about 1/20 of the $\Delta I = \frac{1}{2}$ amplitude, as indicated by the ratio:

$$\left| \frac{A(K^+ \to 2\pi)}{A(K_S^o \to 2\pi} \right| \simeq \frac{1}{20} \quad .$$

Thus we take

$$\left| \frac{A_2}{A_0} \right| \sim \frac{1}{20} \; .$$

If $g \sim 10^{-2}$ the ratio $\mathrm{Im}\, A_2/A_0$ can be as large as 1/2000, i.e.,

$$|\epsilon'| \sim 1/2000$$

to be compared with $\frac{1}{2}|\epsilon + \epsilon'| \sim 1/500$. In this case we have then

$$|\epsilon'| \sim \frac{1}{4} |\epsilon|$$

and one finds in the decay $K_L \to 2\pi$ a possible 25% I = 2 admixture, which is a quite appreciable violation of the $\Delta I = \frac{1}{2}$ rule. The phase of ϵ($\sim$ phase of $1/(\Delta\Gamma + 2i\,\Delta m)$) can be stated in this case, but the phase of ϵ' is unknown as $\delta_2 - \delta_0$ is unknown.

The predictions of a class 1) theory with I spin violation on the K_0 decay are the same as for class 2) theories. For a class 2) theory we have 1% effects on non-radiative weak decays (except K_0, see below), and possibly large effects in radiative weak decays. For instance, the study of $\pi^+\pi^-$ spectra in K_L or $K_S \to \pi^+\pi^-\gamma$ decay would be very interesting. It has been argued by Salzman et al.[29]) that this is not the case if the intermediate vector boson is responsible for the CP violation, due to the heaviness of this object.

The situation in K_L decay is quite complicated. It is not very plausible here to assume that the $\Delta I = 3/2$ or $5/2$ amplitude observed in K^+ decay is of electromagnetic origin, because, barring exceptional cancellations, we would expect then an ϵ' of order 1/20, which is excluded. Thus we take it that in $K^+ \to 2\pi$ the observed $\Delta I = 3/2$ or $5/2$ amplitude is of non-electromagnetic origin, and that the electromagnetic induced A_2 amplitude is of order α. Thus $\epsilon' \sim \alpha$, and of course also $\epsilon \sim \alpha$. Accordingly we can expect in this case an arbitrary mixture of I = 2 and I = 0 final states in $K_L \to 2\pi$. The phase of ϵ is still determined as above, at least if we do neglect contributions of radiative processes to the mass matrix. At this point we may remark that even if the current K_μ is an isoscalar, a violation of the $\Delta I = 1/2$ rule is to be expected, as we are dealing here with second order electromagnetic effects.

Let us now discuss some class 3) models. The first one was the model proposed by Cabibbo[42)], who started by the assumption that the CP violation can be big in weak interactions, and subsequently suggested a particular model for the way in which CP was violated in particular in leptonic decays. The basic assumption is that the baryon current must transform like an SU_3 octet: the consequence is that currents of the first class conserve CP while second class currents violate CP. For baryonic leptonic decays the first and second class currents are respectively (Q = momentum transfer to leptons):

1) class: γ_μ , $\sigma_{\mu\nu} Q_\nu$, $\gamma_\mu \gamma^5$, $i Q_\mu \gamma^5$;

2) class: $i Q_\mu$, $\sigma_{\mu\nu}\gamma^5 Q_\nu$.

Thus, in agreement with the neutron decay experiment, we expect zero phase between vector and axial vector. Obviously everywhere effects are small if Q is small.

In general, however, large effects are expected, in particular in $\Lambda \to N\pi$ or in $K \to 3\pi$ and especially in $K_{\mu 3}$ decay which is very interesting in this theory, Im ξ expected to be quite important. The fact that $K_L \to 2\pi$ is so small is considered an accident in this scheme, and one may therefore, barring a second accident, expect a substantial violation of the $\Delta I = \frac{1}{2}$ rule in the $K_L \to 2\pi$ decay. In view of the present evidence, in particular $K \to 3\pi$, one may be tempted to deduce that although the above model concerning the SU_3 behaviour may be correct, at least the magnitude of the second class currents is substantially smaller than the magnitude of the first class currents.

Another type of model[43] is based on a generalization of the well-known (CP conserving) Cabibbo theory, where currents are obtained by rotation over an angle ϑ around the seventh axis in SU_3 space. The rotation around the sixth axis is not CP conserving, so combining these two rotations one obtains CP violating currents. The simplest idea is to rotate V and A currents in exactly the same way: this leads, however, to one common phase φ for all $\Delta S = 1$ currents that can be absorbed in the definition of strange particles. It is therefore necessary to introduce different rotations for V and A currents. The general form obtained is:

$$\pi_V \cos \vartheta + K_V \sin \vartheta + \pi_A \cos \vartheta + K_A e^{i\varphi} \sin \vartheta ,$$

where $\pi_{V,A}$, $K_{V,A}$ denote components of a current octet behaving as the π and K meson. The phase in K_V can be defined to be zero. The main practical difference with the Cabibbo CP violating model above is the lack of any effect in $K_{\mu 3}$ decay: CP violating effects are due to interference between vector and axial currents, and $K_{\mu 3}$ decay is a purely vector transition.

In both theories there are no $\Delta S = -\Delta Q$ currents. Concerning the magnitude of the CP violation we may consider two extreme cases:

i) φ small. In this case we will have only a small CP violation everywhere in weak interactions. Concerning $K_L \to 2\pi$ decay the situation is similar to what was said before for theories of class 1) with small g.

ii) φ large, and the small rate $K_L \to 2\pi$ is accidental. No effects in $\Delta S = 0$ reactions, appreciable effects in $\Delta S = 1$ reactions (like Λ_β decay), and large effects in non-leptonic decays.

I may finally briefly mention a theory proposed by Zachariasen and Zweig[44]. Starting from considerations based on $\widetilde{U}(12)$ they arrive at a model in which CP is violated in the parity violating part of non-leptonic interactions. On the other hand, Alles[45] has introduced on very different grounds a model giving CP violation in the parity conserving part of the non-leptonic reactions. Clearly, the lack of experimental limitations leaves a wide field for theoretical speculations.

Class 4) theory

Let me now discuss a very interesting model due to Wolfenstein[46] which has the enormous merit of giving very clear experimental predictions.

Suppose there exists an interaction Hamiltonian containing in addition to the usual weak interaction Hamiltonian H_W, with $\Delta S \leq 1$, a CP violating term H_{SW} with $\Delta S = 2$:

$$H = H_0 + H_W + H_{SW}$$

H_0 is the free Hamiltonian, describing also strong and electromagnetic interactions. H_{SW} gives in first order rise to an imaginary term in the off diagonal matrix element of the K_0-$\bar{K}_0$ mass matrix

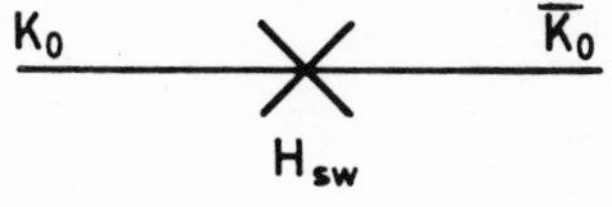

This, in addition to the second order CP conserving contributions of weak interactions, for example:

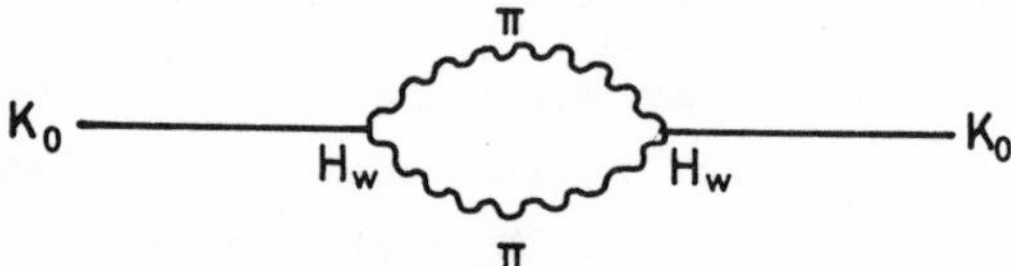

In order to account for the observed effect, the contribution of the H_{sw} induced part must be about 10^{-3} times the second order weak interaction contribution. In other words, H_{sw} describes an interaction weaker than second order weak interaction, reason why we will speak of it as a superweak interaction. The characterizing coupling constant must be of order 10^{-7} - 10^{-8} times the weak interaction coupling constant. We have to do here with a new kind of interaction. In general (with the exception of the K_0, $\bar{K}_0$ system) the contributions of such an interaction to weak processes will be negligible, with CP violating effects of 10^{-7} - 10^{-14} (model dependent). For the K_0, $\bar{K}_0$ system, we have now a one-parameter theory, the only parameter being $|\epsilon|$. In particular, we have $\epsilon' = 0$ (as $\mathrm{Im}\, A_2 = 0$) and a phase δ of ϵ given by:

$$\epsilon = |\epsilon|\, e^{i\delta}, \quad \tan\delta = -2\frac{m_s - m_e}{\Gamma_s - \Gamma_e} .$$

The quantity $|\epsilon|$ is then determined from the experiment of Christenson et al.[1]:

$$|\epsilon| \simeq 4 \cdot 10^{-3} .$$

We note that $\eta_{+-} = \eta_{oo} = \frac{1}{2}\,\epsilon$ thus the K_L branching ratios equal those of K_S

$$\frac{K_L \to 2\pi^0}{K_L \to \pi^+\pi^-} = \frac{K_S \to 2\pi^0}{K_S \to \pi^+\pi^-} .$$

Some other conclusions are:

$$\frac{K_S \to 3\pi^0}{K_L \to 3\pi^0} = |\epsilon|^2$$

$$\frac{K_L \to \pi^+ \ell^- \nu}{K_L \to \pi^- \ell^+ \nu} = 1 + 4 \; |\epsilon| \; \cos \delta \; ,$$

which will probably be measured within reasonable time.

In conclusion we see that the situation for theories of class 3) and 4) is still completely open. For class 1) the PC violation is at most a few percent. In the electromagnetic case the situation is not quite clear but it seems that some symmetries or properties have to be attributed to the C violating part of the electromagnetic current in order to account for the absence of large effects in some processes.

It is a pleasure to acknowledge many fruitful discussions with M. Veltman whose collaboration was essential for the preparation of this report. I am also indebted to J. Bell for interesting remarks and comments.

REFERENCES

1) J.H. Christenson, J.W. Cronin, V.L. Fitch and R. Turlay, Phys.Rev. Letters 13, 138 (1964).

2) J.S. Bell and J.K. Perring, Phys.Rev. Letters 13, 348 (1964).

3) J. Bernstein, N. Cabibbo and T.D. Lee, Physics Letters 12, 146 (1964).

4) X. De Bourad et al., Physics Letters 15, 58 (1965).

5) W. Galbraith et al., Phys.Rev. Letters 14, 383 (1965).

6) M. Lévy and M. Nauenberg, Physics Letters 12, 155 (1964).

7) S. Weinberg, Phys.Rev. Letters 13, 495 (1964).

8) V.L. Fitch et al., Phys.Rev.Letters 15, 73 (1965).

9) A.M.L. Messiah and W. Greenberg, Phys.Rev. 136, 248 (1964);
S.A. Bludman, preprint.

10) B. Laurent and M. Roos, Physics Letters 13, 138 (1964) and CERN preprint 65/1008/5-TH.580.

11) K. Nishijima and M.H. Safouri, Phys.Rev. Letters 14, 205 (1965)

12) J. Prentki and M. Veltman, Physics Letters 15, 88 (1965) and CERN preprint 65/907/5-TH.568;
L.B. Okun, unpublished;
T.D. Lee and L. Wolfenstein, Phys.Rev. 138, B1490 (1965).

13) N. Cabibbo, Phys.Rev. Letters 14, 965 (1965), where further reference on the subject are listed.

14) F. Berends, Physics Letters 16, 118 (1965).

15) V. Soergel, private communication.

16) See, for example:
Y. Fujii and G. Marx, Physics Letters 17, 75 (1965);
S.L. Glashow and C. Sommerfield, Phys.Rev. Letters 15, 78 (1965);
M. Nauenberg, Physics Letters 17, 329 (1965).

17) B. Barrett, M. Jacob, M. Nauenberg and T.N. Truong, preprint, SLAC-pub-123.

18) L. Price and F. Crawford, Phys.Rev. Letters 15, 123 (1965).

19) A. Rittenberg and G.R. Kalbfleisch, preprint.

20) G.R. Kalbfleisch, O.I. Dahl and A. Rittenberg, Phys.Rev. Letters 13, 75 (1964).

21) R.H. Dalitz and D.G. Sutherland, Nuovo Cimento 38, 1945 (1965).

22) R. Armenteros et al., Physics Letters 17, 344 (1965).

23) R. Armenteros et al., private communication.

24) J. Steinberger, private communication.

25) J. Bernstein, G. Feinberg and T.D. Lee, Phys.Rev., to be published.

26) S. Barshay, Physics Letters 17, 78 (1965).

27) T.D. Lee, "Classification of all C non-invariant elementary interactions", preprint; "Minimal elementary interaction and C, T non-invariance", preprint.

28) G. Feinberg, preprint Columbia University.

29) F. Salzman and G. Salzman, Physics Letters 15, 91 (1965) and preprint Northeastern University.

30) T.T. Wu, CERN preprint, to be published in Physics Letters.

31) D. Boulware, preprint Harvard University.

32) R. Jengo and R. Odorice, Physics Letters 16, 168 (1965).

33) M.T. Burgy, V.E. Krohn, T.B. Novey, G.R. Ringo and V.L. Telegdi, Phys.Rev. Letters 1, 324 (1958).

34) J. Cronin and O. Overseth, Phys.Rev. 129, 1795 (1963).

35) U. Camerini et al., Phys.Rev. Letters 14, 989 (1965).

36) M. Ferro-Luzzi et al., Nuovo Cimento 22, 1087 (1961).

37) B. Aubert et al., Physics Letters 17, 59 (1965).

38) M. Baldo-Ceolin et al., Nuovo Cimento, in press.

39) R.G. Sachs, Phys.Rev. Letters 13, 286 (1964).

40) T.T. Wu and C.N. Yang, Phys.Rev. Letters 13, 501 (1964).

41) L. Wolfenstein, CERN preprint 66/1086/5-TH.583.

42) N. Cabibbo, Physics Letters 12, 137 (1964).

43) S.L. Glashow, Phys.Rev. Letters 14, 35 (1965);
A. Morales, R. Nuñez-Lagos and M. Soler, to be published in Nuovo Cimento;
B. d'Espagnat and M.K. Gaillard, Coral Gables conference, 1965.

44) F. Zachariasen and G. Zweig, Phys.Rev. Letters 14, 794 (1965).

45) W. Alles, Physics Letters 14, 348 (1965).

46) L. Wolfenstein, Phys.Rev. Letters 13, 562 (1964);
T.D. Lee and L. Wolfenstein, Phys.Rev. 138, B1490 (1965).

CP VIOLATION AND K DECAY

J. Steinberger,
CERN, Geneva.

I. K DECAYS

1. K-decay modes and branching ratios

Let us start by listing in Table 1 the main decay modes of K mesons.

Table 1

	$K^{\pm}$		K_S		K_L	
Mass width Γ	493.8 ± 0.2 (0.81 ± 0.006) × 10^8		497.7 ± 0.3 (1.24 ± 0.01) × 10^{10}		497.7 ± 0.3 (1.8 ± 0.2) × 10^7	
Decay modes	$\mu^{\pm}\nu$	63%	$\pi^+\pi^-$	68.5 ± 1%	$\pi^{\pm}e^{\mp}\nu$	26 ± 3%
	$\pi^{\pm}\pi^0$	21%	$\pi^0\pi^0$	31.5 ± 1%	$\pi^{\pm}\mu^{\mp}\nu$	35 ± 3%
	$\pi^{\pm}\pi^+\pi^-$	5.5%			$\pi^0\pi^0\pi^0$	25 ± 3%
	$\pi^{\pm}\pi^0\pi^0$	1.7%			$\pi^+\pi^-\pi^0$	14 ± 1%
	$\mu^{\pm}\nu\pi^0$	3.4%			$\pi^+\pi^-$	0.14 ± 0.01%
	$e^{\pm}\nu\pi^0$	4.9%				

K_S and K_L refer, respectively, to the short and to the long-lived neutral K mesons. The table only includes the most important decay rates. For instance it is known that the short-lived component has also all the leptonic decay modes with about the same decay rates as for the long-lived one, but on a fractional basis these rates are very small. For these lectures I will follow closely the review papers given by Bell and myself at the Oxford Conference[1]).

Lecture notes taken by V.P. Henri.

2. Notation and superposition principle

The particles K and $\bar{K}$ produced in strong interactions have definite hypercharge, +1 and -1, respectively. Since the weak interactions do not conserve hypercharge we have transitions between K and $\bar{K}$. Therefore, it will be certain superpositions of K and $\bar{K}$ which have simple exponential decay:

$$
\begin{aligned}
|L\rangle &= p\,|K\rangle + q\,|\bar{K}\rangle \\
|S\rangle &= r\,|K\rangle + s\,|\bar{K}\rangle ,
\end{aligned}
\qquad (I.1)
$$

where p,q,r and s have to be defined. For normalization purposes $|p|^2 + |q|^2 = |r|^2 + |s|^2 = 1$. We are also free to choose the phases of $|L\rangle$, $|S\rangle$, $|K\rangle$ and $|\bar{K}\rangle$. We choose these phases, as Bell has done, in such a way that p,q, and r are real and positive.

In the K rest frame the time dependence of K_L and K_S is exponential:

$$
\begin{aligned}
|L\rangle &\to e^{-iM_L\tau}\,|L\rangle \\
|S\rangle &\to e^{-iM_S\tau}\,|S\rangle
\end{aligned}
$$

M_L and M_S are complex, the real parts are the masses, and the imaginary parts one-half of the K widths:

$$
\begin{aligned}
M_L &= m_L - \frac{i\Gamma_L}{2} \\
M_S &= m_S - \frac{i\Gamma_L}{2} .
\end{aligned}
$$

Let us also define the mass difference Δm as:

$$\Delta m = m_S - m_L .$$

Inverting Eq. (I.1) we obtain for $|K\rangle$ and $|\bar{K}\rangle$ the following expressions

$$|K\rangle = (sp - qr)^{-1} \{s|L\rangle - q|S\rangle\}$$
$$|\bar{K}\rangle = (sp - qr)^{-1} \{-r|L\rangle + p|S\rangle\} . \tag{I.2}$$

In a time τ these states will propagate as follows:

$$\begin{aligned} |K\rangle \underset{\tau}{\rightarrow} & (sp - qr)^{-1} \left\{ s\, e^{-iM_L\tau} |L\rangle - q\, e^{-iM_S\tau} |S\rangle \right\} \\ = & (sp - qr)^{-1} \left\{ \left(sp\, e^{-iM_L\tau} - rq\, e^{-iM_S\tau} \right) |K\rangle \right. \\ & \left. + \left(sq\, e^{-iM_L\tau} - qs\, e^{-iM_S\tau} \right) |\bar{K}\rangle \right\} \end{aligned} \tag{I.3}$$

and

$$\begin{aligned} |\bar{K}\rangle \underset{\tau}{\rightarrow} & (sp - qr)^{-1} \left\{ p\, e^{-iM_S\tau} |S\rangle - r\, e^{-iM_L\tau} |L\rangle \right\} \\ = & (sp - qr)^{-1} \left\{ \left(pr\, e^{-iM_S\tau} - rp\, e^{-iM_L\tau} \right) |K\rangle \right. \\ & \left. + \left(ps\, e^{-iM_S\tau} - rq\, e^{-iM_L\tau} \right) \right\} . \end{aligned} \tag{I.4}$$

In a small interval of time $\delta\tau$

$$|K\rangle \underset{\delta\tau}{\rightarrow} |K\rangle - i\delta\tau \{M|K\rangle + B|\bar{K}\rangle\}$$
$$|K\rangle \underset{\delta\tau}{\rightarrow} |\bar{K}\rangle - i\delta\tau \{A|K\rangle + \bar{M}|\bar{K}\rangle\} . \tag{I.5}$$

Expanding Eqs. (I.3) and (I.4) for small times and comparing with Eq. (I.5), one finds the relation between $M, \bar{M}, A,$ and B and s,p,q and r:

$$\begin{aligned} M &= (M_L\ sp - M_S\ qr)\ (sp - qr)^{-1} \\ \bar{M} &= (M_S\ sp - M_L\ qr)\ (sp - qr)^{-1} \\ A &= (M_S - M_L)\ rp\ (sp - qr)^{-1} \\ B &= (M_L - M_S)\ sq\ (sp - qr)^{-1}\ . \end{aligned} \qquad (I.6)$$

We can now express this in the following way: if a general state Ψ is a mixture of $|K>$ and $|\bar{K}>$

$$\Psi = a|K> + b|\bar{K}>\ .$$

In vector notation $\Psi = \begin{pmatrix} a \\ b \end{pmatrix}$, where the elements a and b refer to the amplitudes of the K and $\bar{K}$ states. The time development of Ψ can now be written in matrix notation:

$$\frac{d}{d\tau}\Psi = -i \begin{pmatrix} M & A \\ B & \bar{M} \end{pmatrix} \Psi\ . \qquad (I.7)$$

The matrix $\begin{pmatrix} M & A \\ B & \bar{M} \end{pmatrix}$ is generally referred to as the mass matrix.

3. CPT invariance

Let us now see what the restrictions are which are a consequence of CPT invariance. CPT invariance requires that the transition amplitudes for the two processes

$$I \rightarrow F \quad \text{and} \quad \bar{F}' \rightarrow \bar{I}'$$

are equal, where the bar denotes interchange between particles and antiparticles, and the prime indicates reversal of spin directions. If this is applied to the transitions

$$K \to K \quad \text{and} \quad \bar{K} \to \bar{K} \ ,$$

the probability amplitudes should be equal, and from the relations (I.3) and (I.4)

$$sp = -rq \, .$$

Since

$$|p|^2 + |q|^2 = |r|^2 + |s|^2 = 1 \ ,$$

we now have

$$s = -q \quad \text{and} \quad r = p \ .$$

Using this last result

$$|L\rangle = p|K\rangle + q|\bar{K}\rangle$$
$$|S\rangle = p|K\rangle - q|\bar{K}\rangle$$

or

$$2p|K\rangle = |L\rangle + |S\rangle$$
$$2q|\bar{K}\rangle = |L\rangle - |S\rangle \ ,$$

and finally

$$\left.\begin{aligned} M = \bar{M} &= \frac{1}{2}\,(M_L + M_S) \\ A &= \frac{1}{2}\,\frac{p}{q}\,(M_L - M_S) \\ B &= \frac{1}{2}\,\frac{q}{p}\,(M_L - M_S) \end{aligned}\right\} . \qquad \text{(I.8)}$$

4. CP invariance

We choose the convention

$$CP|K\rangle = -|\bar{K}\rangle$$
$$CP|\bar{K}\rangle = -|K\rangle .$$

If the decay were CP invariant, then the eigenstates $|S\rangle$ and $|L\rangle$ would be eigenstates of CP. $|S\rangle$ is the state which decays into two pions and is, therefore, CP positive. So

$$CP|S\rangle = |S\rangle, \quad CP|L\rangle = -|L\rangle .$$

In the case of CP invariance therefore

$$p = q = \frac{1}{\sqrt{2}} \quad \text{and} \quad A = B = \frac{1}{2}(M_L - M_S) .$$

5. Experiments on $K \to 2\pi$ decay

Let us now review experiments on CP violation in $K^o \to \pi^+\pi^-$ decay.

$K_L \to \pi^+\pi^-$

This decay was observed by Christenson, Cronin, Fitch and Turlay[2] and almost simultaneously by Abashian et al.[3]. The apparatus of Christenson et al. is shown in Fig. 1. A neutral beam is produced at 30° from an internal target at the AGS at Brookhaven. It goes through a region filled with helium at a distance of about 15 m from the target. The He region is viewed by two telescopes made of spark chambers, bending magnets, spark chambers, scintillation counters and Čerenkov counters. The telescopes are set at about 20° on either side of the beam. A coincidence between the counters in the telescopes triggers the spark chambers and the event is photographed and then measured. The momenta of the two tracks p_+ and p_- are then reconstructed in space and the invariant mass formed:

$$m = \left[(E_+ + E_-)^2 - (\vec{p}_+ + \vec{p}_-)^2 \right]^{1/2} ,$$

where

$$E_{\pm}^2 = p_{\pm}^2 + m_\pi^2 .$$

Figure (2a) shows the observed distribution in m for all events. It corresponds to that expected for the leptonic decays of K_L, which therefore account for most of the observed events. When taking the events in the region about the K mass, 490 MeV < m < 510 MeV, and plotting them as a function of the angle with respect to the beam direction [Fig. (2b)], then again the distribution agrees with that expected for the leptonic decays, except for an excess of 50 events which are in the same direction as the beam. These events are attributed to the $\pi^+\pi^-$ decay of a long-lived particle of mass ~ 500 MeV, presumably the K_L.

The experiment was repeated at Rutherford Laboratory[4)] and at CERN[5)] using beams of higher momentum. Table 2 gives the relative rate

$$\frac{\Gamma_{L,\pi^+\pi^-}}{\Gamma_{L,\text{ all charged}}}$$

for these experiments.

Table 2

Experiment	Decay rate $\Gamma_{L,\pi^+\pi^-}/\Gamma_{L,\text{ all charged}}$	Ref.
Christenson et al.	$(1.87 \pm 0.18) \times 10^{-3}$	2
Abashian et al.	$(2-3) \times 10^{-3}$	3
Galbraith et al.	$(2.08 \pm 0.35) \times 10^{-3}$	4
de Bouard et al.	$(2.24 \pm 0.23) \times 10^{-3}$	5
Average	$(2.04 \pm 0.13) \times 10^{-3}$	

The values in the table correspond to the latest values reported at the Oxford Conference on Elementary Particles (1965). In the case of Ref. 2 and Ref. 5 these differ somewhat from the values reported there.

The corresponding absolute value of the relative amplitude for K_L and K_S decays into $\pi^+\pi^-$,

$$|\eta_{+-}| = \left|\frac{A_{L+-}}{A_{S+-}}\right| = (1.85 \pm 0.19) \times 10^{-3} .$$

II.. INTERFERENCE OF K_S AND K_L IN 2π DECAY

A coherent beam of K_S and K_L will show interference phenomena in common decay modes. The observation of such interference in the 2π decay would be important since it would prove that it is the K_L and not some other particle which decays into pions. Consider the time dependence of $K^o \to 2\pi$ since characteristic oscillatory behaviour should then be observed.

1. $(K_L - K_S)$ mass difference experiments

Before discussing the interference experiments it may be useful, from the experimental point of view, to discuss the experiments on the mass difference between K_L and K_S, since the results of the interference experiments depend so much on the value of Δm.

In one class of experiments one studies the time dependence of the intensity of a given strangeness component of the K^o. In the experiment of Camerini et al.[6]) a K^+ beam interacts in a propane bubble chamber and produces K^o (not $\bar{K}^o$). In time the K^o transforms and the $\bar{K}^o$ admixture becomes appreciable. Neglecting the small CP violating effect (taking $p = q = 1/\sqrt{2}$)

$$K^o \to \frac{1}{2}\,|K>\left[e^{-M_L\tau} + e^{-M_S\tau}\right]$$

$$+ \frac{1}{2}\,|\bar{K}>\left[e^{-M_L\tau} - e^{-M_S\tau}\right].$$

The time dependence of the intensity of the $\bar{K}$ component is, therefore,

$$I_{\bar{K}^o} = \frac{1}{4}\left[e^{-\Gamma_S\tau} + e^{-\Gamma_L\tau} - 2\cos\Delta m\tau \cdot e^{-(\Gamma_L+\Gamma_S)\tau/2}\right]. \qquad \text{(II.1)}$$

In the experiment, the time dependence is studied by looking for the production of hyperons of strangeness -1. The value of Δm which was found, $\Delta m = (1.5 \pm 0.2)\Gamma_S$, is not in agreement with more recent measurements[7].

In the experiments of Aubert et al.[8] and Baldo Ceolin et al.[9] the $\bar{K}_0$ intensity, Eq. (II.1) is studied through the $\bar{K}^0$ decay into $L^- + \pi^+ + \nu$. This experiment will be discussed in more detail in Lecture III in connection with the $\Delta S - \Delta Q$ rule. The mass differences reported are:

Baldo Ceolin et al.[9] $|\Delta m| = 0.15 \pm 0.35$

Aubert et al.[8] $|\Delta m| = 0.47 \pm 0.2$.

In another class of experiments, a beam of K_L mesons is allowed to traverse some matter (the regenerator) and one studies the intensity of the short-lived component K_S.

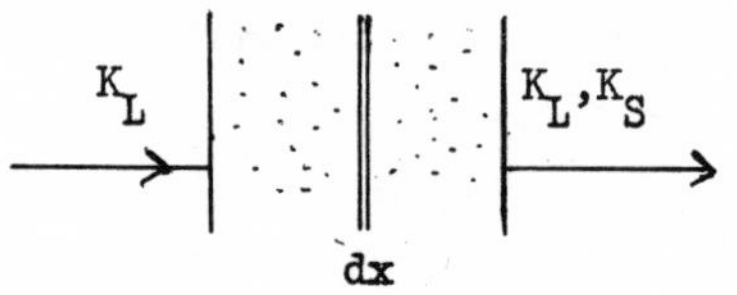

K_L → | regenerator | dx | → K_L, K_S

Consider a slab of thickness dx at a position x in an absorber traversed by an incident K_L beam. Let $\Psi = \begin{pmatrix} a \\ b \end{pmatrix}$ be the state vector of the K at x. In traversing the slab, the K and $\bar{K}$ components will propagate according to the optical theorem, with a refractive index

$$(\epsilon_K - 1) = \frac{2\pi N f(0)}{p^2}$$

and

$$(\epsilon_{\bar{K}} - 1) = \frac{2\pi N \bar{f}(0)}{p^2} ,$$

where $f(0)$ and $\bar{f}(0)$ are the respective complex forward scattering amplitudes, N is the density of nuclei, and p is the K momentum. Equivalently,

$$\frac{d\Psi}{dx} = \frac{2\pi iN}{p} \begin{pmatrix} f(0) & 0 \\ 0 & \bar{f}(0) \end{pmatrix} \begin{pmatrix} a \\ b \end{pmatrix} .$$

Now, however, we would like to consider states of well-defined long-lived or short-lived properties. To do this we rotate, and write

$$\frac{d}{dx}\begin{pmatrix} a_S \\ a_L \end{pmatrix} = \left\{ \frac{\pi iN}{p} \begin{pmatrix} f(0)+\bar{f}(0) & f(0)-\bar{f}(0) \\ f(0)-\bar{f}(0) & f(0)+\bar{f}(0) \end{pmatrix} \begin{pmatrix} a_S \\ a_L \end{pmatrix} \right\} . \quad \text{(II.2)}$$

So far, we have ignored the decay of the K mesons. If this is included, Eq. (II.2) becomes

$$\frac{d}{dx}\begin{pmatrix} a_S \\ a_L \end{pmatrix} = \frac{\pi iN}{p} \begin{pmatrix} f(0)+\bar{f}(0) & f(0)-\bar{f}(0) \\ f(0)-\bar{f}(0) & f(0)+\bar{f}(0) \end{pmatrix} \begin{pmatrix} a_S \\ a_L \end{pmatrix} - \begin{pmatrix} \Lambda_S & 0 \\ 0 & \Lambda_L \end{pmatrix} \begin{pmatrix} a_S \\ a_L \end{pmatrix} , \quad \text{(II.3)}$$

where Λ_S and Λ_L are the free space propagation constants of K_S and K_L, respectively,

$$\Lambda_s = ip_S + \frac{m}{2p}\Gamma_S$$

$$\Lambda_p = ip_L + \frac{m}{2p}\Gamma_L .$$

p_S and p_L are slightly different, since the two K particles have slightly different masses. In the collision, the block of matter absorbs the difference in momentum; however, no energy is exchanged because of its large (macroscopic) mass. Therefore,

$$\sqrt{p_S^2 + m_S^2} = \sqrt{p_L^2 + m_L^2} ;$$

$$p_S = p - \frac{im}{p} \Delta m$$

$$p_L = p + \frac{im}{p} \Delta m$$

and

$$\Lambda_S = ip + \frac{m}{2p} (-i\Delta m + \Gamma_S)$$

$$\Lambda_L = ip + \frac{m}{2p} (+i\Delta m + \Gamma_S) .$$

Expression (II.3) is to be solved for the time dependence of the amplitudes a_S and a_L, where time and space are related by $\tau = xm/p$. The solution is simplified by the recognition that a_S is always small, compared to a_L, for matter available on this earth. Ignoring also the non-zero width of the K_L meson, the solution of (II.3) is easily seen to be:

$$a_L = e^{-im\tau} e^{-N(\sigma + \bar{\sigma})x/2}$$

$$a_S = e^{-im\tau} e^{-N(\sigma + \bar{\sigma})x/2} \qquad \text{(II.4)}$$

$$\times \frac{\pi N[f(0) - \bar{f}(0)]}{\Delta m + i\Gamma_S/2} \left[e^{-(\Gamma_S/2) + i\Delta m)\tau} - 1 \right]$$

when $x = \tau \cdot p/m$.

The experiments which have been done to measure $|\Delta m|$ using regeneration consist chiefly of two types. In the first the regeneration is measured as a function of converter thickness x. In the second method two sheets of matter are introduced in the beam, and the regeneration following the second is measured as a function of their separation x:

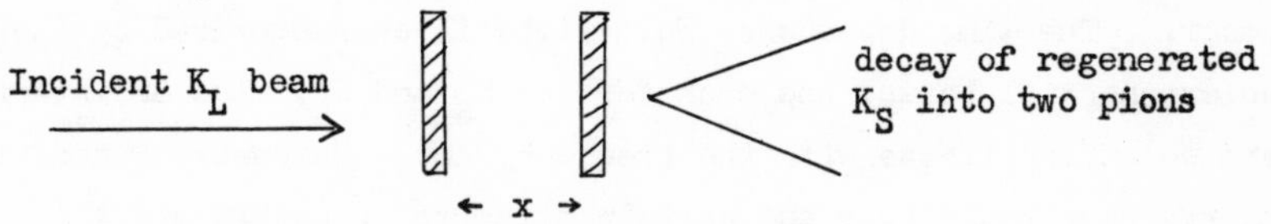

Incident K_L beam → [two plates separated by x] decay of regenerated K_S into two pions

The first experiment using the regeneration technique is that of Good, Piccioni et al.[10] who obtained the value $0.84^{+0.29}_{-0.22}$ in an experiment in which the regeneration was produced in an iron plate, and the detected $K_S \to \pi^+ + \pi^-$ decay occurred in a propane chamber. More recent measurements are due to Fujii et al.[11] who find $|\Delta m|/\Gamma_S = 0.8 \pm 0.2$ and Christenson[12] who finds $0.47^{+0.11}_{-0.13}$, using the "gap" method, and $0.41^{+0.25}_{-0.25}$ and 0.76 ± 0.2 varying the regenerator thickness. The latter two results are obtained from the same data but differ in the analysis. In general, it may be fair to say that the mass difference is, with reasonable certainty, less than 1, but the lower bound is quite insecure.

2. Interference in K_S, K_L 2π decays

Consider a beam for which the K_S amplitude is ρ, and the K_L amplitude is 1 at time t = 0. The $\pi^+\pi^-$ decay per unit time is then

$$\frac{dN_{+-}}{dt} = \Gamma_{S,+-} \left| \rho\, e^{-iM_S\tau} + \eta_{+-}\, e^{-iM_L\tau} \right|^2$$

$$= \Gamma_{S,+-} \left\{ |\rho|^2 e^{-\Gamma_S\tau} + |\eta_{+-}|^2 e^{-\Gamma_L\tau} \right. \qquad \text{(II.5)}$$

$$\left. + 2|\rho||\eta_{+-}| e^{-(\Gamma_S+\Gamma_L)\tau/2} \cos(\Phi_\rho - \Phi_\eta - \Delta m\tau) \right\} .$$

As before $\Delta m = m_S - m_L$; η_{+-} is the ratio of the decay amplitudes $a_L \to \pi^+\pi^- / a_S \to \pi^+\pi^-$ and $\Gamma_{S,+-}$ is the partial width for the $\pi^+\pi^-$ mode.

We found before (Lecture I) that $|\eta| = 1.85 \times 10^{-3}$, a small number. The time dependence Eq. (II.5) is characterized by the sum of two exponential terms, one each for the K_S and K_L, plus an interference term which oscillates with the frequency Δm. Demonstration of this interference would rule out certain attempts to understand the $K_L \to \pi^+\pi^-$ decay without CP violation, and provides a means of studying $\Phi_{\eta+-}$, $\Phi_{\eta oo}$, as well as Δm.

The $K_S - K_L$ mixture may be produced in various ways. In one method a K_L beam is incident on a regenerator. In this case the amplitude ρ is just the regeneration amplitude, which is necessarily small: $|\rho| \gtrsim 0.05$. Since, however, η is so small, there is no problem. The interference will be greatest at times

$$\tau \cong \frac{1}{\Gamma_S} \ln \left|\frac{\rho}{\eta}\right| .$$

Figure 3 shows a representative group of time distributions. Here, $|\Delta m|$ was put equal to $0.7\ \Gamma_S$, $|\rho/\eta|^2$ was put equal to 250, and the distribution (II.5) plotted for four values of the relative phase $\Phi_\rho - \Phi_\eta$. The chief drawback of this method is that the regeneration phase Φ_ρ is not known and although, in principle, measurable, the determination is at best difficult. Since Eq. (II.5) is sensitive only to $\Phi_\rho - \Phi_\eta$, Φ_η will have a corresponding uncertainty.

In another technique, also currently pursued, one begins with a source of K (or $\bar{K}$) mesons, and investigates the interference between the K_S and K_L components of the K. If the zero of time in Eq. (II.5) is the time of production, then $\rho = -1$, and the uncertainty in Φ_ρ falls away. The interference will be most pronounced at times τ

$$\tau \Gamma_S = 2 \ln \left|\frac{1}{\eta}\right| .$$

Yet a third situation has been studied recently by Fitch, Ross, Russ and Vernon[13]. These authors give up the study of the time dependence of the $K \to 2\pi$ intensity, and content themselves with a comparison of the 2π decay rates in, or following, absorbers on which K_L mesons are incident. The intensity of the 2π decay will also reflect the interference between the K_L and K_S two-pion amplitudes.

The apparatus is essentially similar to the previous one. What these authors have done is to measure three things:

i) the decay rate $K_L \to \pi^+\pi^-$ to find the $K_L \to \pi^+\pi^-$ amplitude;

ii) they insert in the sensitive region a piece of dense Be, 8 cm thick;

iii) the region sensitive to the detection of $K \to \pi^+\pi^-$ decays is entirely filled with beryllium of low-average density in the form of a Be-air sandwich.

The spatial extent of the low-density converter is so large that <u>within the converter</u> $\tau\Gamma_S >> 1$, so that expression (II.4) for the regeneration amplitude becomes τ independent:

$$\rho \to \rho_o N = \frac{\pi[f(0) - \bar{f}(0)]}{\Delta m + i\Gamma_S/2} \cdot N \,, \qquad \text{(II.6)}$$

$|f(0) - \bar{f}(0)|$ is determined in part ii) of the experiment. The $\pi^+\pi^-$ decay rate within the converter is then

$$I_{\pi^+\pi^-} = \Gamma_{S,+-} \left\{ |\eta_{+-}|^2 + 2|\eta||\rho_o|N \cos\,(\Phi_\eta - \Phi_\rho) + |\rho_o|^2 N^2 \right\} .$$

The interference term is the one linear in m. The results of parts i) - iii) of the experiment are shown in Fig. 4. In Fig. 5, the data are reduced to values of $\cos\,(\Phi_\eta - \Phi_\rho)$. Because of the mass dependence of Eq. (II.6), this is not unambiguous. If $|\Delta m| \simeq 0.5$ in line with present experiments, then $\cos(\Phi_\eta - \Phi_\rho) \sim 1$, and the interference seems clear. If Δm is nearer to zero, which seems at present not excludable, then $\cos(\Phi_\eta - \Phi_\rho)$ is near to zero, the result which also corresponds to no interference, and the evidence for the interference phenomena is correspondingly weak.

3. K leptonic decay and CP violation

The relative magnitude of the CP violating amplitude in $K \to \pi^+\pi^+$ decay is of the order 10^{-3}. This is also the order of magnitude of the relative rates of other K decay modes. It is conceivable that the entire effect observed in the two pion channels is due to a difference $p - q$ which in turn is the result of a relatively large (complete) CP violation in one or more of the other channels.

Consider then the decay $K \to \pi + L + \nu$. Assume that the lepton current has the usual form; the matrix element has then two terms:

$$M = \frac{G}{\sqrt{2}}\left\{f_+(q^2)(p_K + p_\pi)_\eta + f_-(q^2)(p_K - p_\pi)_\eta\right\} \cdot \bar{\nu} i\gamma_\eta(1 + \gamma_5)L .$$

The second term can be transformed with the help of the Dirac equation and the conservation law

$$p_K - p_\pi = p_\nu + p_L$$

so that

$$M = \frac{Gf_+}{\sqrt{2}}\left\{(p_K + p_\pi)_\eta\, \bar{\nu}\, i\gamma_\eta(1 + \gamma_5)L - \xi m_L \bar{\nu}(1 - \gamma_5)L\right\},$$

where

$$\xi(q^2) = f_-(q^2)/f_+(q^2) .$$

If CP is conserved, ξ is necessarily real. CP violating effects are the result of interference between the first term and the imaginary part of ξ times the second. For electron decay no substantial CP violation effect can be expected because the second term is of order m_L/m_K. In the muon decay one of the expected CP violating results would be a transverse polarization of the muon. Until recently the transverse polarization has not been investigated, however the longitudinal polarization of the muon and the spectrum of both muon and electron had been studied. The data can be fitted without the need of an imaginary component of ξ; in fact, the second term plays quite a small role. All data can be fitted with Re $\xi = 0$ and in any case $|\text{Re } \xi| < 0.5$. The form factor is compatible with being constant, and the electron-muon comparative rates are in agreement with universality.

The problem is now being studied extensively: in particular, the muon transverse polarization. One result which has already appeared is due to a collaboration of Berkeley, Wisconsin and Bari[14]. In a heavy liquid bubble chamber experiment K^+ mesons are stopped; the spectrum and the longitudinal and transverse muon polarization are measured. It is found that $p_{\mu_\perp} = 0.04 \pm 0.35$ so that there is no evidence here of CP violation, however, the limit on Im ξ is poor: Im $\xi = 0.2 \pm 1.8$. From the spectrum and longitudinal polarization, additional information is obtained, all compatible with $\xi = 0$. It is found that

$$\text{Re } \xi = 0.48 \pm 0.32$$

and

$$\text{Im } \xi = 0.8^{+0.6}_{-0.9} \quad .$$

Finally, if electron-muon universality is assumed, and the form factors are taken as constant, the accuracy on Re ξ can be improved by means of the relation:

$$\frac{\Gamma_{K_{\mu 3}}}{\Gamma_{K_{e3}}} = 0.651 + 0.126 \text{ Re } \xi + 0.0189|\xi|^2 \ .$$

Then

$$\text{Re } \xi = 0.28 \pm 0.29 \ .$$

In any case, there is no indication of large CP violation in this channel.

III. CP VIOLATION AND THE $\Delta S = \Delta Q$ RULE IN K^0 LEPTONIC DECAY

Consider now the leptonic K^0 decays and, in particular, their time dependence. We have the following processes and amplitudes:

	Process	Amplitude	Rate	
1)	$K^0 \to L^+ + \nu + \pi^-$	f	N^+	$\Delta S = \Delta Q$
2)	$\bar{K}^0 \to L^- + \bar{\nu} + \pi^+$	f^*	$\bar{N}^-$	
3)	$K^0 \to L^- + \bar{\nu} + \pi^+$	g^*	$\bar{N}^+$	$\Delta S = -\Delta Q$
4)	$\bar{K}^0 \to L^+ + \nu + \pi^-$	g	N^-	

Here, we have assumed CPT conservation in order to write the amplitudes for 2) and 4) in terms of those for 1) and 3). If CP is conserved, the parameter $x = g/f$ must be zero, and conversely, all CP violating effects are proportional to Im x. Since $x = 0$ if the $\Delta S = \Delta Q$ rule is valid, no CP violation is observable in this channel, unless the $\Delta Q = \Delta S$ rule is also violated.

The rate of appearance of L^+ from an initial K^0 can be written as

$$N^+ = (\text{I} + \text{II} - \text{III}) \ ,$$

where

$$I = |1+x|^2\, e^{-\Gamma_L \tau} + |1-x|^2\, e^{-\Gamma_S \tau}$$

$$II = 2(1-|x|^2) \cos \Delta m\tau\, e^{-\frac{1}{2}(\Gamma_S+\Gamma_L)\tau}$$

$$III = 4\, \mathrm{Im}\, x \sin \Delta m\tau\, e^{-\frac{1}{2}(\Gamma_S+\Gamma_L)\tau} .$$

In the same way the rates N^- and $\bar{N}^+$ will be:

$$N^- = (I - II - III)$$

$$\bar{N}^- = (I + II + III)$$

$$\bar{N}^+ = (I - II + III) .$$

We note that if $|x|$ is small compared to 1, i.e. for small violation of $\Delta S = \Delta Q$

$$I \simeq (1+2\,\mathrm{Re}\,x)e^{-\Gamma_L \tau} + (1-2\,\mathrm{Re}\,x)e^{-\Gamma_S \tau}$$

$$II \simeq 2 \cos \Delta m\tau\, e^{-\frac{1}{2}(\Gamma_S+\Gamma_L)\tau} .$$

Term I is, therefore, sensitive to small Re x, term III is sensitive to Im x, that is CP violation, and term II is insensitive to small x but is sensitive to the mass difference Δm.

The terms I, II, and III can be separately studied by forming certain combinations of the N's:

$$\frac{1}{4}(N^{+} + N^{-} + \bar{N}^{-} + \bar{N}^{+}) = I$$

$$\frac{1}{2}(N^{+} - N^{-}) \quad \text{or} \quad \frac{1}{2}(\bar{N}^{-} - \bar{N}^{+}) = II$$

$$\frac{1}{4}[-(N^{+} + N^{-}) + (\bar{N}^{-} + \bar{N}^{+})] = III \ .$$

There are three recent experiments which have been directed towards this question. In the experiments of the Ecole Polytechnique group [8)] and the Padua group[9)] K^0 mesons are generated in the interaction of 800 MeV/c K^+ mesons in a bubble chamber filled with an equal mixture by volume of freon and propane. At the time of production therefore, the $\bar{K}^0$ amplitude is zero. Electron decays are selected by the characteristic radiative behaviour of the electrons, given the name spiralization. Muonic decay is not detected. The experimental problems are chiefly two:

i) the momentum of the K^0, which is necessary to convert the decay path into proper time, cannot be determined individually, but is inferred on the average from observations on $K \to \pi^+\pi^-$ decays;

ii) a non-negligible fraction of the events do not permit unambiguous association.

The third experiment is by the Columbia Rutgers group[15)]. The K^0 are generated in the annihilation of antiprotons stopped in a liquid hydrogen chamber. K^0 and $\bar{K}^0$ are, therefore, produced in equal numbers. In about one half of the cases the strangeness character can be determined, because a K^+ or K^- is seen at the production vertex. The leptonic decays are selected on the basis of kinematics after a measurement of all V's in the pictures. In this way it is possible to find both muonic and electronic decays. The momentum is usually known by combining the information available from the production and decay vertices. The sign of the electron charge is usually not available. The data of the three experiments are presented in Figs. 6 to 10. The Padua data are in agreement with the $\Delta S - \Delta Q$ prediction, except for the case of negative electrons at short times. In the first 2×10^{-10} seconds, six events are observed against an expectation of two. Accommodation of the extra four events requires a large $|x|$, and this is the source of the discrepancy reported by thig group. The Paris data, which are statistically superior, do not confirm this particular discrepancy. Here, the chief deviation from the $\Delta S - \Delta Q$ prediction is in the distribution $N^+ + N^-$ where a small contribution of term III, corresponding to Im x = 0.2, permits a better fit. The fit with $\Delta S = \Delta Q$ seems, however, adequate (see Fig. 8). The Columbia results permit a separation of terms I and III, because of the fact that equal numbers of K^0 and $\bar{K}^0$ are contributing. The data are in good agreement with $\Delta S = \Delta Q$, and therefore CP conservation. This is especially graphic in the difference distribution $\bar{K}^0 - K^0$ of Fig. 10. The solid curve in Fig. 10 is the expectation for a CP violation of the order of $|\text{Im } x| = 0.2$ (and $|\Delta m| = 0.5$), which roughly corresponds to the statistical sensitivity of the experiment.

These experimental data have been analysed by their respective authors, using the maximum likelihood method, to yield best values of Re x, Im x and, in the case of Padua and Paris, also Δm. The results are given in Table 3.

Table 3

Maximum likelihood fits to obtain best values of Re x, Im x and Δm

Experiment	Re x	Im x*)	Δm
Padua	0.06 ± 0.25	$\pm 0.43 \pm 0.25$	$0.15^{+0.35}_{-0.50}$
Paris	$0.04^{+0.11}_{-0.13}$	$-0.21^{+0.15}_{-0.11}$	0.47 ± 0.2
Columbia-Rutgers	0 ± 0.2	$+0.07 \pm 0.20$	0.5 (assumed)

*) The algebraic sign of Im x corresponds to positive Δm.

The combined results give no convincing demonstration of $\Delta S - \Delta Q$ violation, and also therefore not of CP violation. They are probably sufficient to permit the statements:

$$|\text{Re } x| \leq 0.1$$

$$|\text{Im } x| \leq 0.25 \ .$$

It may be useful to append here experimental results available on the question of $\Delta S - \Delta Q$ violation in other reactions.

4. $K \rightarrow 3\pi$

In Table 4 we list four channels, the observed rate Γ_{exp}, Φ the phase space relative to the phase space in channel 1, the expected rate from the $\Delta I = \frac{1}{2}$ rule, and finally the ratio Γ_{exp}/Φ.

Table 4

Channel	Γ_{exp}	Φ	$\Delta I = \frac{1}{2}$ Expected rel. central rate	Γ_{exp}/Φ
$K^+ \rightarrow \pi^+\pi^+\pi^-$	4.85 ± 0.16	1.000	4	4.85 ± 0.16
$K^+ \rightarrow \pi^+\pi^0\pi^0$	1.45 ± 0.11	1.245	1	1.16 ± 0.09
$K_L \rightarrow \pi^0\pi^0\pi^0$	4.85 ± 0.5	1.495	3	3.23 ± 0.3
$K_L \rightarrow \pi^+\pi^-\pi^0$	2.52 ± 0.2	1.225	2	2.06 ± 0.16

From this table we see that $\Delta I = \frac{1}{2}$ is quite good. If CP is conserved $K_S \rightarrow 3\pi^0$ is forbidden, and $K_S \rightarrow \pi^+\pi^-\pi^0$ can only occur in odd orbital states for $\pi^+\pi^-$, so that the rate can be expected to be depressed by a factor of $\sim (Q/m_K)^2 = 1/200$.

If CP were conserved in the $K \to 3\pi$ decay, then a) the decay $K_S \to 3\pi^0$ would be forbidden and b) the decay $K_S \to \pi^+\pi^-\pi^0$ would be inhibited by angular momentum barriers. The $\pi^+\pi^-$ which in K_L decay are dominantly in an S orbit would have to be in a p orbit, and the decay would be expected to be depressed by an angular momentum barrier factor of the order of $(Q/m_K)^2 \simeq 1/200$. Such a low rate is virtually impossible to detect in the large background of K_L three-pion decays. If CP is violated, the $K_S \to 3\pi$ decay can proceed to the extend of the violation. At present no adequate evidence exists for any $K_S \to 3\pi$ decay, and the sensitivity is quite poor. The experimental upper bound for $\Gamma_{K_S \to 3\pi}$ is about equal to the rate $\Gamma_{K_L \to 3\pi}$. The experiment excludes, therefore, CP violating amplitudes larger than the CP conserving ones.

Because of the difficulty in measuring the $K_S \to 3\pi$ decay rates, a remark of Cabibbo[16)] and Gaillard[17)] is useful. These authors point out that if the $\Delta I = \frac{1}{2}$ rule is assumed, the ratio $\Gamma_{+-0}/\Gamma^{+}_{+00}$ can be used to discuss CP violation in the 3π decay. The argument is as follows: let A and $\bar{A}$ be the 3π decay amplitudes of K and $\bar{K}$, respectively, in the S state. Assume CPT, then $|A| = |\bar{A}|$. Let A_L, A_S be the decay amplitudes of K_L and K_S respectively into the three pions. Then neglecting the small $(p - q)$ terms,

$$A_L = \frac{1}{\sqrt{2}}(A + \bar{A})$$

and

$$A_S = \frac{1}{\sqrt{2}}(A - \bar{A}) .$$

As we have already noted, if CP is conserved $\bar{A} = A$, and $A_S = 0$. In general, we can write $A = |A|e^{i\Phi}$ and $\bar{A} = |A|e^{-i\Phi}$, where $\Phi = 0$ corresponds to CP conservation. Then

$$A_L = \cos\Phi\,|A|\,, \quad A_S = \sin\Phi\,|A|$$

and

$$\frac{\Gamma^L_{+-o}}{\Gamma^+_{+oo}} = 2\cos^2\Phi\,.$$

From the values quoted in Table 4 $\cos^2\Phi = 0.89 \pm 0.10$ and $\cos\Phi = 0.94 \pm 0.05$. This is consistent with CP conservation ($\Phi = 0$) and an upper limit can be put on Φ: $|\Phi| \leq 25°$.

The time distribution of $\pi^+\pi^-\pi^0$ decays from an initial K^o is the following:

$$P(t) = \left|\cos\Phi\, e^{-iM_L\tau} - i\sin\Phi\, e^{-iM_S\tau}\right|^2 .$$

Anderson et al.[18] find from an analysis of the time distribution of an 18 event sample

$$|\tan\Phi| = 0.9 \pm 0.5\,.$$

IV. UNITARITY AND $p^2 - q^2$

We now wish to write the conditions imposed by unitarity and CPT on CP violating amplitudes. This has been discussed by Wu and Yang[19], and here I will follow Bell's lecture at Oxford[1].

We write now the requirement of unitarity: let

$$\bar{\Psi} = a\, e^{-iM_L\tau}|L> + b\, e^{-iM_S\tau}|S>\,, \qquad \text{(IV.1)}$$

the total transition rate is

$$- \frac{d}{d\tau} |\bar{\Psi}|^2 = \Sigma | a \langle F|T|L\rangle + b \langle F|T|S\rangle |^2 ,$$

when the sum is over all final states F,

$$= \Sigma[|a|^2 |\langle F|T|L\rangle|^2 + |b|^2 |\langle F|T|S\rangle|^2 + ab^* \langle F|T|S\rangle^* \langle F|T|L\rangle$$
$$+ a^*b \langle F|T|L\rangle^* \langle F|T|S\rangle] . \qquad (IV.2)$$

From Eq. (IV.1) it follows that at $\tau = 0$

$$\frac{d|\Psi|^2}{d\tau} = |a|^2 \Gamma_L + |b|^2 \Gamma_S + i(M_S - M_L)(ab^* \langle S|L\rangle - a^*b \langle L|S\rangle). \qquad (IV.3)$$

Equating (IV.2) and (IV.3)

$$\Gamma_L = \Sigma | \langle F|T|L\rangle |^2$$
$$\Gamma_S = \Sigma | \langle F|T|S\rangle |^2$$

and

$$\langle L|S\rangle = p^2 - q^2 = \frac{1}{i(M_S - M_L)} \Sigma \langle F|T|S\rangle^* \langle F|T|L\rangle . \qquad (IV.4)$$

Let $\Gamma(G)$ be the partial width to a group G of final states:

$$\Gamma_S(G) = \sum_G |\langle F|T|S\rangle|^2$$

and

$$\Gamma_L(G) = \sum_G |\langle F|T|L\rangle|^2 .$$

Then the overlap of the transition amplitudes must be smaller than the square root of the product of the width:

$$\sum_G \langle F|T|S\rangle^* \langle F|T|L\rangle \lesssim \sqrt{\Gamma_L(G)\Gamma_S(G)} \ . \qquad \text{(IV.5)}$$

With the help of (IV.5) and experimental results, it is possible to give upper limits for the contributions of various decay channels to the overlap (IV.5). This was done in the paper of Wu and Yang[19].

Table 5

Channel	$\lvert\sum_G \langle F\lvert T\rvert S\rangle^* \langle F\lvert T\rvert L\rangle\rvert/\Gamma_S$
(1) $\pi^+\pi^-$	$< 1.3\times 10^{-3}$
(2) $\pi^0\pi^0$	$< 5.3\times 10^{-3}$
(3) $\pi L \nu$	$< 0.3\times 10^{-3}$
(4) 3π	$< 0.6\times 10^{-3}$
$\lvert p^2 - q^2\rvert$ $<$	$7.5\times 10^{-3}/\lvert 1 + i\ 2\Delta m/\Gamma_S\rvert$

1. Isospin analysis of the 2π decay mode

We would like to discuss now what measurements are necessary in order that the CP violation in the two-pion decay mode can be completely understood. There are two isospin states, I = 0 and I = 2 (I = 1 is forbidden to two pions of even angular momentum) and, therefore, four amplitudes:

$$\langle 0|T|K\rangle,\ \langle 0|T|\bar{K}\rangle,\ \langle 2|T|K\rangle$$

and

$$\langle 2|T|\bar{K}\rangle .$$

The CPT theorem requires

$$\langle F|T|K\rangle = e^{2i\delta}(\bar{F}'|T|\bar{K})^* ,$$

where F is some final state and $\bar{F}'$ is the charge conjugate state with spins reversed. Using then this CPT requirement, write

$$\langle 0|T|K\rangle = i\,e^{i\delta_0} A_0 ,\ \langle 0|T|\bar{K}\rangle = -i\,e^{i\delta_0} A_0^*$$

$$\langle 2|T|K\rangle = i\,e^{i\delta_2} A_2 ,\ \langle 2|T|\bar{K}\rangle = -i\,e^{i\delta_2} A_2^* .$$

The two states of well-defined charge are, in terms of those of well-defined I spin:

$$+- = \sqrt{\frac{2}{3}}\,(I = 0) + \sqrt{\frac{1}{3}}\,(I = 2)$$

$$0\,0 = \sqrt{\frac{1}{3}}\,(I = 0) - \sqrt{\frac{2}{3}}\,(I = 2) .$$

Then, forgetting $p^2 - q^2$ in comparison with 1:

$$\Gamma_{S,+-} = \frac{2}{3}\left|A_0 + \frac{1}{\sqrt{2}} A_2\, e^{i(\delta_2-\delta_0)}\right|^2$$

$$\Gamma_{S,00} = \frac{1}{3}\left|A_0 - \sqrt{2}\, A_2\, e^{i(\delta_2-\delta_0)}\right|^2 .$$

The experimental ratio $\Gamma_{S,+-}/\Gamma_{S,oo}$ is 2.33 ± 0.05; so barring accidental cancellations,

$$|A_2/A_0| \simeq 0.05 .$$

The relative two-pion decay amplitudes of long and short-lived kaons are:

$$\eta_{+-} = |\eta_{+-}| e^{i\Phi_{\eta+-}} = \frac{\langle +-|T|L\rangle}{\langle +-|T|S\rangle} = \frac{(pA_0 - qA_0^*) + \frac{1}{\sqrt{2}} e^{i(\delta_2-\delta_0)}(pA_2 - qA_2^*)}{(pA_0 + qA_0^*) + \frac{1}{\sqrt{2}} e^{i(\delta_2-\delta_0)}(pA_2 + qA_2^*)} \tag{IV.6}$$

$$\eta_{oo} = |\eta_{oo}| e^{i\Phi_{\eta oo}} = \frac{\langle 00|T|L\rangle}{\langle 00|T|S\rangle} = \frac{(pA_0 - qA_0^*) + \sqrt{2}\, e^{i(\delta_2-\delta_0)}(pA_2 - qA_2^*)}{(pA_0 + qA_0^*) + \sqrt{2}\, e^{i(\delta_2-\delta_0)}(pA_2 + qA_2^*)} , \tag{IV.7}$$

and the unitarity condition reads:

$$0 = \mathrm{Re}\left\{\frac{(p^2 - q^2)(|A_0|^2 + |A_2|^2) - 2pqi\ \mathrm{Im}(A_0^2 + A_2^2) + \gamma\Gamma_S}{\Delta m + i\Gamma_S/2}\right\}, \tag{IV.8}$$

where $\gamma\Gamma_S$ are the contributions of channels other than the two pion channels. We know from table 5 that these are not large enough by themselves to account for a $p^2 - q^2$ which could produced the CP violation observed in the two-pion channel without specific violation in the two-pion transition. Experimentally they may turn out to be very small. Therefore, it is not without interest to explore the possibility that $\gamma \ll 10^{-3}$, and is therefore negligible. Equation (IV.8) then becomes:

$$\frac{2pq \; \mathrm{Im} |A_0|^2 + |A_2|^2}{(p^2 - q^2)(|A_0|^2 + |A_2|^2)} = \frac{2\Delta m}{\Gamma_S} . \qquad \text{(IV.8')}$$

If we know Γ_S and Δm from other measurements, we have the six quantities to determine: Re A_0, Im A_0, Re A_2, Im A_2, $p - q$, and $\delta_2 - \delta_0$ and the following measurements already done: the two partial widths $\Gamma_{S,+-}$ and $\Gamma_{S,oo}$, and the absolute value of η_{+-}. Experiments on the absolute value of η_{oo} are in progress at CERN, as well as interference experiments to find $\Phi_{\eta+-}$. These five measurements, together with (IV.8'), are sufficient to determine all parameters. Checks are possible, for instance, by measuring $\Phi_{\eta oo}$, or by using pion-production experiments to find $\delta_2 - \delta_0$, or by studying the ratio

$$\frac{\Gamma(K_L \to \pi^+ + L^- + \nu)}{\Gamma(K_L \to \pi^- + L^+ + \nu)}$$

for which the ratio $(q/p)^2$ is expected if the $\Delta S - \Delta Q$ rule is valid.

REFERENCES

1) J.S. Bell and J. Steinberger, "Weak interactions of kaons", Oxford International Conference on Elementary Particles (1965).

2) J.H. Christenson, J.W. Cronin, V.L. Fitch and R. Turlay, Phys.Rev. Letters 13, 138 (1964).

3) A. Abashian, R.J. Abrams, D.W. Carpenter, G.P. Fisher, B.M.K. Nefkens and J.H. Smith, Phys.Rev. Letters 13 243 (1964).

4) W. Galbraith, G. Manning, A.E. Taylor, B.D. Jones, J. Malos, A. Astbury, N.H. Lipman and G.T. Walker, Phys.Rev. Letters 14, 383 (1965).

5) X. de Bonard, D. Dekkers, B. Jordan, R. Mermod, T.R. Willitts, K. Winter, P. Scharff, L. Valentin, M. Vivargent and Bolt-Bodenhausen, Physics Letters 15, 58 (1965).

6) U. Camerini, W.F. Fry, J.A. Gaidos, H. Huzita, S.V. Natali, R.B. Willmann, R.W. Birge, R.P. Ely, W.M. Powell and H.S. White, Phys.Rev. 128, 362 (1962).

7) In discussion at the Oxford Conference Frye stated that this value is probably incorrect, that new experiments of this type are in progress, but that the new results are not yet available.

8) B. Aubert, L. Behr, F.L. Canavan, L.M. Chounet, J.P. Lowys, P. Mittner and C. Pascaud, Physics Letters 17, 59 (1965).

9) M. Baldo-Ceolin, E. Calimani, S. Giampolillo, C. Filippi-Filosofo, H. Huzita, F. Mattioli and G. Miari, Nuovo Cimento 38, 684 (1965).

10) R.H. Good, R.P. Matsen, F. Muller, O. Piccioni, W.M. Powell, H.S. White, W.B. Fowler and R.W. Birge, Phys.Rev. 124, 1223 (1961).

11) T. Fujii, J.V. Jovanovitch, F. Turkot and G.T. Zorn, Phys.Rev. Letters 13, 253 (1964).

12) J.H. Christenson, Ph.D. Thesis, Princeton University (1964).

13) V.L. Fitch, R.F. Roth, J.S. Russ and W. Vernon, Phys.Rev. Letters 15, 73 (1965).

14) G.E. Kalmus, W.M. Powell, R.T. Pu and C.L. Sandler, Phys.Rev. Letters 14, 989 (1965).

15) P. Franzini, L. Kirsch, P. Schmidt and J. Steinberger, Phys.Rev. (in press).

16) N. Cabibbo, Proceedings of Erice School (1964) p. 284.

17) M.K. Gaillard, Nuovo Cimento 35, 1225 (1965).

18) J.A. Anderson, F.S. Crawford, Jr., R.L. Golden, D. Stern, T.O. Binford and V.G. Lind, Phys.Rev. Letters 14, 475 (1965).

19) T.T. Wu and C.N. Yang, Phys.Rev. Letters 13, 501 (1964).

FIGURE CAPTIONS

Fig. 1 : Arrangement of $\pi^+\pi^-$ detector in experiment of Christenson et al.

Fig. 2 : a) Invariant mass distribution in the experiment of Christenson et al.
b) Angular distribution of events in the mass interval 490 < m < 510 in the same experiment. The calculated distributions are the results of a Monte Carlo calculation on the expectation for leptonic decays.

Fig. 3 : Expected time distributions for K_L, $K_S \to \pi^+\pi^-$ interferences for several values of the relative phase $\Phi_\rho - \Phi_{\eta+-}$, assuming $\Delta m = 0.7\ \Gamma_S$ and a regeneration $|\rho|^2 = 1.2 \times 10^{-3}$.

Fig. 4 : Results for the three parts of the experiment of Fitch et al. in the form of angular distributions of the events in the mass region 485 < m < 505.

Fig. 5 : Cos $(\Phi_{\rho_o} - \Phi_{\eta_{+-}})$ as a function of assumed mass difference. Experiment of Fitch et al.; $\alpha = \Phi_{\rho_o} - \Phi_{\eta_{+-}}$.

Fig. 6 : Results of Baldo-Ceolin et al. on the time distribution of leptonic K^o decays. The curves are the predictions of the $\Delta S = \Delta Q$ rule.

Fig. 7 : Results of Aubert et al. on the time distribution of positive and negative K^o leptonic decays. The curves are drawn on the basis of Im x = 0.2, Re x = 0.

Fig. 8 : Results of Aubert et al. on the time distribution of all K^o leptonic decays. The solid curve is the prediction of the $\Delta S = \Delta Q$ rule.

Fig. 9 : Results of Franzini et al. on the time distribution of the leptonic decays of an equal mixture of K^0 and $\bar{K}^0$ mesons. The solid curve is the prediction of the $\Delta S = \Delta Q$ rule.

Fig. 10 : Results of Franzini et al. on the time distribution of the difference of K^0 (plotted downwards) and $\bar{K}^0$ (plotted upwards) leptonic decays. The difference (shaded experimental result) is zero if $\Delta S = \Delta Q$. The solid curve is drawn for Im $x = 0.2$.

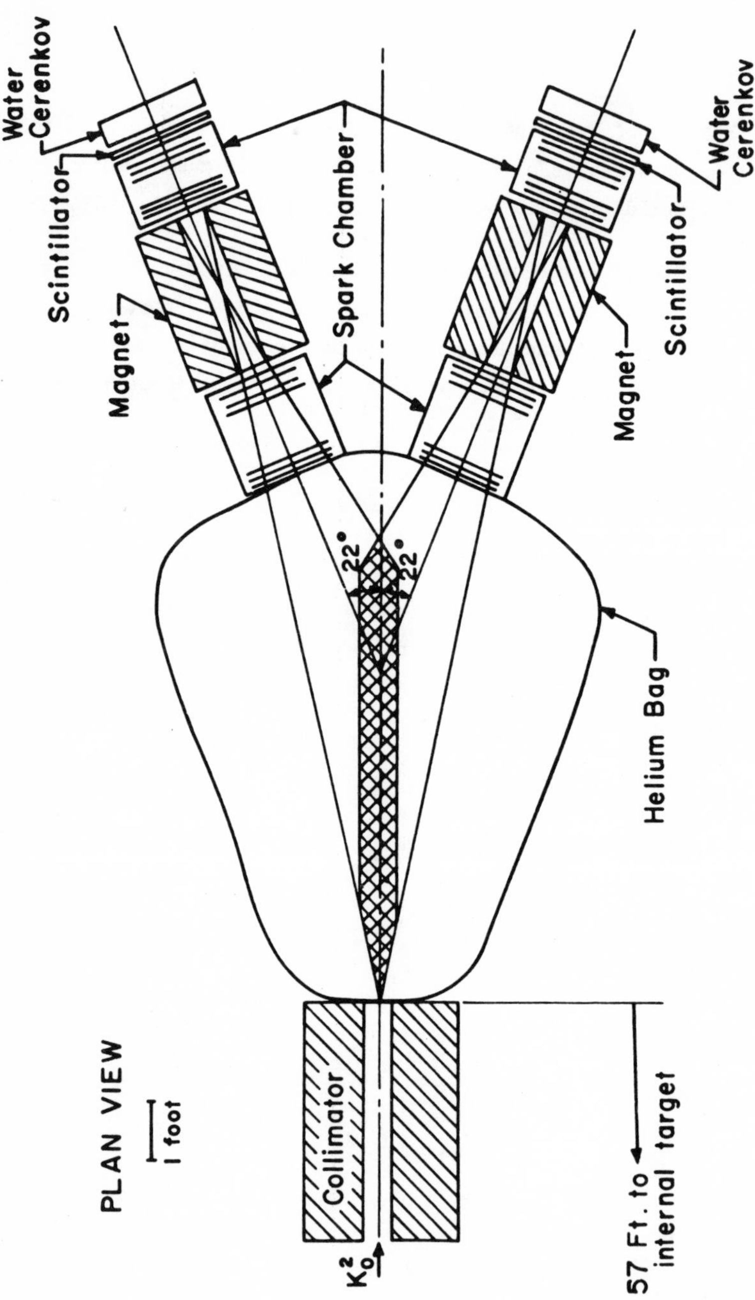
PLAN VIEW
1 foot
Collimator
K_2^o
57 Ft. to internal target
Helium Bag
22°
22°
Spark Chamber
Magnet
Scintillator
Water Cerenkov
Magnet
Scintillator
Water Cerenkov

Fig. 1

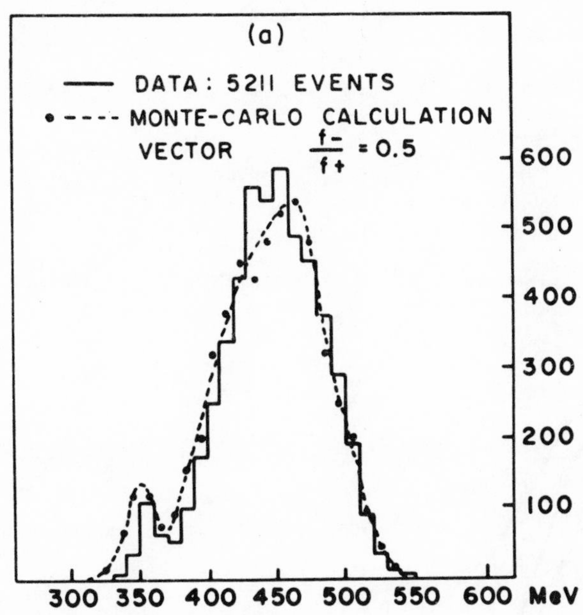
(a)
DATA : 5211 EVENTS
MONTE-CARLO CALCULATION
VECTOR $\frac{f-}{f+}$ = 0.5
600
500
400
300
200
100
300 350 400 450 500 550 600 MeV

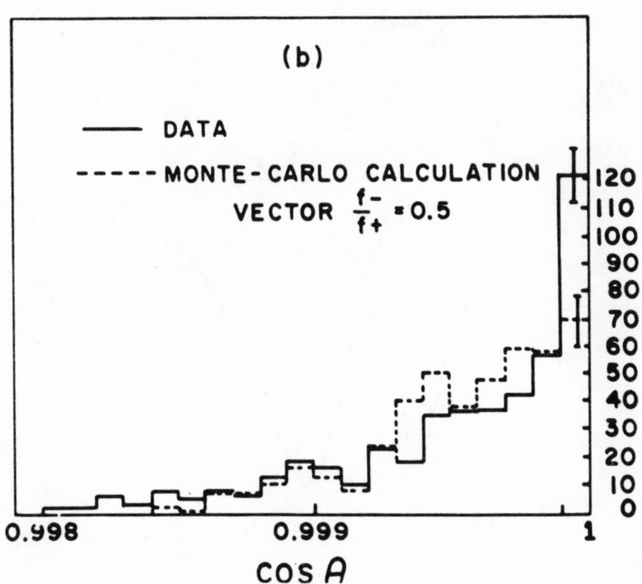
(b)
DATA
MONTE-CARLO CALCULATION
VECTOR $\frac{f-}{f+}$ = 0.5
120
110
100
90
80
70
60
50
40
30
20
10
0
0.998
0.999
1
cos A

Fig. 2

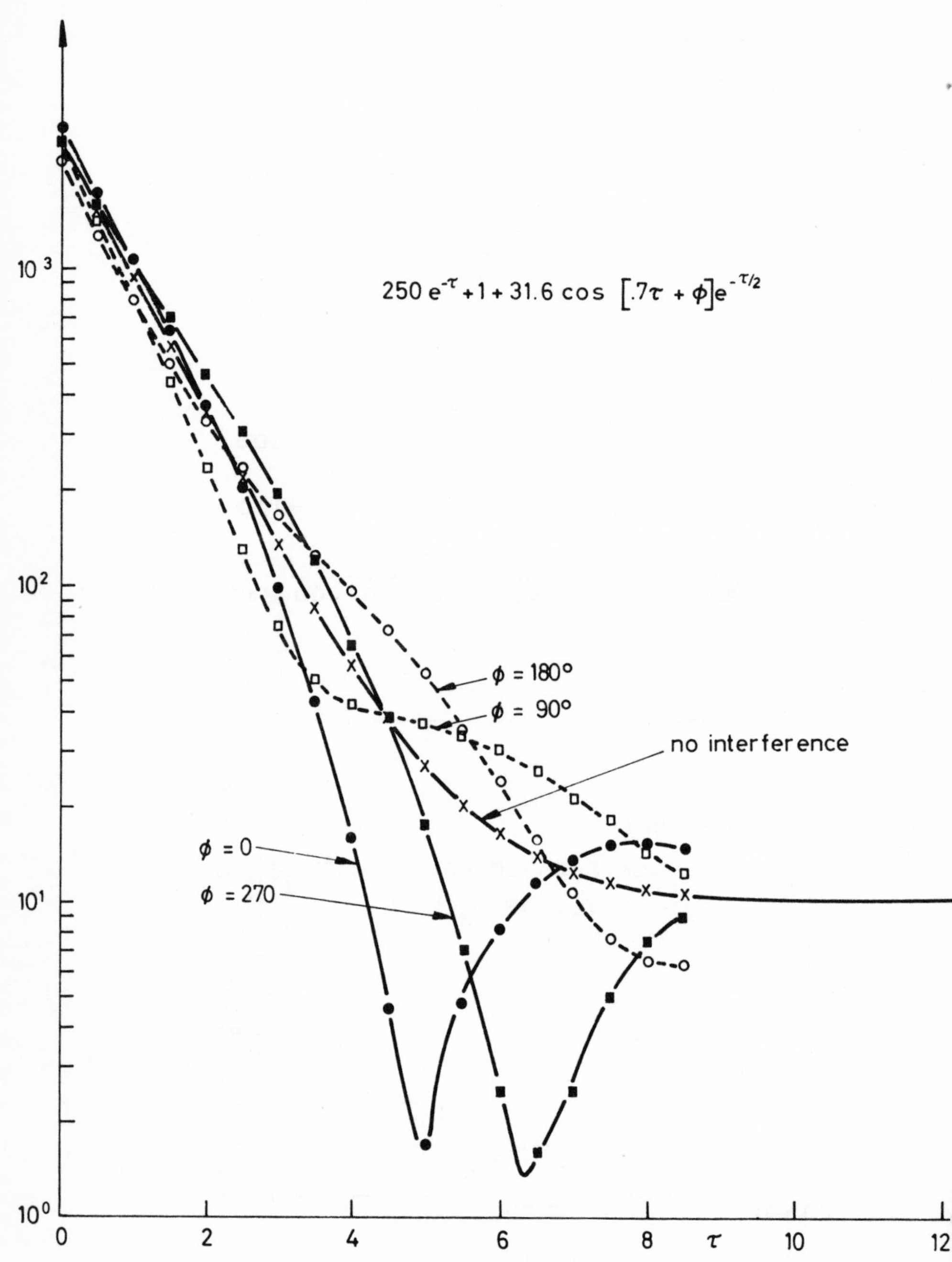

$250\,e^{-\tau}+1+31.6\cos\,[.7\tau+\phi]e^{-\tau/2}$
10^3
10^2
10^1
10^0
0
2
4
6
8
10
12
τ
ϕ = 180°
ϕ = 90°
no interference
ϕ = 0
ϕ = 270

Fig. 3

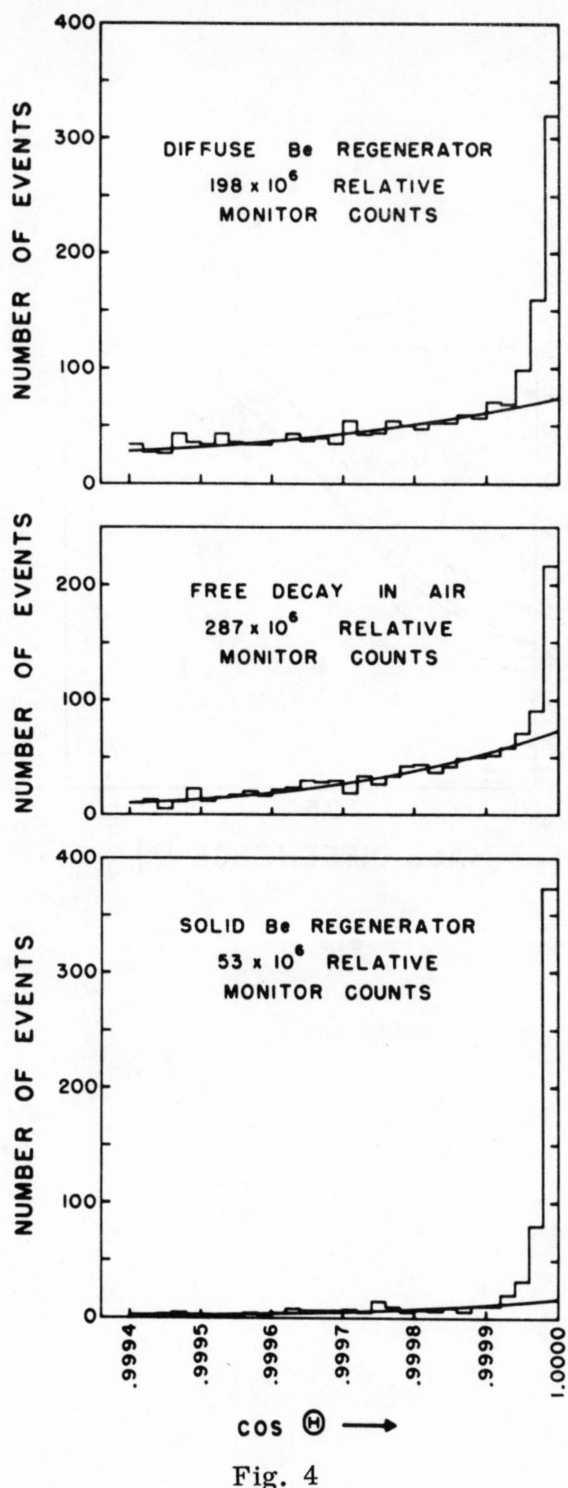

400
300
200
100
0
NUMBER OF EVENTS
DIFFUSE Be REGENERATOR
198 x 10^6 RELATIVE
MONITOR COUNTS
200
100
0
NUMBER OF EVENTS
FREE DECAY IN AIR
287 x 10^6 RELATIVE
MONITOR COUNTS
400
300
200
100
0
NUMBER OF EVENTS
SOLID Be REGENERATOR
53 x 10^6 RELATIVE
MONITOR COUNTS
.9994
.9995
.9996
.9997
.9998
.9999
1.0000
COS Θ

Fig. 4

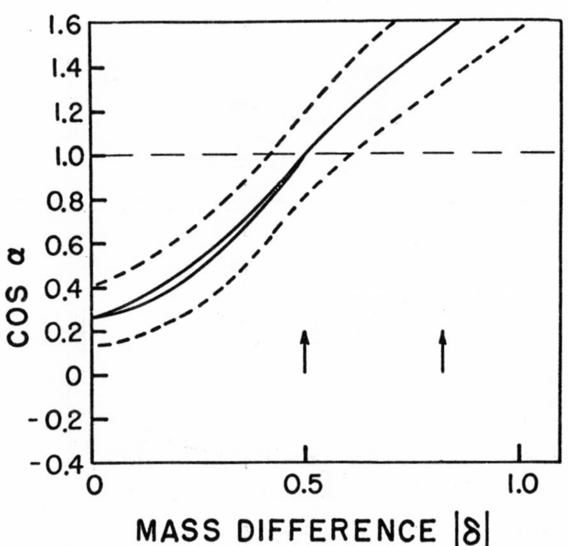

Fig. 5

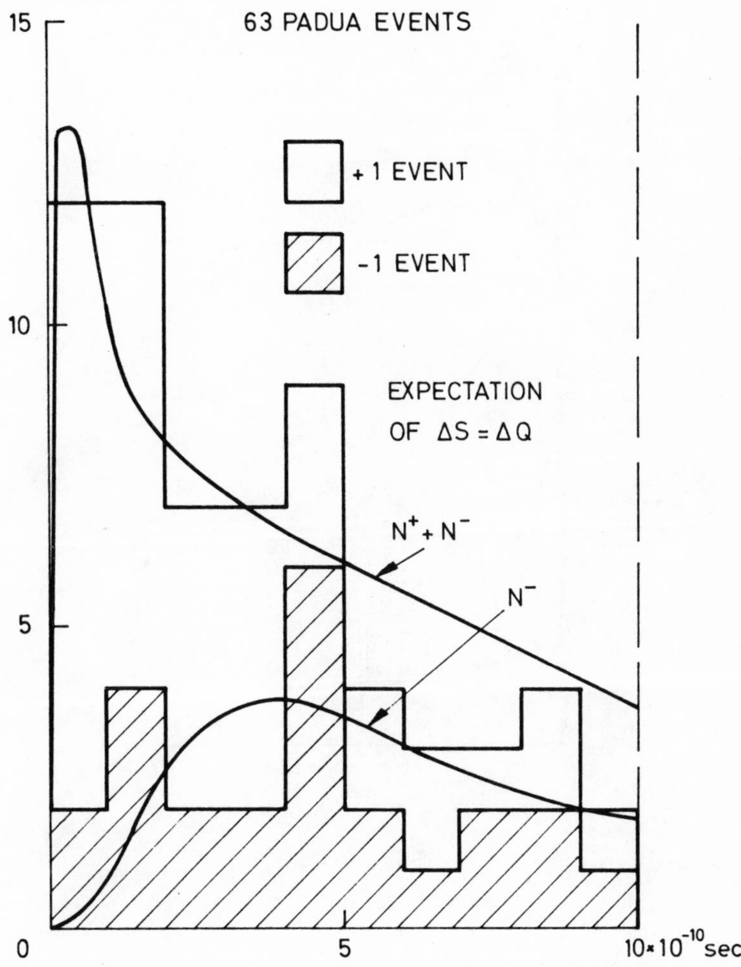
63 PADUA EVENTS
+ 1 EVENT
-1 EVENT
EXPECTATION
OF ΔS = ΔQ
N⁺ + N⁻
N⁻
15
10
5
0
5
10×10⁻¹⁰sec

Fig. 6

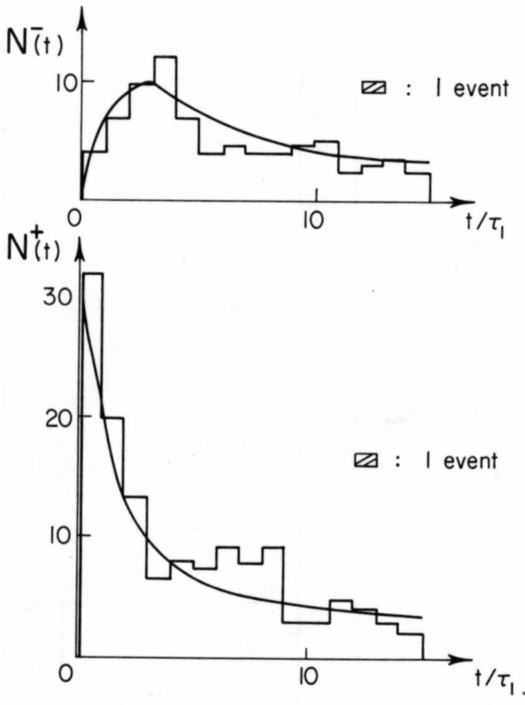

Fig. 7

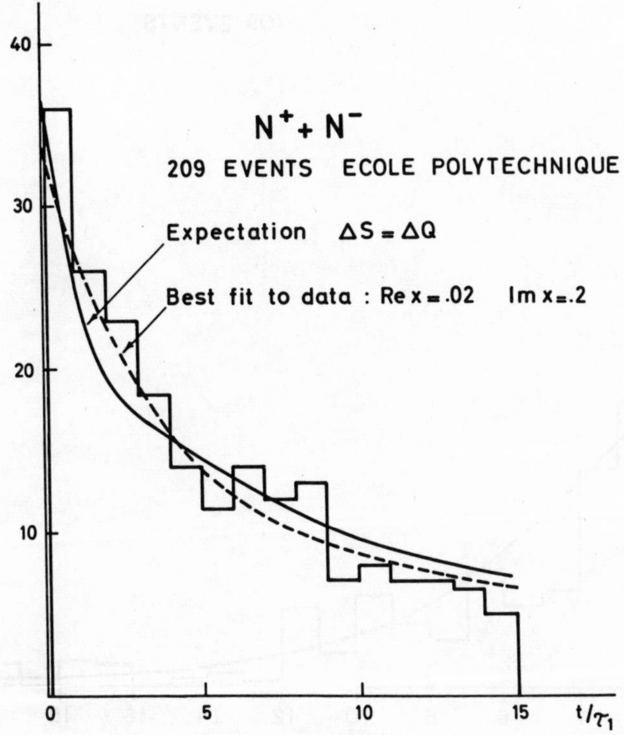

40
30
20
10
0
5
10
15
t/τ_1
$N^+ + N^-$
209 EVENTS ECOLE POLYTECHNIQUE
Expectation $\Delta S = \Delta Q$
Best fit to data : Re x = .02 Im x = .2

Fig. 8

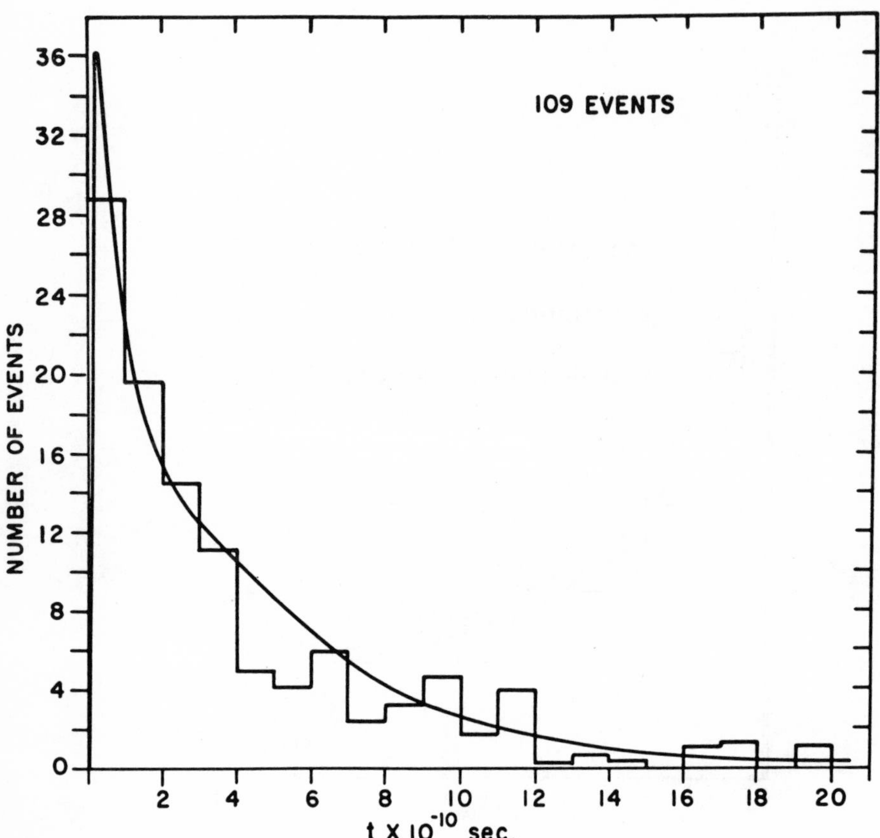
109 EVENTS
NUMBER OF EVENTS
0
4
8
12
16
20
24
28
32
36
2
4
6
8
10
12
14
16
18
20
t X 10^{-10} sec.

Fig. 9

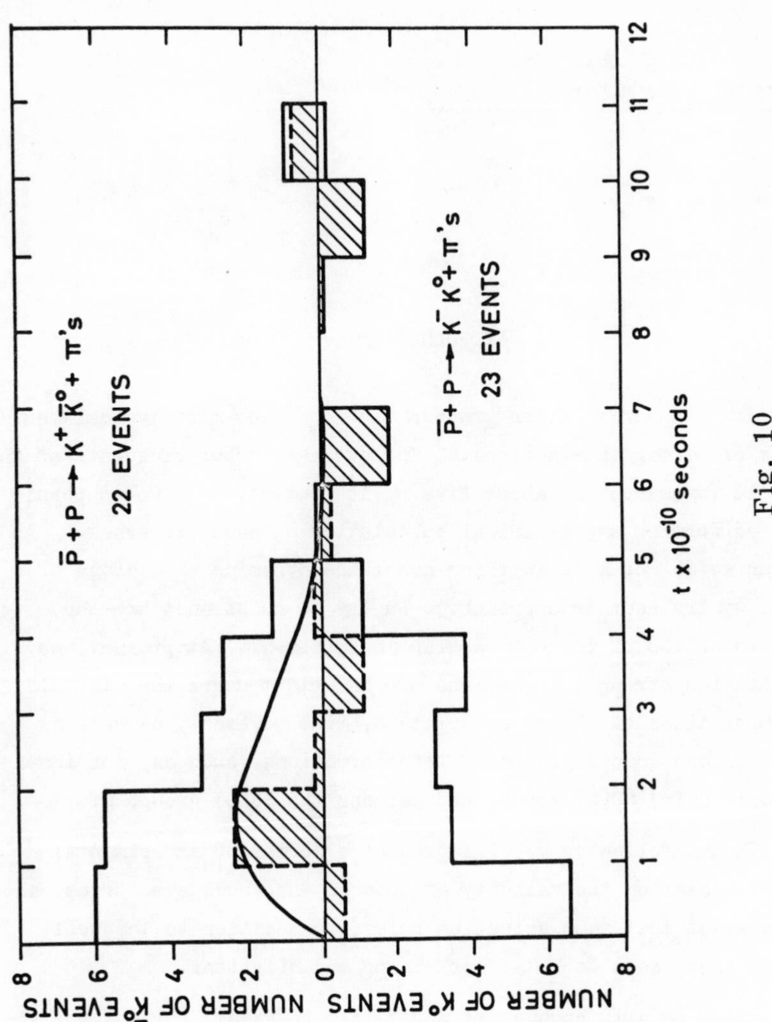

$\bar{P} + P \rightarrow K^+ \bar{K}^0 + \pi$'s
22 EVENTS
$\bar{P} + P \rightarrow K^- K^0 + \pi$'s
23 EVENTS
NUMBER OF $\bar{K}^0$ EVENTS
NUMBER OF K^0 EVENTS
8 6 4 2 0 2 4 6 8
1 2 3 4 5 6 7 8 9 10 11 12
t x 10^{-10} seconds

Fig. 10

PROTON-ANTIPROTON ANNIHILATIONS AT REST*)

P. Franzini,

Columbia University,
New York.

I. INTRODUCTION

The p-$\bar{p}$ annihilation process is one of the most typical manifestations of strong interactions. The average number of quanta of the strong field (π mesons) is about five. At present, we have no possibility of performing any dynamical calculation on such processes. As usual, when we do not know anything about the dynamics of a class of processes, we try another approach. We try to guess what are the symmetry properties of the interaction in question. At present, we believe that the strong interactions are invariant under the discrete symmetry operations of charge conjugation, time reversal, as well as under certain continuous groups of transformations, such as, for instance, the (isotopic spin) SU(2) group, and perhaps the SU(3) group, etc.

In the following we will present a series of experimental results which bear on the validity of some of the above symmetries, and strongly suggest that some selection rules, in addition to the well-established ones, seem to be at work in $\bar{p}$p annihilations.

First of all, however, I would like to recall for you the so-called statistical model, essentially a one parameter phenomenological theory of multipion production in strong interactions.

---

*) Work supported in part by the United States Atomic Energy Commission.

II. THE STATISTICAL MODEL

The statistical model was first proposed by Fermi[1], who assumed that the rates for different processes are proportional to the available phase space. I will outline the formulation of Srivastava and Sudarshan[2] which has the advantage of being clearly covariant.

Consider the reaction

$$\bar{p} + p \rightarrow n\pi \, .$$

The total transition rate can be written as [see Feynman[3] or Schweber[4]]:

$$W_n = \int d^4 p_1 d^4 p_2 \ldots d^4 p_n \delta(p_1 + p_2 + \ldots p_n - P) \times$$
$$\times \, \delta(p_1^2 - \mu^2) \, \delta(p_2^2 - \mu^2) \ldots \delta(p_n - \mu^2) \, f(p, P) \, ,$$

where P is the four momentum of the $\bar{p}p$ initial state, p_j is the four momentum of the jth pion and μ is the pion mass. $f(p, P)$ is given by

$$f(p, P) = \Sigma |M|^2 \, ,$$

and is an invariant function of the four momenta involved. The statistical model is the statement

$$f(p, P) = \text{const} = A \, ,$$

where the constant depends on the number of external meson lines.

In fact, from dimensional arguments one has

$$A = \text{const} \times K^{-2n} \, ,$$

where K has the dimension of a mass. The Compton wave length $\lambda = k^{-1}$ corresponds to the radius of interaction in the Fermi model.

It is to be noted that the dependence on the radius of the interaction appears only if we compare annihilation with a different number of pions.

All features of annihilations into a given number n of pions (like energy spectrum of one pion, mass spectra of m out of n pions, angular distributions between two pions, etc.) are independent of the radius of interaction.

The radius of interaction was determined from the average multiplicity. Experimentally $< n_\pi > \cong 5$ from which $\lambda \cong 3 \times 10^{-13}$ cm, not an unreasonable number, although somewhat large.

Now, we can check the theory in more detail. For instance, compare the experimental mass spectra with the phase-space prediction. It does not work. Sometimes sharp peaks appear: we call them resonances, and almost always the general behaviour of the experimental spectra is different from the phase-space calculations, even if we disregard the peaks.

Another observation was made early in the study of antiproton annihilations. The experimental distribution of the angle between two pions (out of n) of identical charge is different from that of two pions of opposite charge[5)]. Now, in the statistical model, we have forgotten about Bose statistics and angular-momentum conservation.

By weighing the allowed points of phase space by a symmetrical n-pion free particle wave function, Pais et al.[6)] have introduced correlations due to Bose statistics. With this model they indeed succeeded in modifying the phase space predictions, bringing them into closer agreement with experiment. However, the radius of the interaction required to do so turns out to be of the order of 0.5×10^{-13} cm, which is in contradiction with the value deduced from the average multiplicity.

Other discrepancies are connected with the experimental values of the relative rates into two, three etc. pions, and those expected on the basis of the above model.

III. PHENOMENOLOGICAL ANALYSIS OF SOME FINAL STATE IN $p\bar{p}$ ANNIHILATIONS AT REST

While we have no dynamical way to compute the function $f(p,P)$ mentioned above, its form is greatly limited by the various invariance and symmetries, especially if the final state in question has few particles (in practice $n \leq 3$).

In the following we will assume relativistic invariance, conservation of angular momentum and parity, and also isotopic spin conservation and charge conjugation invariance.

We have good reasons to believe that antiprotons at rest are annihilated on protons in the S-wave orbits. This was discussed theoretically by Day, Sucher and Snow[7)] and by Desay[8)], and was proved experimentally, at least for the reaction

$$p + \bar{p} \rightarrow K^0 + \bar{K}^0 ,$$

to an accuracy of 2%[9)].

Thus, the initial $\bar{p}p$ state which we are concerned with in the study of antiproton annihilations at rest is a statistical mixture of ^3S and ^1S states.

In addition, each of these states can have isotopic spin $I = 0$ or $I = 1$. A $\bar{p}p$ system at rest is an eigenstate also of parity, C conjugation and, for a given I, of G parity ($= C \times e^{i\pi I_2}$). We can thus make the following table.

Table 1

State	J^P	C	I	G	2π	3π	4π
^1S	0^-	+	0	+	no (parity)	no	yes
			1	-	no	yes	no
^3S	1^-	-	0	-	no	yes	no
			1	+	yes	no	yes

An n-pion state is an eigenstate of G with eigenvalue $(-1)^n$, thus annihilation into n pions is allowed only for two of the four possible initial states. Moreover, these two states belong one to the triplet and one to the singlet, hence, there is no interference of the two amplitudes.

With all this in mind we can begin examining some particular case. We shall build in the following transition amplitudes, which satisfy all the invariances above, and which are totally symmetric in any two pions, and with the limitation that we shall only keep terms, at most, linear in the momenta (this is equivalent in the non-relativistic limit to considering only the lowest angular momenta.)

1. The reaction $p\bar{p} \rightarrow 3\pi$

Let us look first at the experimental distribution of the invariant mass of two π's, in any of the combinations $\pi^+\pi^-$, $\pi^+\pi^0$, $\pi^-\pi^0$ (see Fig. 1). We see a pronounced peak around 770 MeV which approximately contains one half of the events.

This peak suggests that the ρ meson is produced copiously, hence we will try to analyse the three-pion final state under the assumption that it is reached in two different ways:

$$p\bar{p} \rightarrow \pi^+ + \pi^- + \pi^o \tag{1}$$

$$p\bar{p} \rightarrow \overset{\pm}{\pi^o}\ \overset{\pm}{\rho^o} \rightarrow \pi^+\pi^-\pi^o \ . \tag{2}$$

Looking back to our table, only the states ^3S, I = 0 and ^1S, I = 1 can contribute to the final states of reactions (1) and (2).

Let us first write the amplitudes for reaction (1). Since $P(p\bar{p}) = -1$ and $P(\pi) = -1$, where P is the parity of a particle or a state, we have

$$M_{^1S \rightarrow 3\pi} = 1 \ . \tag{3}$$

For the ^3S case we proceed in the following way. A $J^P = 1^-$ state is described by a polarization four vector S^i_μ (it transforms as a true vector under L transformation), which satisfies the auxiliary condition

$$S_\mu p_\mu = 0 \ , \tag{4}$$

where p_μ is the momentum of the state, and the completeness relation

$$\sum S^i_\mu \, S^i_\nu = \delta_{\mu\nu} - \frac{p_\mu p_\nu}{\mu^2} \ . \tag{5}$$

The transition amplitude, involving any number of external lines corresponding to a 1^- particle or state, is linear in the corresponding polarization vector.

We can now build our amplitude $M_{^3S \rightarrow 3\pi}$. We further notice that 3π's have an intrinsic parity -1, hence our amplitude has to be a pseudoscalar. There is only one which we can build linear in S_μ, and it is given by

$$M_{^3S\to 3\pi} = \epsilon_{\mu\nu\rho\sigma} S_{\mu}^{p\bar{p}} \, p_{\nu}^{+} \, p_{\rho}^{-} \, p_{\sigma}^{o} \, . \tag{6}$$

We can now investigate the symmetry properties of our amplitudes with respect to interchange of two pions. We must remember that the above amplitudes have to be multiplied by the appropriate isospin functions, and then we require the resulting amplitude to be symmetric under simultaneous exchange of space and isospin variables.

$M_{^1S}$ is obviously symmetric in the space part. The corresponding isospin part represents 3π's with $I = 1$. This function has no definite symmetry. We can always separate a symmetric part, and forget it at that, as long as we do not care about branching ratios for different charge modes.

For the ^3S case, the 3π's must have $I = 0$. The isospin wave function is totally antisymmetric in this case and so is the space function which satisfies the requirements of Bose statistics.

Next we have the $\rho\pi$ case. It is easy to see that the triplet amplitude is given by*)

$$M_{^3S\to\,\rho\pi} = S_{\mu}^{p\bar{p}} P_{\mu} \, S_{\nu}^{\rho} \, q_{\nu} \, \frac{1}{m_{\pi\pi} - m_{\rho} - i\,\Gamma/2} \, , \tag{7}$$

where $S^{p\bar{p}}$ is the polarization vector for the p-$\bar{p}$ initial state, P_{μ} is a pseudovector (to account for one negative parity pion in the final state) linear in S^{ρ}, the ρ polarization vector: q_{ν} is the relative momentum of the two pions from the ρ decay. $m_{\pi\pi}$ is the invariant mass of the dipion system, m_{ρ}, and Γ are the ρ mass and width. The only pseudovector that we can form with the variables describing the process is:

---

*) In order to write simpler expressions, we have linearized the propagator $1/q^2 - M^2$, where M is the complex ρ-meson mass.

$$P_{\mu} = \epsilon_{\alpha\beta\gamma\mu} S^{\rho}_{\alpha} p^{\rho}_{\beta} p^{p\bar{p}}_{\gamma} \; . \tag{8}$$

The matrix element is given by

$$M_{^3S\to\rho\pi} = \frac{S^{p\bar{p}}_{\mu} \epsilon_{\alpha\beta\gamma\mu} S^{\rho}_{\alpha} p^{\rho}_{\beta} p^{p\bar{p}}_{\gamma} S_{\nu} q_{\nu}}{m_{\pi\pi} - m_{\rho} - i\,\Gamma/2} = \frac{S^{p\bar{p}}_{\mu} \epsilon_{\alpha\beta\gamma\mu} q_{\alpha} p^{\rho}_{\beta} p^{p\bar{p}}_{\gamma}}{m_{\pi\pi} - m_{\rho} - i\,\Gamma/2} \; . \tag{9}$$

In the p-$\bar{p}$ centre-of-mass, the above matrix element can be finally written as (after summing coherently over the ρ polarization)

$$M_{^3S\to\rho\pi} = \frac{\vec{S}_{p\bar{p}} \cdot \vec{p} \times \vec{q}}{m_{\pi\pi} - m_{\rho} - i\,\Gamma/2} \; , \tag{10}$$

where S, p and q are the space part of the corresponding four vectors, and

$$\sum S^{i}_{j} S^{i}_{k} = \delta_{jk} \; . \tag{11}$$

Finally, we have to account for ρ^+, ρ^- and ρ^o production. With the help of Bose statistics and isospin wave functions, we can write

$$M_{^3S\to\rho\pi} \propto \left(\frac{(\vec{p}_+ - \vec{p}_-) \times \vec{p}_o}{D_{+-}} + \frac{(\vec{p}_- - \vec{p}_o) \times \vec{p}_+}{D_{o-}} + \frac{(\vec{p}_o - \vec{p}_+) \times \vec{p}_-}{D_{o+}} \right) \; , \tag{12}$$

where $\vec{p}_+$, $\vec{p}_-$ and $\vec{p}_o$ are the three momenta for the π^+, π^- and π^o, respectively, and D_{ij} is the function

$$D_{ij} = m_{ij} - m_{\rho} - i\,\frac{\Gamma}{2} \; . \tag{13}$$

For the case of $^1S(p\bar{p}) \to \rho\pi$ we first note that the isospin part of the wave function does not contain $\rho^o\pi^o$. Hence, if the $\rho\pi$ state can be produced by the ^1S and ^3S states in any ratio, the amount of ρ^+, ρ^- and ρ^o must be different. (There is no interference between ^1S and ^3S amplitudes.)

The transition amplitude is given by

$$M_{^1S\to\rho\pi} = \frac{p_\mu^{p\bar{p}}\, s_\mu^\rho\, s_\nu^\rho\, q_\nu}{m_{\pi\pi} - m_\rho - i\,\Gamma/2} \quad , \tag{14}$$

with the same notation as in formula (7). Putting in isotopic spin and using relation (5), and the definition (13), we obtain

$$M_{^1S\to\rho\pi} = \frac{(p_\mu^+ - p_\mu^o)p_\mu^-}{D_{+o}} - \frac{(p_\mu^o - p_\mu^-)p_\mu^+}{D_{o-}} \quad , \tag{15}$$

where the p_μ are the four momenta of the pions. Finally, the distribution function in the Dalitz variables is given by

$$D(E_1\,,\,E_2) = \sum_{\text{spins}} \Big\{ \left| aM_{^3S\to3\pi} + bM_{^3S\to\rho\pi} \right|^2 + \tag{16}$$
$$+ \left| cM_{^1S\to3\pi} + dM_{^1S\to3\pi} \right|^2 \Big\}\, dE_1\, dE_2 \ .$$

In an experiment[10] at Columbia University 823 events, attributed to the reaction $p + \bar{p} \to \pi^+ + \pi^- + \pi^o$, have been compared to the above distribution. A two-dimensional likelihood fit of the above functions to the data, with the complex parameter a,b,c and d free, has proved within the statistical sensitivity of our experiment that the $\rho\pi$ final state is fed only by the ^3S initial state. This result had also been obtained by the Oxford-Padua collaboration[11]. With regard to the three-pion final state, our experiment (again within the statistical accuracy of about 10%) proves that it comes from the ^1S $p\bar{p}$ state. This result also is implicit in the Padua-Oxford experiment, since they state in their paper that the non-resonant background is consistent with a flat distribution. Since, however, there have been confusing statements made recently by some of the authors of that paper, I would like to stress the point that the reaction

$$p + \bar{p} \rightarrow \rho + \pi \rightarrow 3\pi$$

proceeds mostly from the ^3S initial state, while the reaction

$$p + \bar{p} \rightarrow 3\pi$$

proceeds from the ^1S initial state.

The results of our experiment are shown again in the form of a Dalitz plot (Fig. 2). You can see (as in the projection) the 3 ρ bands. Independently from the fitting, we can count the number of events in the ρ^+, ρ^- and ρ^0 bands. We obtain, respectively, 175, 184 and 186, which confirms the above result that only the ^3S state goes into $\rho\pi$. Finally, the best way of displaying the uniformity of the 3π density is to fold the Dalitz plot over six times, and divide it into ten equal area portions, as in Fig. 3. The density of points approaches a constant value at the boundary instead of dropping to zero, as required by the ^3S matrix element for the 3π final state.

2. The reaction $p\bar{p} \rightarrow \omega\pi^+\pi^-$

As soon as the number of particles in the final state becomes larger than 3, it becomes a formidable task to write all the possible matrix elements and especially to perform meaningful fits to the data. Because of the narrow width of the ω meson, and because of its copious production in $p\bar{p}$ annihilations, it is possible to treat this channel as a true three-body one and to collect enough events to make the analysis meaningful.

At the time when this reaction was studied, it was also of interest, with respect to the SU(6,6) predictions[12,13], to compare the production of 3 PS mesons with the production of one vector and 2 PS mesons in $\bar{p}$-p annihilations.

The G parity of $\omega\pi\pi$ is negative, hence only the states $^1S(p\bar{p})$ with I = 1 and $^3S(p\bar{p})$ with I = 0 can contribute to the process. The simplest Lorentz invariant amplitudes are

$$M_{^1S\to\omega\pi\pi} = \epsilon_{\alpha\beta\gamma\delta} S^{\omega}_{\alpha}\, p^{\omega}_{\beta}\, p^{+}_{\gamma}\, p^{-}_{\delta} \qquad (17)$$

$$M_{^3S\to\omega\pi\pi} = S^{(p\bar{p})}_{\alpha} S^{\omega}_{\beta} A_{\alpha\beta} \quad , \qquad (18)$$

where S_{μ} is the polarization four vector for a $J_p = 1^-$ state, and p_{μ} is the four momentum.

$A_{\alpha\beta}$ is a tensor, symmetric in the π^+, π^- variables since the two pions in the reaction $^3S(p\bar{p}) \to \omega^o\pi^+\pi^-$ are in an I = 0 state. Up to terms linear in the pion momenta, the tensor $A_{\alpha\beta}$ can be written as

$$A_{\alpha\beta} = a\delta_{\alpha\beta} + b(p^{+}_{\alpha}p^{-}_{\beta} + p^{-}_{\alpha}p^{+}_{\beta}) \; , \qquad (19)$$

where a and b are arbitrary complex coefficients.

Since there is evidence for the presence of the ρ resonance in the $\omega^o\pi^+\pi^-$ events, we also have to consider the amplitude for $\bar{p}p \to \rho^o\omega^o$. C invariance restricts this reaction to the $^1S(p\bar{p})$ state. The corresponding amplitude is

$$M_{^1S\to\omega^o\rho^o} = \epsilon_{\alpha\beta\gamma\delta} p^{(p\bar{p})}_{\alpha}\, p^{\omega}_{\beta}\, S^{\omega}_{\gamma}\, S^{\rho}_{\delta}\, \frac{1}{D_{+-}}\, S^{\rho}_{\mu}(p^{+}_{\mu} - p^{-}_{\mu}) \; , \qquad (20)$$

which summing over the ρ polarization, reduces to

$$M_{^1S\to\omega^o\rho^o} = \epsilon_{\alpha\beta\gamma\delta} p_\alpha^{(p\bar{p})} p_\beta^\omega S_\gamma^\omega (p_\delta^+ - p_\delta^-) \frac{1}{D_{+-}} , \qquad (21)$$

where D_{+-} is defined in Eq. (13).

In all the above, the fact that the ω meson has a finite (~ 10 MeV) width has been neglected, and the ω has been treated as a stable particle.

Experimentally, the $\omega^o\pi^+\pi^-$ final state is observed among the annihilations into $\pi^+\pi^-\pi^+\pi^-\pi^o$. The above matrix elements can be used in comparison with the experimental data, only if we can single out of the five-pion events those in which an ω^o is produced, and if we can single out which three of the five pions form an omega. This is only partly true in our experiment[14].

In Fig. 4 the invariant mass of a π^+, π^- and a π^o out of the five-pion final state is plotted on a very expanded scale. A very prominent peak is observed at 785 MeV, corresponding to the ω^o mass. If we take a narrow region in mass around 785 MeV, the events contained there correspond to ω production plus a background coming from other reactions, as well as from the three other (+- o) combinations which are possible for the true ω^o events. In our case, approximately 55% of the events are true $\omega\pi\pi$ and to be compared to the distribution functions obtained by squaring and summing over all spins the amplitudes given in Eqs. (17), (18), (20) and (21). To take care of the remaining background, we assume that its distribution is identical at the immediate right and left of the ω peak. Hence, a subtraction procedure is possible.

The results of a two-dimensional fit to the Dalitz plot density for the $\omega\pi\pi$ events were the following:

i) the channel $\omega^o + \pi^+ + \pi^-$ contains both non-resonant pions and pions resonating as ρ. The non-resonant state dominates; $\omega^o\rho^o$ production accounts for (15 ± 6)% of the channel.

ii) The experimental distribution of points in the Dalitz plot for the $\omega\pi\pi$ events is in agreement with that due to ^3S annihilation into $\omega\pi^+\pi^-$, and ^1S annihilation into $\rho^0\omega^0$.

iii) The amount of ^1S annihilation into $\omega\pi^+\pi^-$, with the two pions not resonating as ρ, is less than 10% of the ^3S annihilation into $\omega\pi\pi$.

3. The reaction $p\bar{p} \to K + \bar{K} + \pi$

This reaction is similar to the three-pion case but for one complication. All four initial states of the $p\bar{p}$ system can contribute in the reaction. This makes the analysis somewhat more complicated and full details of it, together with the results of an experiment performed by the Columbia group, are given in Ref. 15.

I will only mention the main features of the experimental data and the conclusion which can be drawn.

i) The K*(880) resonance is observed in the final state $K_1^0 K_2^0 \pi^0$ and not in the state $K_1^0 K_1^0 \pi^0$, hence, we conclude that the reaction $p\bar{p} \to K^*K$ goes only from the $^3S(\bar{p}p)$ state $[C(K_1 K_2 \pi^0) = -1, C(^3S) = -1]$.

ii) In the final state $K_1^0 \pi^\pm K^\mp$, negative interference is observed at the crossing of the two K* bands in the Dalitz plot. As shown in Ref. 15 this favours the ^3S, I = 1 state with respect to the ^3S, I = 0 state.

iii) Experimentally, the boundary of the Dalitz plot is highly populated.

This fact, together with point i), indicates that in contrast to the reaction $p\bar{p} \to K^*K$, the reaction $p + \bar{p} \to K + \bar{K} + \pi$ proceeds mainly from the ^1S proton-antiproton system.

4. Summary

In the following table, I will summarize the results.

Table 2

Dominant state in some $p\bar{p}$ annihilation reactions at rest

Final State	Initial state
$\pi^+\pi^-\pi^0$	^1S
$\rho^{\pm,0}\,\pi^{\mp,0}$	^3S
$\omega^0\pi^+\pi^-$	^3S
$\omega^0\rho^0$	^1S (from C invariance)
K^*K	^3S, I = 1
$KK\pi$	^1S

As you see, for five particular final states we have found a predominance of one amplitude among the possible uses. This fact might well be the result of some symmetry being at work. In fact, you have heard from Professor Pais a very definite prediction in this regard.

IV. PARTIAL RATES FOR THE ANNIHILATIONS INTO TWO AND THREE MESONS

From the analysis of the reactions discussed above, the partial rates were determined for the various channels involved. There are, however, many other reactions and, correspondingly, more two-body and three-body final states whose rates were determined in our study of anitproton annihilations at rest.

They are relevant to the various symmetry schemes proposed hence, I will include a table presenting all our results on this subject[9,10,14-16])

Table 3

Two and three body partial annihilation rates

Channel	Partial rate ($\times 10^{-2}$)
$\pi^+\pi^-$	0.31 ± 0.03
K^+K^-	0.11 ± 0.01
$K_1^o K_2^o$	0.06 ± 0.009
$\pi^{\pm}_{o}\rho^{\mp}_{o}$	4.0 ± 0.4
$\rho^o\eta^o$	0.22 ± 0.17
$\rho^o\omega^o$	0.7 ± 0.3
$\rho^o\rho^o$	0.38 ± 0.3
K^oK^{o*}	0.12 ± 0.02
$K^{\pm}K^{\mp*}$	0.09 ± 0.02
$K^{\pm*}K^{\mp*}$	0.13 ± 0.05
$K^{o*}\overline{K^{o}}^{*}$	0.29 ± 0.05
$\pi^+\pi^-\pi^o$	7.3 ± 0.9 (includes $\rho\pi$ rate above)
$\rho^o\pi^+\pi^-$	5.8 ± 1.0
$\omega^o\pi^+\pi^-$	3.9 ± 0.5
$\eta^o\pi^+\pi^-$	1.2 ± 0.3

V. A DIRECT TEST OF CHARGE CONJUGATION INVARIANCE IN PROTON-ANTIPROTON ANNIHILATION AT REST

The observation by Christenson et al.[17]) that the long-lived K^o meson decays into two charged pions has proved that CP invariance is violated in this decay. Several authors have suggested that this violation might be the consequence of the violation of charge conjugation (C) invariance in the strong interactions[18-20]). No precise experimental verification of C invariance in strong interactions has been performed up to now in a direct way. It is possible, however, to relate time reversal (T) and charge conjugation if CPT invariance is assumed, because parity is known to be conserved to a high degree of accuracy in strong interactions.

Experiments on proton-proton triple scattering[21]) and on detail balance[22]) have proved that the T-violating amplitude should be no more than 2 - 3% of the T-conserving amplitude. By the argument above, the same limits apply to the C-violating amplitude.

In the case of $p\bar{p}$ annihilation at rest, it is possible to directly test C invariance by comparing the energy and invariant mass distributions of charge-conjugate channels[23]). The two states of the $\bar{p}p$ system at rest are eigenstates of C. These states do not interfere if the final states are integrated over the spatial orientation, since the two states have different angular momentum. If C is conserved, therefore, the distributions in the momentum of the annihilation products must be invariant with respect to the interchange of positive and negative particles. If C is not conserved, CPT invariance would still ensure the equality of the above distributions in the absence of final-state interactions. However, the final-state interactions are strong, and the equality is expected to be broken by C non-invariance.

The experimental sample[23] which was tested consisted of 34,811 pionic annihilations and 4,663 annihilations in which one or more K^0 mesons are emitted (kaonic annihilations). The test consists of first making histograms for the corresponding distributions in various dynamical variables, such as the kinetic energies of the π^+ and π^- mesons, or the invariant mass combination $K^+\pi^-$ and $K^-\pi^+$, etc. and then forming the sum

$$\chi^2 = \sum_{i=1}^{N_B} \frac{(N_i^+ - N_i^-)^2}{(N_i^+ - N_i^-)}$$

which is the χ^2 function for the difference in the two distributions, assuming no C violation. N_i is the number of events in the ith bin. The χ^2 value in the case of no C violation is expected to be equal to the number of bins N_B, with a variance equal to $\sqrt{2N_B}$. It should be mentioned that in several cases a particular annihilation reaction furnishes more than one distribution; for instance, in the channel $K^0 K^\pm K^\mp$ we compare p_{K^+} to p_{K^-}, p_{π^+} to p_{π^-}, and $M_{K^+\pi^-}$ to $M_{K^-\pi^+}$. These distributions are essentially independent for the purpose of this test.

The following tables list the tests which were performed, the number of counts for the test, the number of bins, and the χ^2 for pionic and kaonic annihilations. There is no indication of any statistically significant deviation from the χ^2 value expected if C is conserved.

It remains to relate this null result to a limitation on the relative amplitude of a possible C violating part of the annihilation interaction. For the case of a small violation this limit is given by:

$$\alpha \approx \frac{|\text{C violating amplitude}|}{|\text{C conserving amplitude}|} \lesssim \left[\frac{\sqrt{2N_B}}{N_{counts}}\right]^{1/2} .$$

Table 4

Summary of pionic annihilations

Annihilation channel	No. of counts	Distributions which are compared	Interval size (MeV)	$\sum \frac{(N_i^+ - N_i^-)^2}{N_i^+ + N_i^-}$	No. of interval N_B
$\pi^+\pi^-\pi^0$	1,830	$E_+ \leftrightarrow E_-$	100	6.95	9
$\pi^+\pi^- X^0$ $m_{X^0} > m_{\pi^0}$	5,099	$E_+ \leftrightarrow E_-$	100	13.2	9
$\pi^+\pi^+\pi^-\pi^-$	7,324	$E_+ \leftrightarrow E_-$	100	3.7	8
	3,662	$m_{++} \leftrightarrow m_{--}$	100	13.3	13
$\pi^+\pi^+\pi^-\pi^-\pi^0$	27,992	$E_+ \leftrightarrow E_-$	100	5.3	8
	13,996	$m_{++} \leftrightarrow m_{--}$	100	18.6	12
	27,992	$m_{++-} \leftrightarrow m_{+--}$	100	7.16	12
$\pi^+\pi^+\pi^-\pi^- X^0$ $m_{X^0} > m_{\pi^0}$	20,508	$E_+ \leftrightarrow E_-$	100	6.16	7
	10,254	$m_{++} \leftrightarrow m_{--}$	100	4.35	9
	20,508	$m_{++-} \leftrightarrow m_{+--}$	100	6.68	12
All pionic events	139,165			85.4	99

Table 5

Summary of kaonic annihilations

Annihilation channel	No. of counts	Distributions which are compared	Interval size (MeV)	$\sum \frac{(N_i^+ - N_i^-)^2}{(N_i^+ + N_i^-)}$	No. of interval N_B
$K^0 K^+ \pi^-$ (413 events)	851	$P_{\pi^+} \leftarrow \rightarrow P_{\pi^-}$	80	2.2	7
$\bar{K}^0 K^- \pi^+$ (438 events)	851	$P_{K^+} \leftarrow \rightarrow P_{K^-}$	80	8.86	8
	851	$m_{K^+\pi^-} \leftarrow \rightarrow m_{K^-\pi^+}$	80	4.97	8
$K^0 \bar{K}^0 \pi^+ \pi^-$	1,909	$P_{\pi^+} \leftarrow \rightarrow P_{\pi^-}$	80	7.74	6
$\bar{K}^0 K^+ \pi^- \pi^0$ (957 events)	1,910	$P_{\pi^+} \leftarrow \rightarrow P_{\pi^-}$	80	7.99	6
$K^0 K^- \pi^+ \pi^0$ (953 events)	1,910	$P_{K^+} \leftarrow \rightarrow P_{K^-}$	80	6.17	7
	1,910	$m_{K^+\pi^-} \leftarrow \rightarrow m_{K^-\pi^+}$	80	6.81	6
	1,910	$m_{\pi^+\pi^0} \leftarrow \rightarrow m_{\pi^-\pi^0}$	80	5.49	8
	1,910	$m_{K^+\pi^-\pi^0} \leftarrow \rightarrow m_{K^-\pi^+\pi^0}$	80	7.56	7
All kaonic events	14,012			57.8	63

Using this relation we obtain

$$\alpha_{\text{pionic}} \lesssim 0.01$$

$$\alpha_{\text{kaonic}} \lesssim 0.03 .$$

In addition, a simpler test is possible in the case of the annihilations in which kaons of opposite charges are produced. Within statistics, these numbers are the same (see Table 5), and we conclude from this that $\alpha_{\text{kaonic}} \lesssim 0.02$.

We conclude that this search for possible C violation in $\bar{p}$-p annihilation at rest gives no indication of violation either in pionic or kaonic annihilations, and that a reasonable upper limit for C-violation amplitudes in the two cases is 1% and 2% respectively. Tests of C invariance in $\bar{p}$p annihilation in flight have been carried out previously[24)], but with more limited statistics.

REFERENCES

1) E. Fermi, Progr.Theoret.Phys.(Kyoto) 5, 570 (1950).

2) P.P. Srivastava and G. Sudarshan, Phys.Rev. 110, 765 (1958).

3) R.P. Feynman, Theory of fundamental processes, (Benjamin, New York).

4) S.S. Schweber, An introduction to relativistic quantum field theory, (Harper and Row, New York) p. 484.

5) G. Goldhaber, W.B. Fowler, S. Goldhaber, T.F. Hoang, T.E. Kalageropoulos and W.M. Powell, Phys.Rev.Letters 3, 181 (1959).

6) G. Goldhaber, S. Goldhaber, W. Lee and A. Pais, Phys.Rev. 120, 300 (1960).

7) T.B. Day, G.A. Snow and J. Sucher, Phys.Rev. Letters 3, 61 (1959).

8) B. Desay, Phys.Rev. 119, 1385 (1960).

9) C. Baltay, N. Barash, P. Franzini, N. Gelfand, L. Kirsch, G. Lütjens, D. Miller, J.C. Severiens, J. Steinberger, T.H. Tan, D. Tycko, D. Zanello, R. Goldberg and R.J. Plano, Phys.Rev. Letters 15, 532 (1965).

10) C. Baltay, P. Franzini, N. Gelfand, G. Lütjens, J.C. Severiens, J. Steinberger, D. Tycko and D. Zanello, Phys.Rev. 140, B1039 (1965).

11) G.B. Chadwick, W.T. Davies, M. Derrick, C.J.B. Hawkins, J.H. Mulvey, D. Radojicik, C.A. Wilkinson, M. Cresti, S. Limentani and R. Santangelo, Phys.Rev. Letters 10, 62 (1963).

12) N.P. Chang and J.M. Shpiz, Phys.Rev. Letters 14, 617 (1965).

13) R. Delbourgo, Y.C. Leung, M.A. Rashid and J. Strathdee, Phys.Rev. Letters 14, 609 (1965).

14) C. Baltay, P. Franzini, G. Lütjens, J.C. Severiens, J. Steinberger, D. Tycko and D. Zanello, Phys.Rev. 140, B1042 (1965).

15) N. Barash, P. Franzini, L. Kirsch, D. Miller, J. Steinberger, T.H. Tan, R. Plano and P. Yaeger, Phys.Rev. 139, B1659 (1965).

16) C. Baltay, P. Franzini, G. Lütjens, J.C. Severiens, D. Tycko and D. Zanello, to be published.

17) J.H. Christenson, J.W. Cronin, V.L. Fitch and R. Turlay, Phys.Rev. Letters 13, 138 (1964).

18) J. Prentki and M. Veltman, Physics Letters 15, 88 (1965).

19) T.D. Lee and L. Wolfenstein, Phys.Rev. 138, B1490 (1965).

20) L.B. Okun (preprint).

21) P.E. Hillman, A. Johansson, G. Tibell, Phys.Rev. 110, 1218 (1958); A. Abashian, and E.M. Hafner, Phys.Rev. Letters 1, 255 (1958).

22) L. Rosen, J.E. Brolley, Jr., Phys.Rev. Letters 2, 98 (1959); D. Bodansky, S.F. Eccles, G.W. Farwell, M.E. Rickey and P.C. Robinson, Phys.Rev. Letters 2, 101 (1959).

23) C. Baltay, N. Barash, P. Franzini, N. Gelfand, L. Kirsch, G. Lütjens, J.C. Severiens, J. Steinberger, D. Tycko and D. Zanello, Phys.Rev. Letters 15, 591 (1965).

24) N.H. Xuong, G.R. Lynch, C.K. Hinrichs, Phys.Rev. 124, 575 (1961); B.C. Maglić, G.R. Kalbfleisch and M.L. Stevenson, Phys.Rev. Letters 7, 137 (1961).

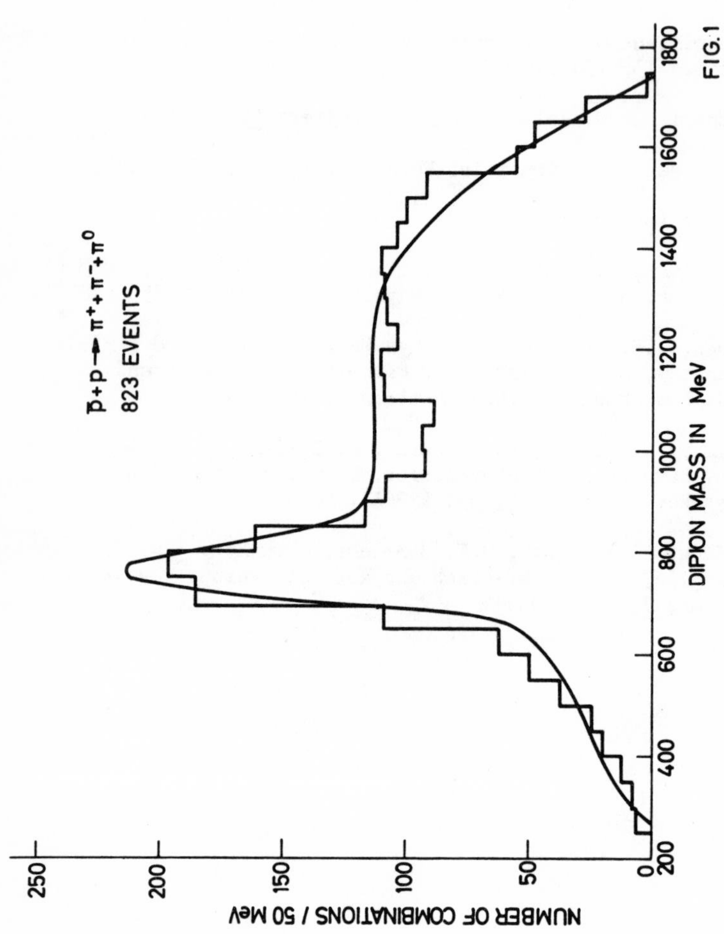
$\bar{p}+p \rightarrow \pi^{+}+\pi^{-}+\pi^{0}$
823 EVENTS
NUMBER OF COMBINATIONS / 50 MeV
0
50
100
150
200
250
200
400
600
800
1000
1200
1400
1600
1800
DIPION MASS IN MeV

FIG.1

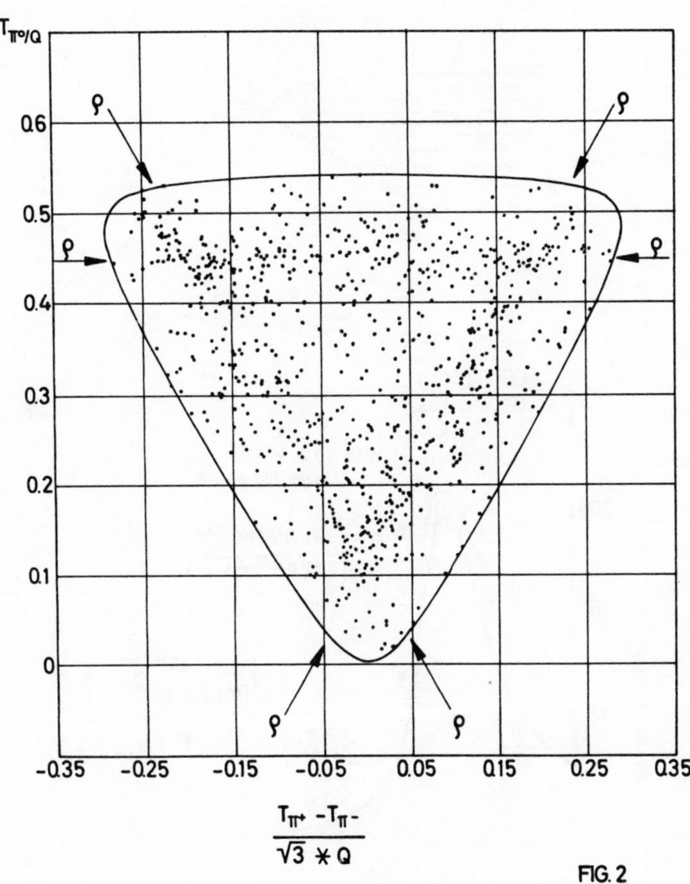
T_{π^0}/Q
0.6
0.5
0.4
0.3
0.2
0.1
0
ρ
ρ
ρ
ρ
ρ
ρ
-0.35
-0.25
-0.15
-0.05
0.05
0.15
0.25
0.35
$\frac{T_{\pi^+} - T_{\pi^-}}{\sqrt{3} * Q}$

FIG. 2

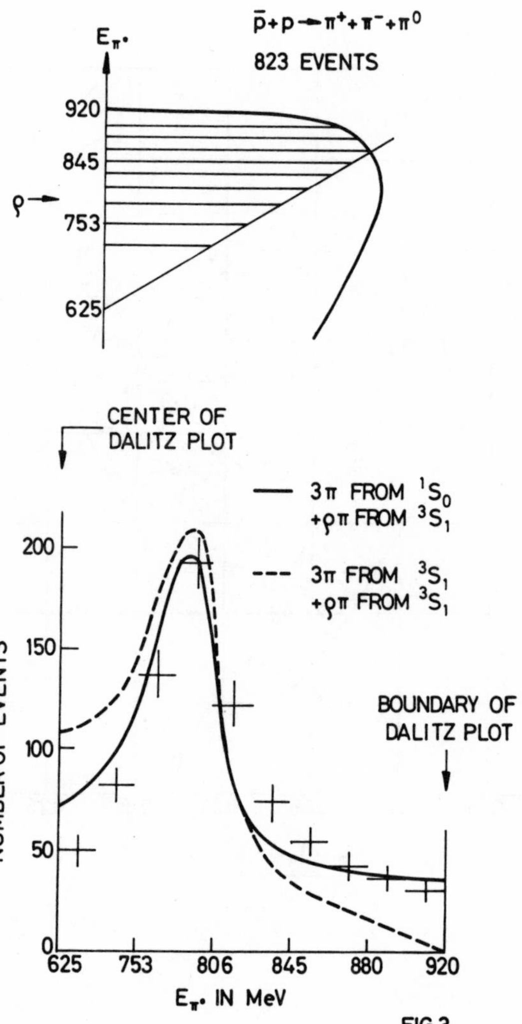
$\bar{p}+p \rightarrow \pi^{+}+\pi^{-}+\pi^{0}$
823 EVENTS
E_{π^0}
920
845
ρ
753
625
CENTER OF DALITZ PLOT
3π FROM $^1S_0$ +ρπ FROM $^3S_1$
3π FROM $^3S_1$ +ρπ FROM $^3S_1$
BOUNDARY OF DALITZ PLOT
NUMBER OF EVENTS
200
150
100
50
0
625
753
806
845
880
920
E_{π^0} IN MeV

FIG.3

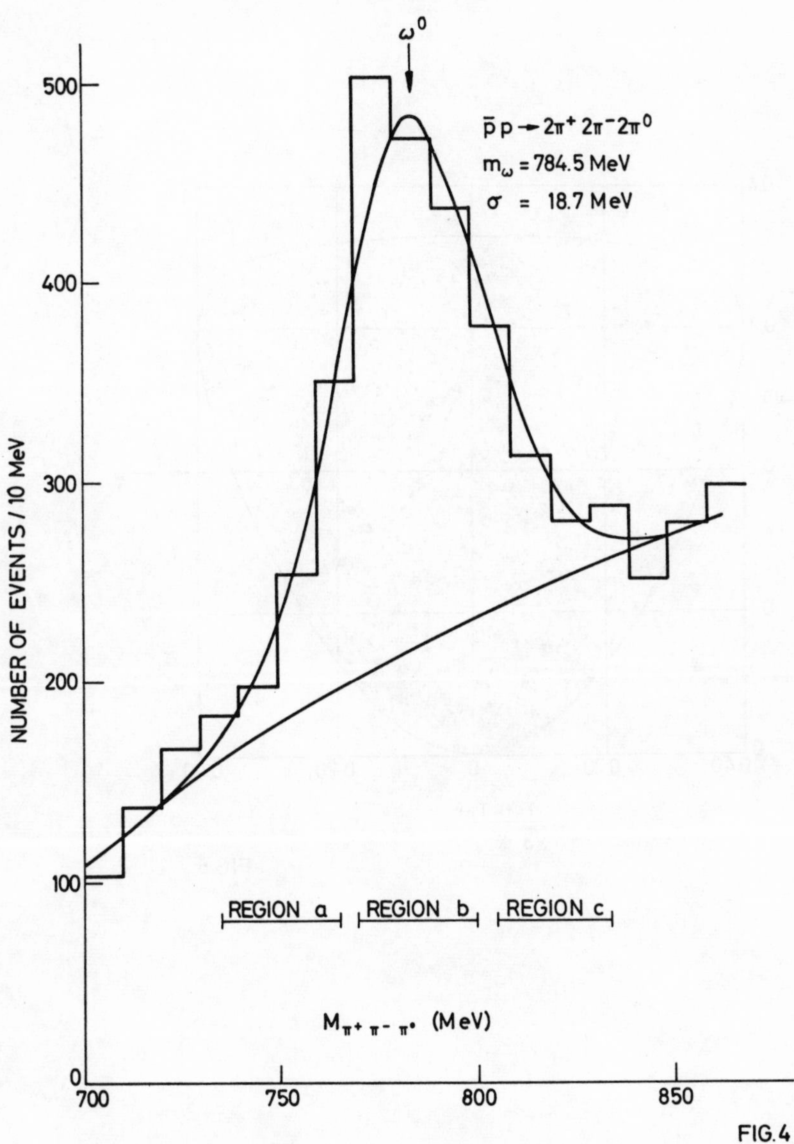
ω^0
$\bar{p}p \to 2\pi^+ 2\pi^- 2\pi^0$
m_ω = 784.5 MeV
σ = 18.7 MeV
NUMBER OF EVENTS / 10 MeV
500
400
300
200
100
0
REGION a
REGION b
REGION c
$M_{\pi^+ \pi^- \pi^0}$ (MeV)
700
750
800
850

FIG.4

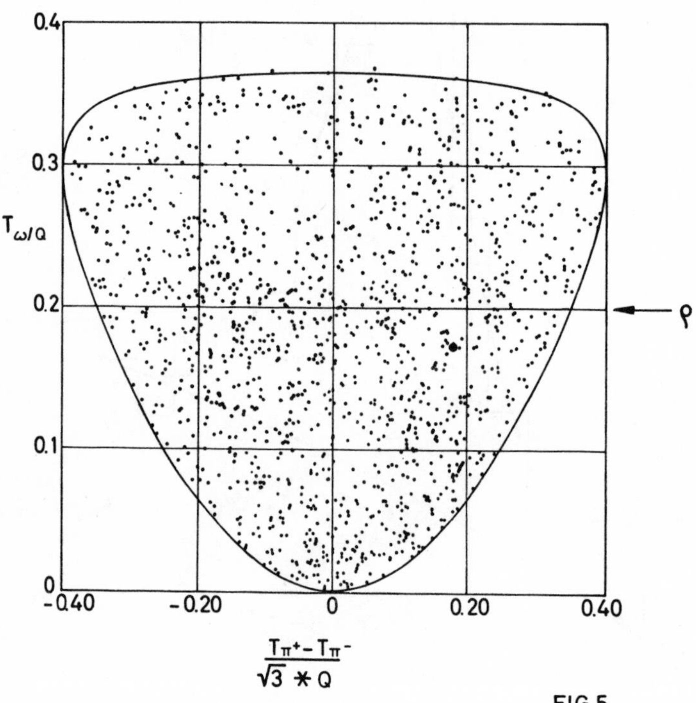

FIG.5

CLOSING LECTURE

THE SIGNIFICANCE OF INTERNAL SYMMETRIES

L.A. Radicati,
Scuola Normale Superiore,
Pisa.

1. The hierarchy of internal symmetries

Group theory was first applied to quantum mechanics to analyse the implications of geometrical symmetries, i.e. symmetries which depend upon the properties of space time. The most significant result of that work was, according to von Laue[1)], "the recognition that almost all rules of spectroscopy follow from the symmetry of the problem". Today we should add, as equally significant, the understanding of the relation between the spin of a particle and its behaviour under space-time transformations.

The importance and the generality of these results depend upon the fact that space-time symmetries are believed to be exact and therefore lead to absolute conservation laws. However, from the very beginning it was recognized that under special dynamical situations one could also deduce approximate conservation laws. For example, in many cases the generators of space rotations $\vec{J}$ can be split into two parts $\vec{L}$ and $\vec{S}$, which are separately approximately conserved. In atoms, a necessary, though not sufficient condition for this is

$$v \sim \sqrt{B/m} \ll 1 ,$$

where v, m and B are the velocity, the mass and the binding energy of the electron. One can thus represent the Hilbert space $\mathcal{H}$ of the atom by the tensor product

$$\mathcal{H} = \mathcal{R} \otimes \mathcal{S}$$

of two spaces which are independently transformed by $\vec{L}$ and $\vec{S}$. Of course, for an exact theory which takes into account relativity, this splitting is no longer correct and the only constant of the motion that remains is the total angular momentum.

Today, the main interest in group theory has shifted from the study of geometrical to that of internal symmetries, which depend upon the behaviour of the state vector under transformations acting on the internal variables.

Strictly speaking, we should call internal variables those operators (excluding the mass and the spin) which are left invariant by all space-time transformations. Internal symmetry transformations would then commute with the geometrical ones. The only variables satisfying this definition, and thus exactly conserved, are the baryon and lepton numbers B and L, and the charge Q.

There are however other internal variables which, though not exactly conserved can, nevertheless, be used for an approximate classification of states and for deducing approximate selection rules in the same way as the variables L and S have been used in atomic spectroscopy.

These variables are usually defined within the frame of a specific dynamical model based on perturbation theory. The total energy operator P_0 is assumed to be the sum of four terms:

$$P_0 = P_0^W + P_0^E + P_0^M + P_0^V , \tag{1}$$

which represent the contributions from weak, electromagnetic, medium strong and very strong interactions. The contribution to P_0 from gravitation is neglected.

In terms of this model we define:

i) two commuting operators Y and T_3 related to the charge Q by

$$Q = \tfrac{1}{2} Y + T_3$$

and such that

$$[P_0 - P_0^W , T_3] = [P_0 - P_0^W , Y] = 0 ; \qquad (2)$$

ii) two operators T_1 and T_2 which complete with T_3 the algebra of SU(2)

$$[T_i , T_j] = i \epsilon_{ijk} T_k \qquad (i,j,k = 1,2,3)$$

and such that

$$[T_i , Y] = 0$$

$$[P_0 - P_0^W - P_0^E , T_i] = 0 \quad ; \qquad (3)$$

iii) eight operators F^A ($A = 1 \dots 8$) satisfying the algebra of SU(3)

$$[F^A , F^B] = i f^{ABC} F^C$$

and the relations

$$F^3 = T_3$$

$$F^8 = \frac{\sqrt{3}}{2} Y \qquad (4)$$

$$[P_0 - P_0^W - P_0^E - P_0^M , F^A] = 0 \; .$$

The internal symmetry associated with the various parts of the total energy operator is thus seen to increase with the strength of the interaction.

All these symmetries are only approximate. The most badly violated is the SU(3) symmetry associated with the very strong interactions, the separation within an SU(3) multiplet being of the order of a few hundred MeV. The SU(2) symmetry of the medium strong interactions is about one hundred times more accurate, its violation being measured by a splitting of the isotopic-spin multiplets of a few MeV. The corresponding parameter for the violation of hypercharge conservation is probably of the order of 10 eV.

2. The well-ordered violation of internal symmetries

An importand and, a priori, not obvious property of the internal symmetries is that even though they are only approximate, they are violated in a well-defined way. This makes it possible to deduce from internal symmetries a number of consequences even for transitions which violate them. It also explains the applicability of various mass formulae which give the separation between the states of each representation of the internal symmetry group caused by the interactions that have been neglected in order to define the group itself.

Thus, the hypercharge conserving weak interactions transform like the adjoint representation of SU(2). Those which violate hypercharge conservation obey the selection rule $\Delta Y = 1$ and, together with the former, complete the set of operators of the adjoint representation of SU(3) capable of inducing transitions with $\Delta T_3 \neq 0$. Similarly, the electromagnetic interaction which violates both SU(2) and SU(3) have well-defined transformation properties under both groups. From this it follows that the separation between the $(2T+1)$ states of an isotopic spin multiplet is[2]:

$$\Delta M\ (\alpha\,,\ T\,,\ T_3) = A(\alpha\,,\ T)T_3 + B(\alpha\,,\ T)\,T_3^2\ ,$$

where T_3 is the quantum number which labels the states within the representation T, and α is the set of the other quantum numbers necessary to completely specify the state. This expression for ΔM can now be tested experimentally in light nuclei where isotopic quartets and quintets are identifiable[3]. A similar argument leads to the Coleman-Glashow sum rule for the electromagnetic self energies of the baryon octet[4]. Finally, the Okubo-Gell-Mann mass formula for SU(3) follows from the fact that medium strong interactions transform like an octet[5].

The fact that isospin conservation is violated in a well-defined way had been noticed long ago both for the weak[6] and the electromagnetic interactions[7]. In both cases, however, this appeared to have its origin more on a special feature of nuclear models than on a general property of the symmetry itself. Two recent developments have helped to realize the generality of the result, and have given a new meaning to it.

i) The discovery of the conservation of the vector current of weak interactions[8]. This has given a model independent proof of Wigner's selection rule for Fermi transitions and, what is more important, has clarified the connection between the currents associated with symmetry violating interactions, and the generators of a symmetry group.

ii) The discovery of the SU(3) symmetry[9]. Since in this group there is one isovector and one isoscalar generator we can now understand the presence in the conserved electric current of the two components with $T = 0$ and $T = 1$. Moreover, the hypercharge violating currents find their natural place in this scheme[10].

Though badly violated, SU(3) provides a unifying scheme for the four kinds of interactions which appear in Eq. (1). From the fact that $P_0 - P_0^W$ commutes with both T_3 and Y, we can conclude that the largest possible internal symmetry group is of rank two. Amongst all Lie groups of rank two, Nature has chosen SU(3), i.e. a specific relation between the scale of charge and hypercharge[11]. All hadronic currents known at present transform like the various components of the adjoint representation of SU(3)[9], a fact which allows one to unify all mass formulae and selection rules for all interactions.

3. The relation of internal and geometrical symmetries: Wigner's theory of nuclear supermultiplets

It is clear, however, that a complete understanding of the hierarchy of interactions requires a better knowledge of the relations between internal and geometrical symmetries. Indeed, at one point in the scale of weakness, the violation of an internal symmetry group is accompanied by violation of symmetry under reflection. This anomalous behaviour of the weak interactions spoils the simple connection between currents and generators of internal symmetry groups, which we discussed in the previous section. Indeed the axial vector current, though transforming like an SU(3) octet, cannot find its place among the generators of SU(3).

A specific relation between the two kinds of symmetry has been considered long ago by Wigner[12)] in connection with a nuclear model invariant under a group of transformations acting on both geometrical and internal variables. The model is based on the assumption - valid, as we have seen, only in the non-relativistic approximation - that the intrinsic spin S and the orbital angular momentum L are independent constants of the motion, and that Coulomb forces are negligible. In this case one can consider unitary transformations in the product space

$$\mathcal{S} \otimes \mathcal{I}$$

spanned by the four orientations of the spin and isotopic spin of each nucleon.

Invariance under SU(4) transformations in the space $\mathcal{S} \otimes \mathcal{I}$ obtains, only if the nucleon-nucleon interaction is of the Wigner or Majorana type, undoubtedly a rather crude approximation to reality. This is why Wigner's theory cannot explain finer and, from the point of view of nuclear structure, important features of nuclear spectra.

However, it makes a number of predictions about the dependence of level spacing on the symmetry which are in surprisingly good agreement with experiment[13]. It even appears that one of the main objections against SU(4) invariance, namely, the large difference between the singlet and triplet scattering lengths, can be explained as due to the accidental presence of the triplet n-p bound state[14]. At energies sufficiently removed from this bound state, the singlet and triplet phase shifts for n-p scattering show indeed a remarkable similarity.

An important feature of Wigner's theory is that it provides a relation between the Fermi and the Gamow-Teller allowed transitions[6]. Both are induced by the appropriate generators of the symmetry group and, therefore, both can occur only between states belonging to the same SU(4) representation, a prediction which is reasonably well borne out by experiment.

One more feature of Wigner's model should be pointed out. In a strictly SU(4) invariant theory all states of a given representation are degenerate. The degeneracy is removed by spin-dependent forces and by the electromagnetic interactions. The latter cause a separation of the isotopic multiplets, whereas the former split the states with different spin S and isotopic spin T. It turns out that at least in light nuclei where the model is expected to be more reliable, the two splittings are of comparable order, namely, a few MeV. Therefore, Wigner's SU(4) symmetry is expected to have more or less the same validity as the SU(2) symmetry connected with isospin.

4. The combination of internal and geometrical symmetries in the physics of elementary particles

If one disregards relativistic effects, the extension of Wigner's theory to elementary particles[15] is almost obvious. Indeed, we find here a situation which is reminiscent of the one discussed in the preceding paragraph: the separation between states of the known

SU(3) multiplets is of the same order of magnitude as the distance between the centre-of-mass of different multiplets. For example, for the $J = \frac{1}{2}$ octet and $J = \frac{3}{2}$ decuplet, we find

$$\delta_8 = (\Xi - N) = 382 \text{ MeV} \qquad \delta_{10} = (\Omega - N^*) = 438 \text{ MeV}$$

$$\Delta = M_{10} - M_8 = 342 \text{ MeV} ,$$

where M_{10} and M_8 are the average masses of the two multiplets. Thus, we are led to consider the group SU(6), instead of SU(4) acting in spin and isotopic spin space, in order to include along with the spin the SU(3) degrees of freedom.

A number of encouraging results are obtained by considering this symmetry group:

i) The correlation between the spin and the representations of SU(3) is explained in a natural way[15].

ii) The origin of the $\omega - \Phi$ mixing is qualitatively understood[16]. One can make a quantitative prediction by assuming a suitable, and rather plausible, behaviour of the physical particles under a subgroup of SU(6).

iii) The coincidence between the coefficients of the Okubo-Gell-Mann mass formula for the octet and the decuplet, and a similar relation between the mass differences in the $J = 1^-$ and $J = 0^-$ octets, follows naturally from SU(6) invariance[15].

Even more interesting results are obtained by studying the coupling of baryons to the meson, the electromagnetic and the lepton field. For a rigorous treatment one would need a relativistic definition of SU(6) invariance: however, by assuming a specific statis coupling one obtains the following predictions:

i) $$D/F = \frac{3}{2}$$

for the ratio between D and F coupling[16];

ii)

$$G_A/G_V = -\ ^5\!/_3$$

for the ratio of the Gamow-Teller and Fermi coupling constants[16];

iii)

$$\mu_n/\mu_p = -\ ^2\!/_3$$

for the ratio of the neutron and proton-magnetic moments[17].

Whereas the first two values are about 30% off the experimental ones, the third is surprisingly accurate.

In view of the encouraging results obtained from SU(6), one would like to have a consistent justification of the scheme. In the case of Wigner's theory there is no need for a relativistic formulation, since relativistic effects play a negligible role in nuclear physics. On the other hand, in many phenomena in particle physics the momentum transfer is such as to require a fully relativistic treatment. This is particularly necessary to explain the constancy of the ratio of the magnetic form factors

$$\frac{G_{MN}(q^2)}{G_{MP}(q^2)} = -\frac{2}{3}$$

from $q^2 = 0$ to several $(\mathrm{GeV})^2$. Similarly, the neutron charge form factor remains zero up to very large momentum transfers.

Unfortunately, a consistent relativistic extension of SU(6) faces a number of serious difficulties due to quite general theorems on group extension established by Michel[18] and by McGlinn[19]. Even the less ambitious schemes which have been proposed[20] lead to violation of the unitary condition[21].

A different approach to justify the results deduced from a static model in SU(6) has been advocated by Lee and Dashen and Gell-Mann[22]. It is based on the study of the equal time commutation relations satisfied by the current densities, which are of the type:

$$[C_i(\vec{x}), C_j(\vec{y})] = i\, g_{ijk}\, C_k(\vec{x})\, \delta(\vec{x} - \vec{y}) + \dots , \tag{5}$$

where the dots in the right-hand side stand for terms containing the gradient of the δ function. The $C_i(\vec{x})$ are vector or pseudovector currents (and, perhaps, also tensor, scalar and pseudoscalar densities), and g_{ijk} are the structure constants of an appropriate Lie algebra. The commutation relations (5) are satisfied in the quark model where the only contribution to the current comes from a spinor field transforming like a triplet under SU(3). In general, they are assumed to hold an assumption that cannot obviously be proved a priori.

From Eq. (5) it follows that the integral Q_i of the C's over the whole space satisfies the Lie algebra

$$[Q_i , Q_j] = i\, g_{ijk}\, Q_k . \tag{6}$$

It is easy to see that by including only the vector and axial vector currents one obtains the algebra of $U(6) \otimes U(6)$. It is possible to deduce from this algebra all the results obtained from the static model by taking the matrix element of (6) between physical states, and by limiting the sum over the intermediate states in the left-hand side to one-particle states.

The advantage of this approach is that it does not seem to require that the Q's be constant of the motion and, therefore, be interpreted as the generators of a symmetry group. The advantage is, however, only illusory as the approximation used in calculating the sum over the intermediate states in Eq. (6) is essentially equivalent to assuming the constancy of the Q's[23]). Now, among the 72 Q's which define the algebra of $U(6) \otimes U(6)$ only nine at most, the integrals over the fourth component of the vector currents, are constant of the motion, at least in the approximation of disregarding the mass differences in an SU(3) multiplet. Except in the non-relativistic limit, no theory is known for which the integral over the space components of the axial vector currents are constant.

5. An outlook

It is too early to judge the significance of the predictions of the SU(6) symmetry. The agreement with experiment is certainly encouraging, but we need to wait for a consistent justification before considering these results as well established. There seems little doubt however that for a complete understanding of internal symmetries and, therefore, of the mutual relations of the various interactions, the geometrical symmetries have to be brought into the picture. I believe that the results obtained from SU(6) should indeed be considered as evidence that this is the case, even though the precise way to accomplish this aim still eludes us.

Since the time when internal symmetries were first introduced in the physics of strong interactions[24)] our knowledge of their consequences has vastly increased, and their significance for the classification of the various interactions has clearly emerged. It is perhaps not too optimistic to hope that one day they will give one of the more important clues for the understanding of the whole field of particle physics.

REFERENCES

1) Quoted by E.P. Wigner, Group theory, p. V, Academic Press, New York (1959).

2) The formula was derived and discussed for nuclei by E.P. Wigner and E. Feenberg, Rept.Progr.Phys. 8, 274 (1941);
See also W.M. MacDonald, Phys.Rev. 98, 60 (1955); 100, 51 (1955); 101, 271 (1956).
A discussion which is independent of the model has been given by S.B. Treiman and S. Weinberg, Phys.Rev. 116, 465 (1959).

3) J. Cerny, R.H. Pehl, F.S. Goudling and D.S. Landis, Phys.Rev. Letters 13, 726 (1964).

4) S. Coleman and S.L. Glashow, Phys.Rev. Letters 6, 423 (1961).

5) S. Okubo, Progr.Theor.Phys.(Kyoto) 21, 232 (1963).

6) E.P. Wigner, Phys.Rev. 56, 519 (1939).

7) L.A. Radicati, Phys.Rev. 87, 521 (1952).

8) S.S. Gershtein and Ya.Zel'dovich, JETP 2, 576 (1956);
R.P. Feynman and M. Gell-Mann, Phys.Rev. 109, 193 (1958);
The best experimental test of the conservation of the isovector current has been performed by Y.K. Lee, L.W. Mo and C.S. Wu, Phys.Rev. Letters 10, 253 (1963).

9) M. Gell-Mann, Phys.Rev. 125, 1067 (1962);
Y. Ne'eman, Nuclear Phys. 26, 222 (1961).

10) N. Cabibbo, Phys.Rev. Letters 10, 531 (1963).

11) This point has been emphasized by D. Speiser in a seminar at the Institute for Advanced Study (1961);
See also D. Speiser and T. Tarski, J.Math.Phys. 4, 588 (1963).

12) E.P. Wigner, Phys.Rev. 51, 106 (1937).

13) P. Franzini and L.A. Radicati, Physics Letters 6, 322 (1963).

14) S.M. Bilenki, Yu M. Kazarinov, L.I. Lapidus, and R.M. Ryndin, Preprint, Dubna (1965).

15) F. Gursey and L.A. Radicati, Phys.Rev. Letters 13, 173 (1964).

16) F. Gursey, A. Pais and L.A. Radicati, Phys.Rev. Letters 13, 299 (1964).

17) M.A. Beg, B.W. Lee and A. Pais, Phys.Rev. Letters 13, 514 (1964).

18) L. Michel and B. Sakita, Annales de l'Institut Poincaré, to be published.

19) W.B. McGlinn, Phys.Rev. Letters 12, 467 (1964).

20) M.A. Beg and A. Pais, Phys.Rev. Letters 14, 261 (1965);
R. Delbourgo, A. Salam and J. Strathdee, Proc.Roy.Soc. A284, 146 (1965);
B. Sakita and W. Wali, Phys.Rev. Letters 14, 404 (1965);
W. Ruhl, Physics Letters 14, 346 (1965);
T. Fulton and J. Wess, Physics Letters 14, 334 (1965).
This list is far from exhaustive: I apologize to all authors whose work is not quoted here.

21) M.A. Beg and A. Pais, Phys.Rev. Letters 14, 509 (1965);
R. Blanckenbecker, M.L. Goldberger, K. Johnson and S.B. Treiman, Phys.Rev. Letters 14, 518 (1965);
See also S. Weinberg, Phys.Rev. 139, B 597 (1965).

22) B.W. Lee, Phys.Rev. Letters 14, 676 (1965);
R.F. Dashen and M. Gell-Mann, Physics Letters 17, 142 and 145 (1965).

23) S. Coleman, CERN preprint (1965).

24) It is difficult to decide when the concept of internal symmetry was first introduced in physics. To my knowledge the concept of invariance under isotopic spin SU(2) transformations was introduced, for the first time, and its consequences fully explored, in nuclear physics by G. Breit and E. Feenberg, Phys.Rev. 50, 850 (1936) and by B. Cassen and E.U. Condon, ibid. 50, 846 (1936). The first generalization to meson theory was given by N. Kemmer, Proc.Cambridge Phil.Soc. 34, 354 (1938).

LOW-ENERGY HYPERON-PROTON INTERACTIONS

G.A. Snow*),

University of Rome,
Italy.

I would like to present a general, qualitative report on the low-energy hyperon-proton interactions from the experimental point of view, with some theoretical comments à la SU(3) and SU(6) liberally mixed in. This subject is probably as good as any to illustrate the difficulties of comparing these symmetry schemes with experiments dealing with two-particle scattering.

In the past two years a substantial amount of data has been accumulated on the subject by groups at Heidelberg[1], Rehovoth[2], Berkeley[3] and Maryland[4] using bubble chamber film with K^- mesons stopping in a hydrogen chamber. In addition, some information on hyperon-proton interactions has come from the study of $(Yn)K^+$ final states in pp collisions by the Rochester-BNL[5] and Princeton[6] groups. Nevertheless, the statistical accuracy of the data in some channels, such as the charged Σ-hyperon interactions, is still rather poor. At least, the number of good events has exceeded the number of theoretical papers on the subject, which was not true a few years ago.

The kinds of interactions that are amenable for study in stopping (K^-,p) film are:

$$\begin{aligned} &\Lambda + p \to \Lambda + p && T_\Lambda < 37\text{ MeV} \\ &\Sigma^- + p \to \begin{pmatrix}\Lambda \\ \Sigma^0\end{pmatrix} + n && \multirow{2}{*}{} \\ &\Sigma^\pm + p \to \Sigma^\pm + p && \left.\right\} T_\Sigma < 13\text{ MeV}, \end{aligned} \qquad (1)$$

*) Fulbright and Guggenheim Fellow, 1965-66, on sabbatical leave from the University of Maryland, College Park, Maryland.

where the hyperons are produced in limited low-energy regions predominently from the (K^-,p) absorption reaction

$$K^- + p \rightarrow Y + \pi \ . \tag{2}$$

Since these hyperons belong to the same SU(3) octet as the neutrons and protons, it is natural to compare these hyperon-proton reactions with (n,p) and (p,p) reactions, and to inquire whether such a comparison can shed any light on the usefulness of SU(3), SU(6), ... for these low energy baryon-baryon interactions. Speculations along these lines have been made by Oakes[7]), de Souza, Meshkov and Snow[8]), Iwao[9]), Akyeampong and Delbourgo[10]), Kantor et al.[11]), A. Goldhaber and Socolow[12]), Dyson and Xuong[13]) and many others.

If one considers the $\ell = 0$ state of two baryons with Y = +2 or Y = +1 in pure SU(3) symmetry, one finds for singlet and triplet-spin states, using the generalized Pauli principle, the following independent amplitudes.

Table 1

Baryon-baryon S-wave scattering amplitudes in SU(3)

Baryon-baryon	Spin singlet	Spin triplet
(n,p)	$\underline{27}$	$\underline{10}^*$
(Σ^+,p)	$\underline{27}$	$\underline{10}$
(Σ^-,p)	$\underline{27}$, $\underline{8}_s$	$\underline{10}^*$, $\underline{10}$, $\underline{8}_a$
(Λ,p)	27, $\underline{8}_s$	$\underline{10}^*$ $\underline{8}_a$

I have not included in this table the more esoteric Y = 0 and Y = -1 baryon-baryon states, since they are not easily available for experimental study, particularly in $\ell = 0$ angular momentum states. Note that the triplet state has three independent SU(3) amplitudes while the singlet state has two. The theoretical situation becomes much "simpler" in SU(6) where only two independent amplitudes suffice for both singlet and triplet states. This is so since there are only two SU(6) amplitudes [490 and 1050] that are antisymmetric in spin and space combined. As a result of this simplicity, there are a few SU(6) predictions which can be tested and, in fact, they are not well satisfied experimentally.

The Y = 2 system has a triplet-bound state, the deuteron, in a pure 10* representation and an almost bound singlet state in a pure 27. Low-energy (n,p) scattering can be parameterized by four parameters a_s, r_{os}, a_t and r_{ot} respectively, the singlet and triplet scattering length and effective ranges defined by the usual expansion

$$k \cot \delta = -\frac{1}{a} + \frac{1}{2} r k^2 + \dots .$$

In terms of these parameters the total n-p cross-section is given by:

$$\sigma = \tfrac{1}{4}\,\sigma_s + \tfrac{3}{4}\,\sigma_t$$

$$\sigma = \frac{\pi}{k^2 + \left(-\frac{1}{a_s} + \frac{1}{2} r_{os} k^2\right)^2} + \frac{3\pi}{k^2 + \left(-\frac{1}{a_t} + \frac{1}{2} r_{ot} k^2\right)^2} \qquad (3)$$

where

$$a_s = -23.7 \text{ fermi} \qquad a_t = 5.4 \text{ fermi}$$

$$r_{os} = 2.5 \text{ fermi} \qquad r_{ot} = 1.7 \text{ fermi} .$$

The existence of the deuteron bound state and the availability of thermal and sub-thermal beams of neutrons have made possible the experimental determination of all four of these S-wave parameters. We are clearly in a much less favourable situation with respect to hyperon-proton S-wave interactions. SU(6) predicts[10,11] that $\sigma_s = \sigma_t$ or $a_s = a_t$, which is not very well satisfied by the n-p scattering data.

Consider first the (Λ,p) interaction. The study of hyperfragments has shown that no (Λ,p) bound state exists, neither in the singlet nor the triplet spin state. As is well known a substantial amount of information about the strengths of S wave singlet and triplet (Λ,p) interaction potentials has been extracted theoretically from the experimental binding energies of the 3,4 and 5 body Λ hyperfragments. The singlet interaction is more attractive than the triplet interaction. Several different theoretical estimates for the four S wave (Λ,p) parameters are listed in Table 2.

Table 2

(Λ,p) scattering lengths and effective ranges as deduced from hyperfragments binding energies

a_s (fermi)	r_{os} (fermi)	a_t (fermi)	r_{ot} (fermi)	Refs.
$-3.6\ {}^{+3.6}_{-1.8}$	≈ 2	-0.53 ± 0.12	≈ 5	(14)
$-2.89\ {}^{+0.59}_{-0.41}$	1.94 ± 0.08	-0.71 ± 0.06	3.75 ± 2.2	(15)
-2.3 ± 0.5	2.2 ± 0.2	-0.90 ± 0.15	3.55 ± 0.4	(16)

These results depend, somewhat, on the assumptions made with respect to the shape of the two-body potentials and the role, if any, ascribed to three-body forces.

A comparison of these results with our most recent experimental (Λ,p) scattering data indicates that the central values of the theoretical fits in Table 2 are of the right order of magnitude but systematically too small by $\sim$ 30%. Parameters which fit $\sigma_{el}(\Lambda p \rightarrow \Lambda p)$ better are obtained by increasing either a_s or a_t, or both, from the central values listed in Table 2 (for example, $a_s = -5.0$, $r_{os} = 2.0$, $a_t = -1.0$, $r_{ot} = 3.0$ fits the latest Maryland data). There is no indication of any resonant behaviour in the low-energy (Λ,p) scattering data. The simplest interpretation of the bump found by Melissinos et al.[5)] in the (Λ,p) mass distribution in the reaction $pp \rightarrow \Lambda p K^+$, is that it is a direct reflection of the rather strong $^1S_0$ attractive potential for the Λp system. It would appear to be difficult to separate the dynamics of the production reaction from the final state Λp interaction in order to obtain more detailed information about the Λp interaction. The general conclusion is that 3,4 and 5 body hyperfragments analyses did lead to the correct qualitative description of the S wave singlet and triplet (Λ,p) interactions. Before leaving (Λ,p) scattering there is one more qualitative feature that I would like to point out. The differential cross-section indicates a pronounced forward peaking in the energy region $E_{lab} = 25-37$ MeV. At the same energy n-p scattering is more isotropic with a tendency toward slight backward and forward scattering. The lack of backward peaking in (Λ,p) scattering indicates that K^+-meson exchange does not play an appreciable role. Again SU(6) predicts[10,11)] spin independence for the (Λ,p) interaction, i.e. $\sigma_s(\Lambda p) = \sigma_t(\Lambda p)$ or $a_s = a_t$; and there does not appear to be any significant agreement with experiment.

Let us now consider the (Σ^+,p) interactions. Here again only the elastic scattering channel is open at low energies, and $\sigma_{el}(\Sigma^+,p)$ is found to be about $\frac{2}{3}\,\pi\lambdabar^2$ in the region $p_{\Sigma^+}^{lab} \sim 160$ MeV/c. This small cross-section compared to the maximum possible value of $4\pi\lambdabar^2$ shows that the triplet (Σ^+,p) interaction is weak. In the language of SU(3) this implies that the <u>10</u> amplitude is weak. The singlet (Σ^+,p) amplitude is a pure <u>27</u> just as is the (np) or (pp) singlet amplitude. In fact, the observed (Σ^+p) cross-section is of the same order of magnitude as (pp) scattering (after correction for symmetrization) at the same centre-of-mass momentum. If the triplet (Σ^+p) scattering is very small then this data is consistent with SU(3).

Returning to the SU(3) prediction that

$$\delta({}^1S_0,\Sigma^+p) = \delta({}^1S_0,pp)\ ,$$

Mr. de Souza at the University of Maryland has been investigating the effect of pseudoscalar boson-mass splitting on this prediction, assuming that the coupling constants are SU(3) invariant. He uses a simple version of single boson exchange as the source of the NN and YN potentials. He includes a singlet scalar meson plus the pseudoscalar octet and a repulsive core. (Adding any more mesons would introduce more parameters than the data he is fitting.)

For a given F/D ratio in the $\bar{B}BP$ coupling, and a given repulsive core ~ 0.5 f, he adjusts the mass and the coupling constant of the scalar meson so as to fit the np singlet scattering length a_s, and effective range r_{os}. Then, with these parameters he can calculate the potential for the Σ^-n system and its corresponding cross-section or phase shifts. The real pseudoscalar meson and hyperon masses are used so that the theory is no longer SU(3) invariant. What he finds with this simple prescription is that the singlet (Σ^-n) system becomes bound with a tiny binding energy (small enough so that the Σ^+p system would not be bound due to Coulomb repulsion)[17]. Except at $T_\Sigma \leq 1$ MeV, the singlet (Σ^-n) cross-section is very close to π/k^2, and hence is about 30% larger than the (n,p) singlet cross-section. It is quite insensitive to F/D or the repulsive size. Apparently it is the Σ-n mass difference that causes the bound state to appear since it reduces the kinetic energy term in the Schrödinger equation. The dominant attraction in the $^1S_0$ state for n-n comes from the assumed scalar meson, not from the pseudoscalar meson exchange potential. I leave it to you to judge whether this calculation is encouraging or discouraging with respect to the sensitivity of SU(3) symmetry predictions to particle mass splittings.

Finally, let us consider the Σ^-p system, which has two open absorption channels, Σ^0n and Λn, besides the elastic scattering channel. The absorption reaction at low energies ($p_{\Sigma^-}^{lab} \approx 150$ MeV/c) is very strong, i.e. $\sigma_{abs} \sim \pi\lambda^2$. (In fact, some of the experimental results give cross-sections even greater than $\pi\lambda^2$, although there is no convincing evidence for any appreciable p-wave contribution.) Clearly the triplet state must contribute heavily in this channel since the $\ell = 0$, singlet state absorption cross-section, is bounded by $\frac{1}{4}\,\pi\lambda^2$. The Σ^-p elastic scattering is difficult to measure and analyse, but it is found to be slightly less than that of the order of $\pi\lambda^2$. One can also observe the ratio

$$r = \frac{\Sigma^0 n}{(\Sigma^0 n) + (\Lambda n)}$$

for K^-p interactions at rest, r_r and in flight r_f. Experimentally, r_r is found to be between 0.4 and 0.46 and in flight r_f is of the order of 0.45- 0.6. It is clear that the (Σ^0n) channel, despite its phase space handicap, competes well with the Λn channel. A theoretical analysis of this data along the lines carried out a few years ago by de Swart and collaborators[18] using only one and two π-meson exchange potentials, would yield the conclusion that $f_{\Sigma\Sigma\pi} \approx f_{\Sigma\Lambda\pi}$, contrary to their original conclusion. The theoretical situation is, however, very complicated (many channels, many bosons to be exchanged), so this qualitative conclusion is not compelling.

In conclusion, we see that the $^1S_0$ scattering interaction is very strong in the pp or np channel and probably in the Σ^+p or Σ^-n channels, and moderately strong in the Λp elastic scattering channel. The Σ^-p S-wave interaction is dominated by very strong absorption reactions almost equally divided between Λn and $\Sigma^0 n$.

The SU(6) as well as $SU(6)_W$ prediction for $\ell = 0$ hyperon-proton and nucleon-nucleon scattering, namely, spin independence, is not borne out by the experimental data. SU(3) symmetry predicts much less, but from the theoretical point of view even those few predictions are expected to suffer deviations due to effects of mass splittings.

REFERENCES

1) H.G. Dosch, R. Englemann, H. Filthuth, V. Hepp, E. Kluge and A. Minguzzi-Ranzi, Physics Letters 14, 162 (1965).
R. Englemann, H. Filthuth, A. Fridman and G. Alexander et al., Phys.Rev. Letters 13, 484 (1964).

2) G. Alexander, U. Karshon, A. Shapira and G. Yekutieli et al., Phys.Rev. Letters 13, 484 (1964).

3) J. Schultz, W. Chinowsky, R. Kinsey and N. Rybicki, Bulletin APS 10, 589 (1965).

4) R.A. Burnstein, Invited paper, Washington APS Meeting (1965).
B. Sechi-Zorn, R.A. Burnstein, T.B. Day, B. Kehoe and G.A. Snow, Bulletin APS 10, 589 and 467 (1965); Proceedings of International Conference for High-Energy Physics at Dubna (1964); and Phys.Rev. Letters 13, 282 (1964).

5) A.C. Melissinos, N.W. Reay, J.J. Reed, T. Yamanouchi, E. Sacharides, S.J. Lindenbaum, S. Ozaki and L.C.L. Yuan, Phys.Rev. Letters 14, 604 (1965).

6) P.A. Piroue, Physics Letters 11, 164 (1964).
W.J. Hogan, P.F. Kunz, A. Lemonick, P.A. Piroue and A.J.S. Smith, Bulletin APS 10, 517 (1965).

7) R.J. Oakes, Physical Review 131, 2239 (1963).

8) P. de Souza, G.A. Snow and S. Meshkov, Physical Review 135, B565 (1965).

9) S. Iwao, Nuovo Cimento 34, 1167 (1964).

10) D.A. Akyeampong and R. Delbourgo, Trieste preprint IC/65/65.

11) P.B. Kantor, T.K. Kuo, R.F. Peierls and T.L. Trueman, BNL preprint (1965).

12) R. Socolow and A. Goldhaber, Perkeley preprint (May, 1965).

13) F.J. Dyson and N. Xuong, Phys.Rev. Letters 10, 815 (1964).

14) J.J. de Swart and C. Dullemond, Ann.Phys. (N.Y.) 19, 458 (1962).

15) R.C. Herndon, Y.C. Tang and E.W. Schmid, Physical Review 137, B294 (1965).

16) A. Bodmer, Invited paper, Washington APS Meeting (1965) and private communication.

17) G. Snow, Physical Review 110, 1192 (1958).

18) J.J. de Swart, Proceedings of International Conference on Hyperfragments at St. Cergue, p. 191 (1963).

STRANGE RESONANCES

S. Focardi,
Bologne University.

I. INTRODUCTION

I will report here the new results, referring to strange resonances, published after the Dubna Conference.

In describing the methods which are generally used in the analysis of experimental data, I will use some of these results as examples.

In Table 1 the present situation is summarized. A reference is given in the last column to identify the origin of the data. In this table the resonant states have been subdivided according to quantum numbers: B (baryonic number), S (strangeness) and I (isotopic spin). All experiments reported here were done in the hydrogen bubble chamber, unless otherwise specified.

1. Mesonic resonances

In this group we classify the states with B = 0. The experimental method most frequently used in order to discover the existence of resonant states is the following. Suppose we have a hypothetical reaction

$$A + B \rightarrow C + D + E^{+} \ldots \quad (1)$$

to study the CD system of particles. We plot un unbiased sample of type 1 events versus the invariant mass (or squared mass) of the CD system. This experimental distribution must be compared with a phase space relative to the same particles.

We will see several examples of this method. The determination of resonant state quantum numbers will be discussed in detail for K*(1400).

1.1 K*(1400)

Figure 1 shows the mass distribution[1] of the $K\pi$ system in the reaction

$$K^- + p \rightarrow \bar{K}^o + \pi^- + p$$

at 3.5 GeV/c. This is the first evidence of K*(1400) existence. It is clear that the experimental distribution cannot be explained only by K*(890) superimposed on the phase space.

The same evidence was obtained by Hardy et al.[2] by analysing the $\Lambda^o \pi^- K^+$ and $\Sigma^o \pi^- K^+$ systems obtained from the initial state $\pi^- + p$ at 3.9 and 4.2 GeV/c. Figure 2 shows the mass distribution reported by these authors.

As we can see in Fig. 3 we have, finally, a third evidence of this K* existence obtained by studying[3] the reactions

$$\begin{aligned} K^- + p &\rightarrow K^- + \pi^+ + n \\ &\rightarrow K^- + \pi^o + p \\ &\rightarrow \bar{K}^o + \pi^- + p \end{aligned}$$

at 3 GeV/c.

The mass values and the full widths at half height obtained in these three experiments are reported in Table 2.

The K*(1400) isotopic spin can be obtained from the branching ratio R in $\bar{K}^o + \pi^-$ and $K^- + \pi^o$. The isotopic spin conservation predicts a branching ratio R = 2 or R = 0.5 for isotopic spin I = ½ or I = 3/2 respectively. The experimental ratio R = 1.6 ± 0.8 favours the assignment I = ½.

Because the π and the K are pseudoscalars, the possible spins and parities of this resonance are 0^+, 1^-, 2^+ ... etc. The spin determination can be made by analysing the angular distribution of K^- in the K^* centre-of-mass. Following Gottfried and Jackson[4], we define the polar and the azimuthal angles ϑ and φ with respect to the incident K direction and to the production plane. Depending upon whether the spin of the resonance is 0, 1 or 2, the following distribution functions are expected:

$$\text{spin 0} \quad F_0(\cos\vartheta,\varphi) = \frac{1}{4\pi}$$

$$\text{spin 1} \quad F_1(\cos\vartheta,\varphi) = \frac{3}{4\pi}\left[\rho_{00}\cos^2\vartheta + \rho_{11}\sin^2\vartheta - \rho_{1,-1}\sin^2\vartheta\cos 2\varphi\right]$$

$$\text{spin 2} \quad F_2(\cos\vartheta,\varphi) = \frac{15}{16\pi}\left[3\rho_{00}(\cos^2\vartheta - 1/3)^2 + 4\sin^2\vartheta\cos^2\vartheta(\rho_{11} - \rho_{1,-1}\cos 2\varphi)\right. \tag{2}$$

These formulae are valid in a one-particle exchange model if the spin of exchanged particles is < 2. $\rho_{j,j}$ are the density matrix elements in the K^* spin space.

In Fig. 4 the experimental angular distributions are compared with formulae (2). The assignment 2^+ is favoured with respect to 1^-. 0^+ can be excluded.

This K^* might be the partner of the other known states 2^+ (f_0 and A_2) within the scheme of SU(3).

1.2 The K**(1320)

Almeida et al.[5] found evidence of a resonant state in the reaction

$$K^+ + p \rightarrow K^+ + \pi^+ + \pi^- + p$$

at 5 GeV/c by observing the $K^+\pi^+\pi^-$ mass plot. Figure 5 shows this mass plot. The shaded area is obtained after removing the events with the (proton π^+) mass between 1150 - 1340 MeV ($N^*_{3/2}$ band). This technique of removing events in the bands of other resonances is frequently used in order to exclude the possibility that the observed effect is associated with the other resonant states. The mass and the full width at half height of this resonance are $M = 1320 \pm 25$ MeV and $\Gamma = 60 \pm 20$ MeV, respectively. At present no information is available on the spin and isospin of this system. The dominant decay is $K^{**} \to K^{*}\pi^{+}$.

2. Mesonic resonances with strangeness +2

Mesonic resonances with strangeness > 1 are more easily observed in reactions induced by K mesons. In fact, with pion beams these states can be formed only if the final state of the reaction contains four strange particles or Ξ particles. The cross-sections for these processes are so small that good statistics cannot be obtained. The first evidence of an $S = +2$ resonance was given by a CERN group[6] who analysed the reactions

$$\begin{aligned} K^{+} + p &\to K^{+} + K^{+} + \Lambda^{o} \\ &\to K^{+} + K^{+} + \Sigma^{o} \\ &\to K^{+} + K^{o} + \Sigma^{+} \\ K^{+} + p &\to K^{+} + K^{+} + \Lambda^{o} + \pi^{o} \\ &\to K^{+} + K^{o} + \Lambda^{o} + \pi^{+} \end{aligned}$$

at 3, 3.5 and 5 GeV/c. In this energy region, the cross-sections are ~ 20 μbarn.

The mass spectrum of the KK system is shown in Fig. 6. A fit of experimental data with phase space plus two Breit-Wigner functions gives the following results:

M_1 resonance $M = 1280 \pm 20$ MeV $\Gamma = 110 \pm 40$ MeV

M_2 resonance $M = 1055 \pm 20$ MeV $\Gamma = 60 \pm 25$ MeV .

The two resonances have isotopic spin I ≐ 1 because they are observed in the state $I_z = 1$. The distribution of the events in the M_1 mass region, as a function of momentum transfer, is peaked at low values of momentum transfer. The M_2 peak could also be explained by a strong scattering in the s-wave interaction of the KK system.

3. Baryonic resonances

3.1 $Y_1^*(1942)$, $Y^*(2097)$ and $Y^*(2299)$

Considering now systems with baryonic number $B \neq 0$, we find other resonant states which have recently been observed. A CERN-Saclay collaboration[7] in which $\bar{p} + p$ interactions at 5.7 GeV/c were analysed, gives evidence of three new Y^*'s. They are observed in a mass plot of the systems $(KN)^0$ and $(KN\pi)^{\pm}_{0}$ (Fig. 7). The characteristics of these Y^*'s are given in Table 3.

The two highest mass states might be the same as those observed in a photoproduction experiment by a group at Yale University[8] in the reaction

$$\gamma + p \rightarrow K^+ + \dots \quad . \qquad (3)$$

The momentum and velocity of the K^+ are measured by means of bending magnets, Čerenkov counters and scintillation counters. The authors observe that the cross-section for reaction (3) shows enhancements which could be explained by two Y^*'s, with

$$M = 2022 \pm 20 \text{ MeV} \qquad \Gamma = 120 \text{ MeV}$$
$$M = 2245 \pm 25 \text{ MeV} \qquad \Gamma = 150 \text{ MeV} .$$

3.2 $\Xi^*(1933)$

Another Ξ^* resonance has recently been observed by the Ecole Polytechnique Paris-Saclay-Amsterdam collaboration[9]. The reaction studied was

$$K^- + p \to \Xi^- + K^{\overset{0}{+}} + \pi^{\overset{+}{0}}$$

at 3 GeV/c. By plotting the $\Xi^-\pi^+$ invariant squared mass, the $\Xi^*(1530)$ was observed to be produced abundantly. As Fig. 8 shows, the mass spectrum cannot be easily explained using only phase space and the $\Xi^*(1530)$. The accumulation of events in the region of 3.8 GeV^2 also remains after subtraction of events which lie in the $K^*(890)$ band. The mass and width of this Ξ^* are respectively

$$M = 1933 \pm 16 \text{ MeV} \qquad \Gamma = 140 \pm 35 \text{ MeV} .$$

3.3 <u>A Y* with I = 2</u>

A $\Sigma\pi$ resonant state with isotopic spin I = 2 has recently been observed by Pan and Ely[10]. The experiment was made with a propane bubble chamber exposed in a 1.15 GeV/c K^- beam. The authors studied the reaction

$$K^- + n \to \Sigma^- + \pi^+ + \pi^-$$

and observed an enhancement at M = 1415 ± 16 MeV in the $\Sigma^-\pi^-$ mass plot (Fig. 9). The width cannot be determined because of the large background. The $\Sigma^-\pi^-$ system has $I_z = -2$. The assignment I = 2 follows for this Y*.

Table 1

The present situation on strange resonances.
More details can be found in Ref. 11.

B	Symbol	Isotopic spin	Mass (MeV)	Width (MeV)	J^P	Decay modes	Refs.
Mesons	κ	½	725 ± 2	< 12	?	$K\pi$	(11)
	K^*	½	891 ± 1	50 ± 2	1^-	$K\pi$	(11)
	C	≤ 3/2	1215 ± 15	60 ± 10	?	$K\rho, K^*\pi$	(11)
	K^{**}	?	1320 ± 25	60 ± 20	?	$K^*\pi$	(5)
	K^*	½	1410 ± 10	100 ± 20	2^+	$K\pi$	(11)
	M_2	1	1055 ± 20	60 ± 25	?	KK	(6)
	M_1	1	1280 ± 20	110 ± 40	?	KK	(6)
Baryons	$Y_0^*(1405)$	0	1405	35 ± 5	$1/2^-$	$\Sigma\pi, \Lambda\pi\pi$	(11)
	$Y_0^*(1520)$	0	1518.9 ± 1.5	16 ± 2	$3/2^-$	$\Sigma\pi, KN, \Lambda\pi\pi$	(11)
	$Y_0^*(1815)$	0	1815	70	$5/2^+$	$KN, \Sigma\pi, \Lambda\pi\pi, \Lambda\eta$	(11)
	$Y_1^*(1385)$	1	1382.1 ± 0.9	51 ± 2	$3/2^+$	$\Lambda\pi, \Sigma\pi$	(11)
	$Y_1^*(1660)$	1	1660 ± 10	44 ± 5	≥ 3/2	$KN, \Sigma\pi, \Lambda\pi, \Sigma\pi\pi, \Lambda\pi\pi$	(11)
	$Y_1^*(1765)$	1	1765 ± 10	60 ± 10	$5/2^-$	$KN, \Lambda\pi, \Sigma\pi, \Lambda\pi\pi$	(11)
	$Y_1^*(1942)$	1	1942 ± 9	43 ± 18	?	$KN\pi$	(7)
	$Y^*(2097)$	<2	2097 ± 6	34 ± 11	?	$KN, KN\pi$	(7)
	$Y^*(2299)$	<2	2299 ± 6	35 ± 12	?	$KN\pi$	(7)
	$Y_2^*(1415)$	2	1415 ± 16	?	?	$\Sigma\pi$	(10)
	$\Xi^*(1530)$	½	1529.7 ± 0.9	7.5 ± 1.7	$3/2^+$	$\Xi\pi$	(11)
	$\Xi^*(1818)$	½	1818 ± 5	~ 60	$3/2^-$	$\Xi^*\pi, \Lambda\bar{K}, \Xi\pi, \Xi\pi\pi$	(11)
	$\Xi^*(1933)$	?	1933 ± 16	140 ± 35	?	$\Xi\pi$	(9)

Table 2

Mass and width values of K*(1400) as given in Refs. 1,2 and 3

Experiment	Mass	Width	Ref.
$K^- + p$; 3.5 GeV/c	1400 ± 10	~ 160	(1)
$\pi^- + p$; 3.9 and 4.2 GeV/c	1430 ± 20	100 ± 20	(2)
$K^- + p$; 3 GeV/c	1404 ± 14	99 ± 26	(3)

Table 3

Mass, width, isotopic spin and decay modes of
$Y^*(1942)$, $Y^*(2097)$ and $Y^*(2299)$

Mass	Width	Isotopic spin	Main decay modes
2097 ± 6	34 ± 11	0 or 1	$(KN)^0$, $(KN\pi)^0$
2299 ± 6	35 ± 12	0 or 1	$(KN\pi)^0$
1942 ± 9	43 ± 18	1	$(KN\pi)^{\pm}_{0}$

REFERENCES

1) N. Haque, D. Scotter, B. Musgrave, W.M. Blair, A.L. Grant, I.S. Hughes, P.J. Negus, R.H. Turnbull, A.A.Z. Ahmad, S. Baker, L. Celnikier, S. Misbahuddin, H.J. Sherman, I.O. Skillicorn, A.R. Atherton, G.B. Chadwick, W.T. Davies, J.H. Field, P.M.D. Gray, D.E. Lawrence, J.G. Loken, L. Lyons, J.H. Mulvey, A. Oxley, C.A. Wilkinson, C.M. Fisher, E. Pickup, L.K. Rangan, J.M. Scarr and A.M. Segar, Physics Letters 14, 338 (1965).

2) L.M. Hardy, S.U. Chung, O.I. Dahl, R.I. Hess, J. Kirz and D.H. Miller, Phys.Rev. Letters 14, 401 (1965).

3) S. Focardi, A. Minguzzi Ranzi, P. Serra, L. Monari, S. Herrier and A. Verglas, Physics Letters 16, 351 (1965).

4) K. Gottfried and J.D. Jackson, Physics Letters 8, 144 (1964).

5) S.P. Almeida, H.W. Atherton, T.A. Byier, P.J. Dornan, A.G. Forson, J.H. Scharenguival, D.M. Sendall and B.A. Westwood, Physics Letters 16, 184 (1965).

6) M. Ferro-Luzzi, R. George, Y. Goldschmidt-Clermont, V.P. Henry, B. Jongejans, D.W.G. Leith, G.R. Lynch, F. Muller and J.M. Perreau, Physics Letters 17, 155 (1965).

7) R.K. Bock, W.A. Cooper, B.R. French, J.B. Kinson, R. Levi-Setti, D. Revel, B. Tallini and S. Zylberajch, Physics Letters 17, 166 (1965).

8) W.A. Blanpied, J.S. Greenberg, V.W. Hughes, P. Kitching, D.C. Lu and R.C. Minehart, Phys.Rev. Letters 18, 741 (1965).

9) J. Badier, M. Demoulin, J. Goldberg, B.P. Gregory, C. Pelletier, R. Barloutaud, A. Leveque, C.L. Oudec, J. Meyer, P. Schlein, A. Verglas, D.J. Holthuizen, W. Hoogland and A.G. Tenner, Physics Letters 16, 171 (1965).

10) Y.L. Pan and R.P. Ely, Phys.Rev. Letters 13, 277 (1964).

11) A.H. Rosenfeld, A. Barbaro Galtieri, W.H. Barkas, P.L. Bastien, J. Kirz and M. Roos, UCRL 8030, Part 1, March 1965 edition.

FIGURE CAPTIONS

Figure 1 : $\bar{K}^0\pi^-$ invariant mass plot in the reaction $K^- + p \rightarrow \bar{K}^0 + \pi^- + p$ (Ref. 1). The curves represent fits to K*(890) + phase space and K*(890) + K*(1400) + phase space.

Figure 2 : Mass distribution for $K\pi$ system in the reactions $\pi^- + p \rightarrow \Lambda^0 + \pi^+ + K^+$ and $\pi^- + p \rightarrow \Sigma^0 + \pi^- + K^+$ at 3.9 and 4.2 GeV/c (Ref. 2).

Figure 3 : The $K\pi$ invariant mass plot in the reactions $K^- + p \rightarrow K^- + \pi^+ + n$, $K^- + p \rightarrow \bar{K}^0 + \pi^- + p$, $K^- + p \rightarrow K^- + \pi^0 + p$ at 3 GeV/c (Ref. 3).

Figure 4 : The angular distribution relative to K*(1400) as given in Ref. 2. Upper histograms show events in the peak and background on either side.

Figure 5 : The invariant mass plot of the $K^+\pi^-\pi^+$ system. The shaded area is the distribution obtained by removing events containing the $N^*_{3/2}$ (1150 < $p\pi^+$ < 1340 MeV) events (Ref. 5)

Figure 6 : The KK mass spectrum given in Ref. 6.

Figure 7 : Effective mass distribution of the KN and $KN\pi$ combinations in $\bar{p} + p$ at 5.7 GeV/c (Ref. 7).

Figure 8 : The squared mass plot of system $\Xi^-\pi^+$ as given in Ref. 9.

Figure 9 : The $\Sigma^-\pi^-$ invariant mass plot (Ref. 10) obtained in the reaction $K^- + n \rightarrow \Sigma^- + \pi^- + \pi^+$.

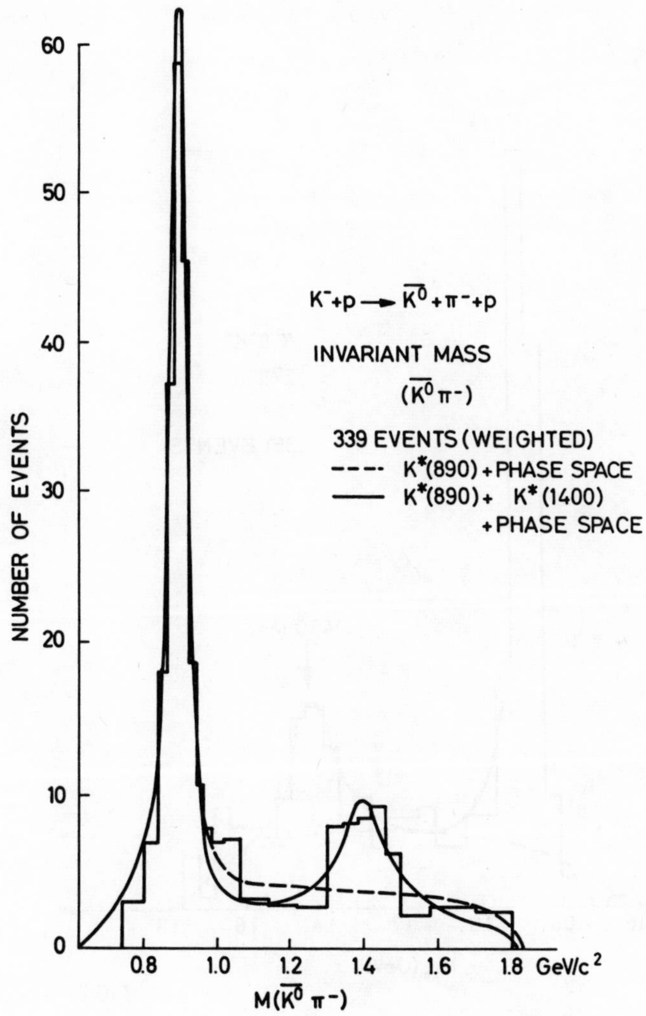
60
50
40
30
20
10
0
NUMBER OF EVENTS
$K^- + p \to \overline{K^0} + \pi^- + p$
INVARIANT MASS
$(\overline{K^0}\pi^-)$
339 EVENTS (WEIGHTED)
---- K*(890) + PHASE SPACE
—— K*(890) + K*(1400) + PHASE SPACE
0.8 1.0 1.2 1.4 1.6 1.8 GeV/c²
$M(\overline{K^0}\pi^-)$

FIG.1

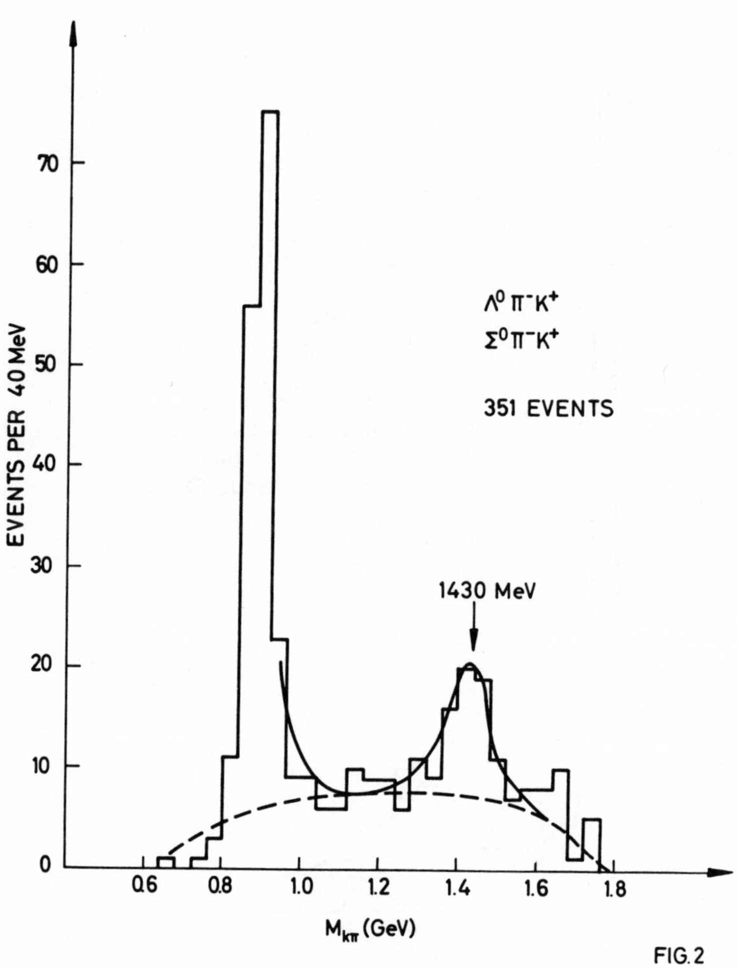
EVENTS PER 40 MeV
70
60
50
40
30
20
10
0
0.6
0.8
1.0
1.2
1.4
1.6
1.8
$M_{K\pi}$ (GeV)
$\Lambda^0 \pi^- K^+$
$\Sigma^0 \pi^- K^+$
351 EVENTS
1430 MeV

FIG. 2

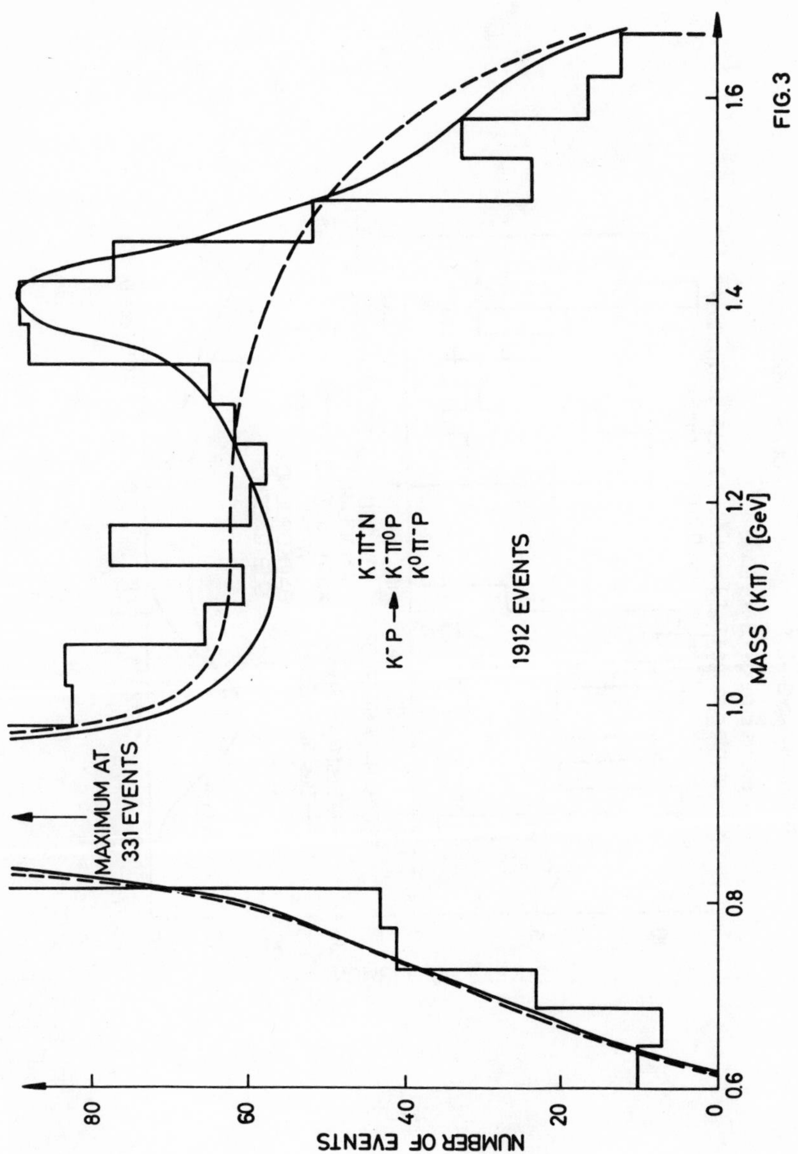
NUMBER OF EVENTS
0
20
40
60
80
0.6
0.8
1.0
1.2
1.4
1.6
MASS (Kπ) [GeV]
MAXIMUM AT
331 EVENTS
K⁻P → K⁻π⁺N
K⁻π⁰P
K⁰π⁻P
1912 EVENTS

FIG.3

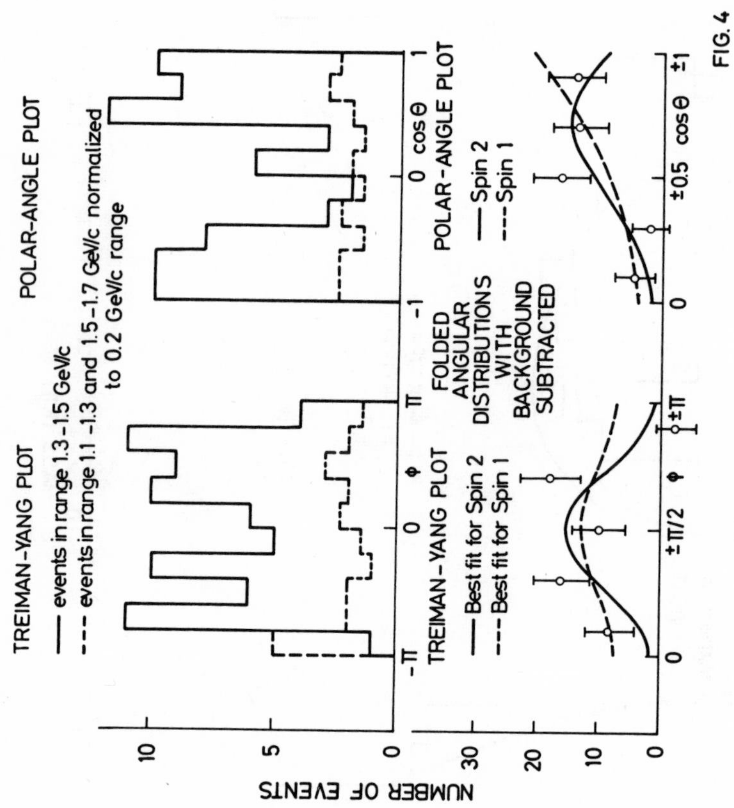

TREIMAN-YANG PLOT
POLAR-ANGLE PLOT
— events in range 1.3 - 1.5 GeV/c
--- events in range 1.1 - 1.3 and 1.5 - 1.7 GeV/c normalized to 0.2 GeV/c range
NUMBER OF EVENTS
10
5
0
30
20
10
0
-π
0
φ
π
-1
0
cos θ
1
TREIMAN-YANG PLOT
— Best fit for Spin 2
--- Best fit for Spin 1
FOLDED ANGULAR DISTRIBUTIONS WITH BACKGROUND SUBTRACTED
POLAR-ANGLE PLOT
— Spin 2
--- Spin 1
0
± π/2
φ
± π
0
± 0.5
cos θ
± 1

FIG. 4

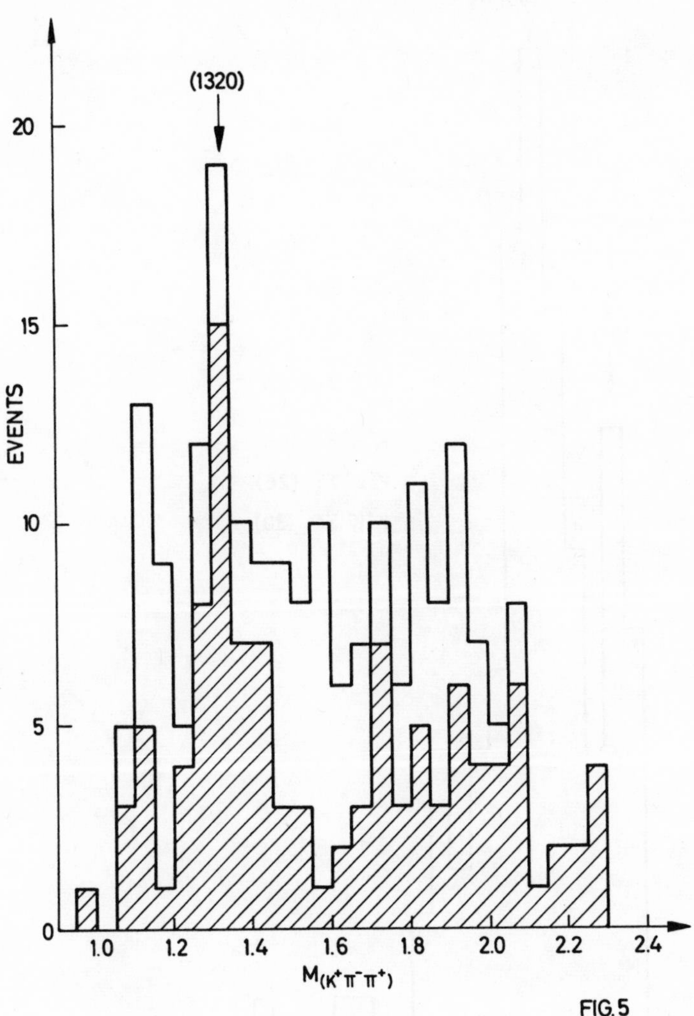

FIG. 5

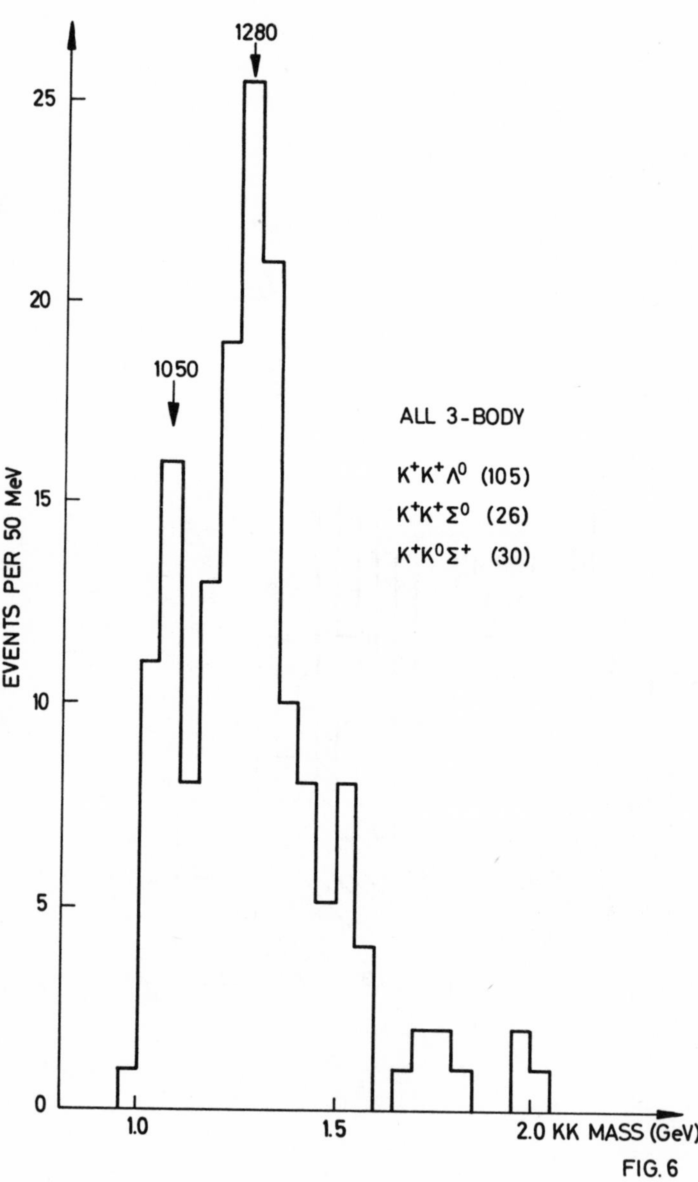
1280
1050
25
20
15
10
5
0
EVENTS PER 50 MeV
ALL 3-BODY
K⁺K⁺Λ⁰ (105)
K⁺K⁺Σ⁰ (26)
K⁺K⁰Σ⁺ (30)
1.0
1.5
2.0 KK MASS (GeV)

FIG. 6

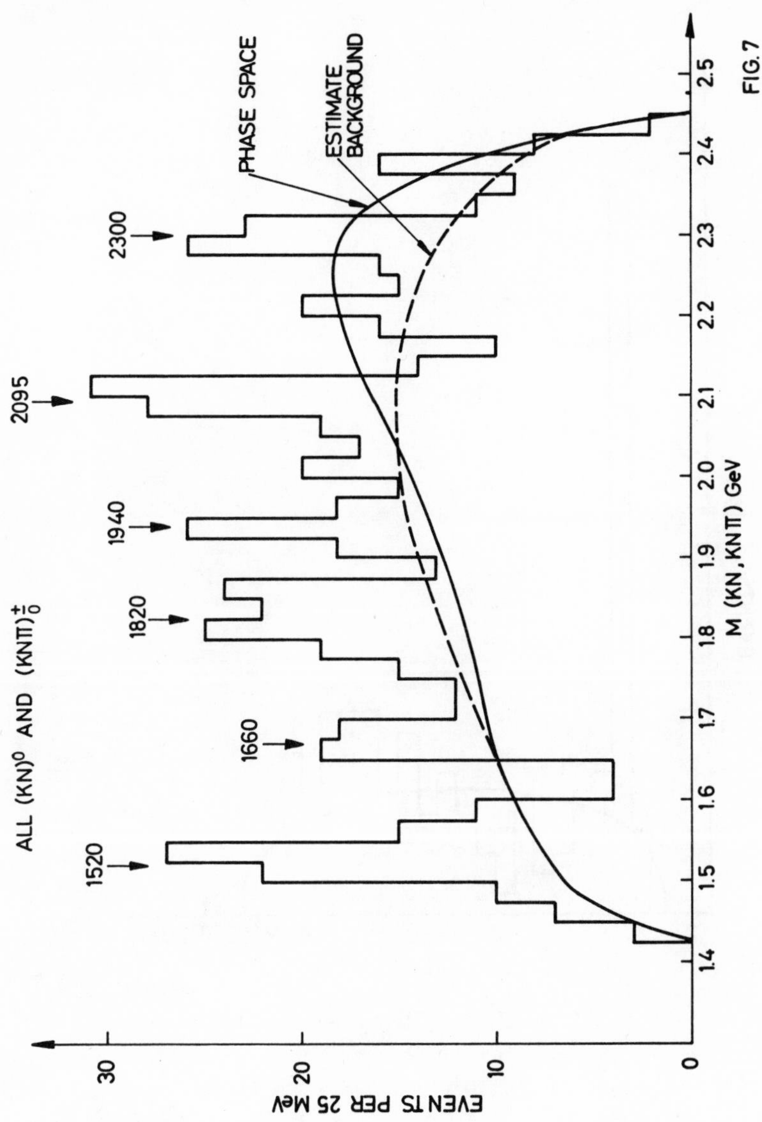

ALL (KN)^0 AND (KNπ)$^{\pm}_{0}$
1520
1660
1820
1940
2095
2300
PHASE SPACE
ESTIMATE
BACKGROUND
EVENTS PER 25 MeV
0
10
20
30
1.4
1.5
1.6
1.7
1.8
1.9
2.0
2.1
2.2
2.3
2.4
2.5
M (KN, KNπ) GeV

FIG.7

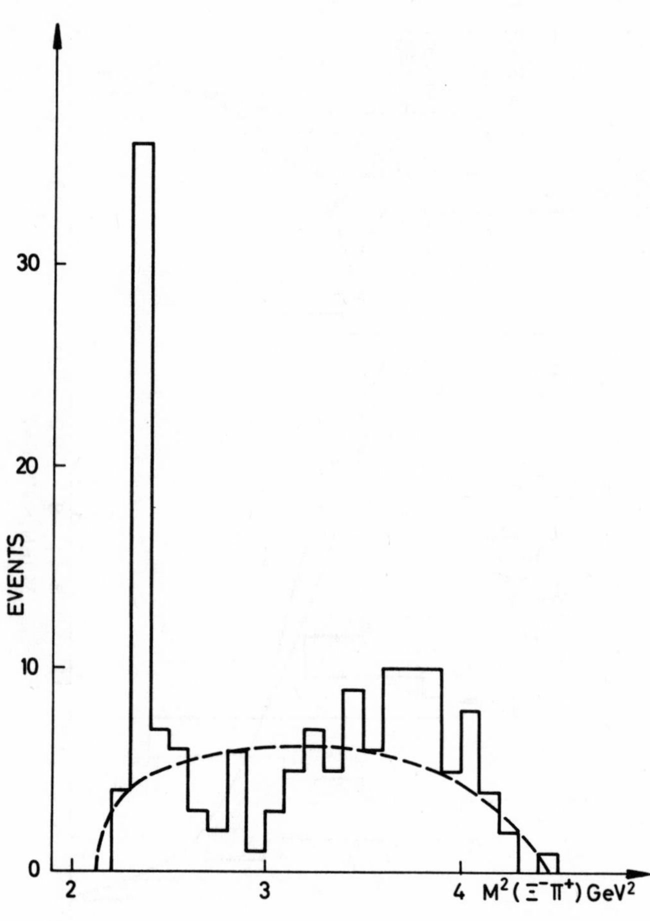

FIG. 8

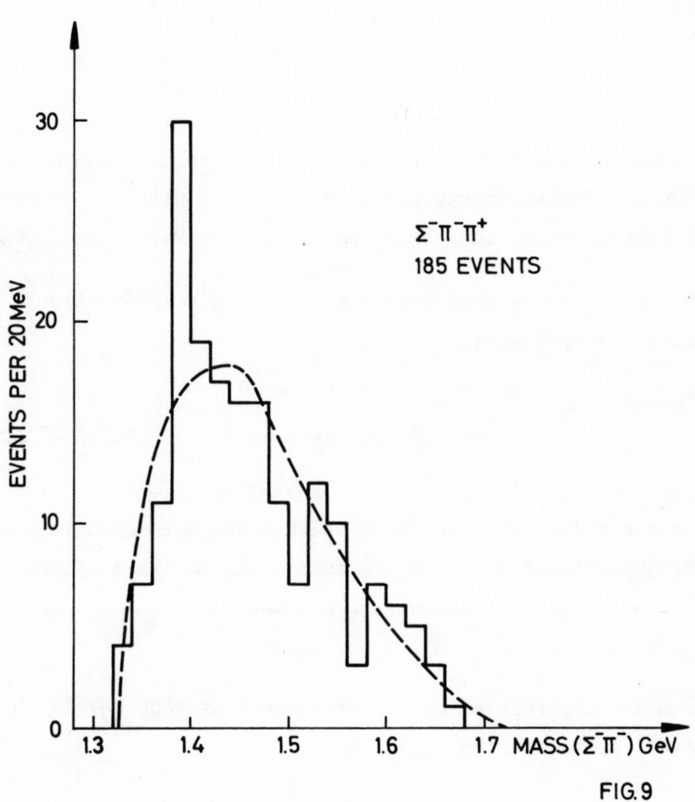
30
20
10
0
EVENTS PER 20 MeV
$\Sigma^-\pi^-\pi^+$
185 EVENTS
1.3
1.4
1.5
1.6
1.7
MASS ($\Sigma^-\pi^-$) GeV

FIG. 9

INVITED DISCUSSION FOLLOWING
THE FOCARDI LECTURE

V.P. Henri,
CERN.

I would like to make a few remarks concerning the K*(1400) and the $K^{**}(1320)$. These remarks refer to recent results[12,13] presented at the Oxford International Conference on Elementary Particles (1965).

1) The K*(1400) has now also been observed in K^+p interactions. The reactions involved are:

a) $$\left.\begin{aligned} K^+p &\to K^0\pi^+p \\ &\to K^+\pi^0 p \end{aligned}\right\} \text{(with } K^0 \text{ seen or unseen) at 3.5 and 5 GeV/c.}$$

b) $K^+p \to K^+\pi^-p\pi^+$ at 5 GeV/c in which a sizeable amount (~ 45%) of K*(890) + N*(1238) events are seen. An analysis of the K*(1400) N*(1238) channel again seems to favour a spin of 2^+ for K*(1400).

c) From the observation of an enhancement at 1400 MeV in the $(K\pi\pi)^0$ system in the reaction

$$K^+p \to (K\pi\pi)^0 \; N^{*++} \text{ at 5 GeV/c ,}$$

and from reaction b) selecting the K*(1400) N^{*++} events, we have determined the decay branching ratio

$$K^*(1400) \to \frac{K\pi}{K^*\pi} = 1.4 \pm 0.3 \;.$$

This result is in good agreement with the theoretical predictions, 1.63 of Glashow and Socolow[14] based on SU(3).

2) Concerning the K**(1320) peak reported by the Cambridge group[5] while investigating the reaction $K^+p \to K^+\pi^-p\pi^+$ at 5 GeV/c, I would like to show our own results based on four to five times higher statistics.

In Fig. 10 we show the Dalitz plots for the reaction $K^+p \to K^*p\pi^+$ with $K^* \to K^+\pi^-$ at 3, 3.5 and 5 GeV/c where K* refers to K*(890). The abscissae correspond to $M^2(K^+\pi^+\pi^-)$ and the ordinate to $M^2(p\pi^+)$.

For the analysis of the $K^+\pi^+\pi^-$ system we remove the events with a mass of the $p\pi^+$ system smaller than 1.4 GeV, thereby reducing the possibility of N*(1238) contamination. The region of the cut is represented in Fig. 10 by the line across the Dalitz plots.

In Fig. 11 the projections on the $K^+\pi^+\pi^-$ axis are shown. The solid curves correspond to phase space, while the dashed curves correspond to a modification of phase space using a one-pion exchange model in which the following graph is considered.

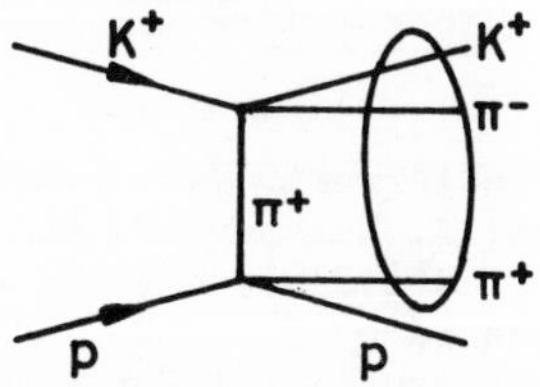

At the nucleon vertex the experimental values for the π^+p elastic scattering cross-section is introduced. This is analogous to the calculation of Moar and O'Halloran[15].

The results indicate that

i) the observed peak in the $K^+\pi^+\pi^-$ system seems to shift while going from 3 to 5 GeV/c;

ii) the peak observed at 5 GeV/c is not centred at 1320 MeV but more likely at ~ 1270 MeV;

iii) the one-pion exchange model considered is unable to account for the size of the effect at least at 3.5 and 5 GeV/c;

iv) from an analysis of the $K\pi\pi$ systems in the reactions

$$K^+p \rightarrow K^o\pi^o\pi^+p$$

and

$$K^+p \rightarrow K^o\pi^o\pi^+n$$

at 5 GeV/c, the isotopic spin I = ½ is favoured over I = 3/2.

REFERENCES

12) R. George, Y. Goldschmidt-Clermont, V.P. Henri, B. Jongejans, W. Koch, D.W.G. Leith, G. Lynch, A. Moissev, F. Muller, J.M. Perreau and V. Iarba, "Formation of N*(1238), K*(890) and K*(1400) in the three-body events produced by 3.5 and 5.0 GeV/c K^+", paper presented at the Oxford International Conference on Elementary Particles (1965), and to be published.

13) W. De Baere, J. Debaisieux, P. Dufour, F. Grard, J. Heughebaert, L. Pape, P. Peeters, F. Verbeure, R. Windmolders, T.A. Fillipas, R. George, Y. Goldschmidt-Clermont, V.P. Henri, B. Jongejans, W. Koch, G.R. Lynch, D.W.G. Leith, F. Muller and J.M. Perreau, "The enhancement ($K\pi\pi$) around 1270 MeV/c^2 in the reaction $K^+p \rightarrow KN\pi\pi$ at 3.0, 3.5 and 5 GeV/c". Paper presented at the Oxford International Conference on Elementary Particles (1965), and to be published.

14) S.L. Glashow and R.H. Socolow, Phys.Rev. Letters 15, 329 (1965).

15) U. Moar and T.A. O'Halloran, Jr., Physics Letters 15, 281 (1965).

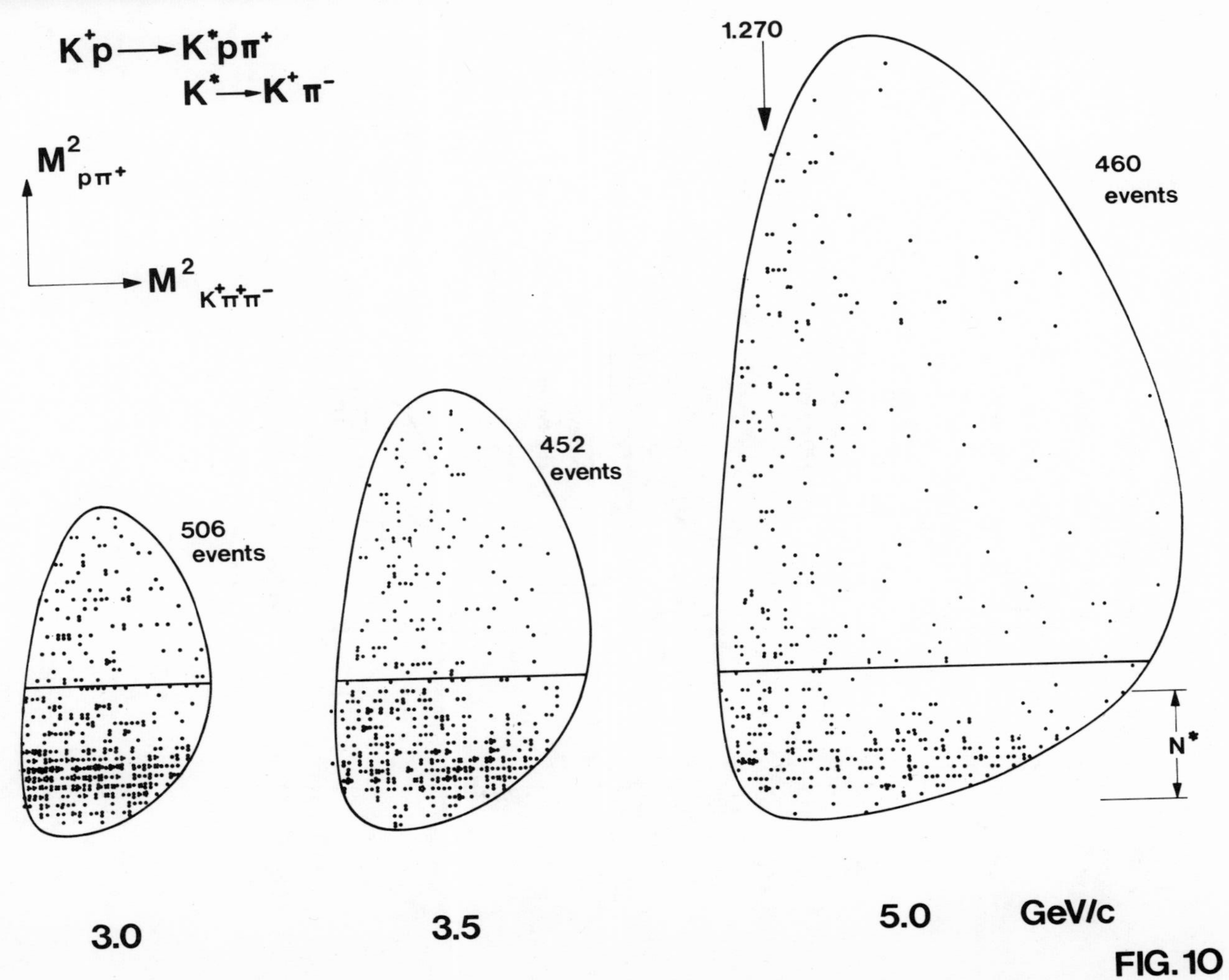
$K^+p \longrightarrow K^*p\pi^+$
$K^* \longrightarrow K^+\pi^-$
$M^2_{p\pi^+}$
$M^2_{K^+\pi^+\pi^-}$
1.270
460 events
452 events
506 events
N^*
3.0
3.5
5.0
GeV/c

FIG.10

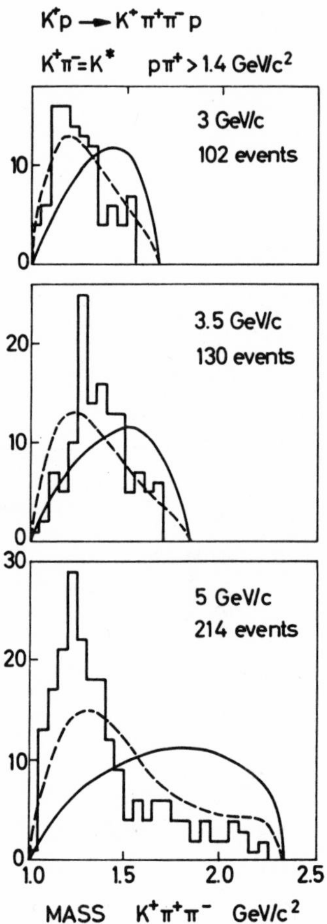
$K^+p \rightarrow K^+\pi^+\pi^- p$
$K^+\pi^- = K^*$ $p\pi^+ > 1.4$ GeV/c²
3 GeV/c
102 events
3.5 GeV/c
130 events
5 GeV/c
214 events
0
10
20
30
1.0
1.5
2.0
2.5
MASS $K^+\pi^+\pi^-$ GeV/c²

FIG. 11

CURRENT EXPERIMENTS AT DESY

U. Meyer-Berkhout,

Deutsches Elektronen-Synchrotron
DESY, Hamburg.

INTRODUCTION

My lecture will be divided into two parts. In the first part I shall try to sketch the current experimental programme at the 6 GeV electron accelerator in Hamburg. In the second part I shall attempt to summarize some of the results obtained during the first year of machine operation.

I. CURRENT EXPERIMENTAL PROGRAMME

Table 1 contains a list of the present experiments at DESY. The experiments are divided into four subgroups: bubble-chamber work, electron scattering-experiments, investigations of photoproduction processes, and "other" experiments. In the experiments marked with one star, some results have already been obtained. The "two-star" experiments are beginning to run, but results cannot yet be released. The preparations for the experiments marked with three stars are more or less advanced, but these experiments are not yet running.

Experiment No. 1 on the list is the work with the 80 cm hydrogen bubble chamber. At present, the main interest is directed towards photoproduction of pions, strange particles, and resonances in the photon-energy range between 0.3 and 5.5 GeV. Thus far, 400,000 pictures have been taken.

Table 1

Current experiments at DESY

Exp. No.	
	A. Bubble-chamber experiments
1*	Photon interactions in a hydrogen bubble chamber in the energy range 0.3- 5.5 GeV.
	B. Electron-scattering experiments
	I. Internal beam
2*	Elastic e-p and e-d scattering
3*	Search for a heavy electron
	II. External beam
4**	Elastic and inelastic e-p scattering at small angles and high energies
5***	Inelastic e-d scattering
	C. Photoproduction experiments
6**	Electroproduction of pions and photoproduction of dipions
7**	Photoproduction of π^0 mesons at small angles
8**	Photoproduction of charged mesons at small angles
9***	Coherent production of η mesons in the Coulomb field of complex nuclei (Primakoff effect)
10***	Photoproduction of wide angle electron-positron pairs
	D. Other experiments
11*	Investigation of Bremsstrahlung spectra with a pair spectrometer
12*	Production of coherent Bremsstrahlung by electrons in a diamond crystal (Überall effect)
13**	Experiments with synchrotron radiation in the vacuum UV region

The main purpose of the e-p and e-d scattering work, listed as experiment No. 2 in Table 1, is, of course, the determination of nucleon form factors. Thus far, only proton form factors in the q^2 range between 0.8 and 1.6 $(GeV/c)^2$ have been measured through detection of e-p coincidences. The results are in reasonable agreement with those obtained at CEA, Cornell and Stanford.

Experiment No. 3 is concerned with a systematic search for the heavy electron in the mass range between 0.5 and 1.0 GeV. Low[1)] has suggested that the Pipkin anomaly[2)] observed in wide angle photoproduction of electron-positron pairs may be due to the existence of a heavy electron, e*, which decays into a photon and an ordinary electron. If such an electron-photon resonance exists it should be possible to produce the heavy electron by inelastic e-p scattering. The e* might be observed as a sharp missing-mass peak in the recoil proton momentum or angular distribution. For improvement of the e*-detection sensitivity, e-p coincidences were registered in this particular experiment.

The external electron beam is also being used for electron scattering work. The ejection efficiency is as high as 70%. The external beam has excellent properties: intensity at present is between 3×10^{11} and 1×10^{12} electrons per second, emittance approximately 1 mm × mrad, energy spread approximately 0.3% for a duty cycle of 2%. The first experiment in the external beam, listed as No. 4 in Table 1, is concerned with elastic and inelastic e-p scattering at fairly small angles ($8^\circ - 40^\circ$) and high energies ($E_0 \leq 7.5$ GeV). Among other things, one hopes to obtain from this experiment more precise electric form factor values than are available so far at q^2 values above 1 $(GeV/c)^2$. Furthermore, in small angle high-energy scattering experiments contributions from two-photon exchange terms to the cross-section may possibly show up. In addition, a systematic study of nucleon isobar excitation by inelastic e-p scattering is intended. A fairly complex magnetic spectrometer, built out of standard DESY lenses and magnets, is used to analyse the momentum spectrum of scattered electrons with an intrinsic momentum resolution of about 0.5% up to scattered electron momenta of 6 GeV.

Experiment No. 5 on inelastic e-d scattering, not yet running, is a spark-chamber experiment which is being prepared by a visiting team from Karlsruhe. Wire chambers will be used. It is intended to detect scattered electrons in coincidence with recoil protons. The experiment may yield more precise neutron form-factor values than are available thus far up to squared four-momentum transfers of 2 $(GeV/c)^2$. In the electron leg, the scattering angle and the momentum of the electron will be measured. Furthermore, the proton recoil angle, and possibly its momentum, will be determined.

The first three photoproduction experiments (Nos. 6,7, and 8) should increase our understanding of the mechanism by which pions, pionic resonances, and possibly kaons are produced in high-energy photon interactions. Experiment No. 6 represents an attempt to study electroproduction of pions and photoproduction of resonances (e.g. the ρ meson) with spark chambers. Experiment No. 7, performed by a visiting team from Bonn University, is concerned with the study of π^0 photoproduction in the energy range between 1.3 and 3 GeV and π^0-production angles between 0 and 60° in the centre-of-mass system. The conventional technique of making use of two total absorption Čerenkov counters located symmetrically above and below the production plane is employed. They serve to detect the two gamma rays from π^0 decay and thus to measure π^0-production angle and momentum. The minimum pion energy to be detected is determined by the maximum angle subtended by the apertures of the two counters. The maximum energy is given by the end-point energy of the brems-spectrum. Except for the smallest π^0-production angles where the recoil-proton energy is too small, the proton is detected in coincidence with the neutral pion. This reduces the undesired background from multi-pion production events.

In experiment No. 8 the photoproduction of charged mesons at small angles will be investigated by the use of a fairly complex magnetic spectrometer. The experimental arrangement is quite versatile and should allow the investigation of single π^+ production, the study of the Drell effect, and an investigation of photoproduction of charged kaons down to small angles. The layout is such that, in principle, production angles down to 0° can be covered.

The coherent production of η mesons in the Coulomb field of a heavy nucleus like Pb, listed as No. 9 in Table 1, is an extension of the corresponding Frascati experiment[3] on π^0 production through the Primakoff effect, from which the lifetime of the π^0 meson was determined. The experiment, which seems at present to be the only feasible way to measure the η lifetime, is being prepared as a collaboration between the Pisa group and a visiting team from Bonn University. The main difficulty in the experiment is to disentangle the Coulomb field contribution to the differential η production cross-section, which peaks at $\frac{1}{2}^\circ$ for $E_0 \approx 4$ GeV, from the contributions of competing nuclear processes, such as coherent and incoherent nuclear photoproduction of η mesons. Separation should be feasible at 4 or 5 GeV, provided an η-detection system with an extremely good angular resolution ($\approx \frac{1}{2}^\circ$) is available. As in the corresponding π^0-production experiment eight total absorption lead glass Čerenkov counters, which are run in various combinations of twofold coincidences, will be used for the detection and energy determination of the two gamma rays from η decay.

Experiment No. 10 is a modified version of the Pipkin experiment[4] on wide angle photoproduction of electron-positron pairs. The modification consists mainly of replacing the two half quadrupoles which were used in the CEA experiment by two more sophisticated magnetic spectrometers. This should provide better resolution in the squared four momentum of the virtual electron and, at the same time, should allow one to use an extended liquid hydrogen target. A repetition of the Pipkin experiment is considered to be highly important, since the results of the experiment, if correct, would imply a deviation from simple q.e.d in e^-e^+ pair production.

In the fourth group some other experiments are listed. A pair spectrometer was built to measure the shape of bremsstrahlung intensity spectra, the knowledge of which is required for the interpretation of almost all photoproduction experiments. This is systematically done in experiment No. 11.

The production of coherent bremsstrahlung by electrons in a single crystal of diamond (experiment No. 12), first observed by Diambrini et al.[5)] in a series of beautiful experiments at the Frascati Synchrotron, is also being investigated at DESY. In fact, Dr. Bologna from the Diambrini group is participating in this work. The pair spectrometer again serves to analyse the intensity distribution of the coherent bremsstrahlung.

For completeness I must mention that a group of spectroscopists at DESY is currently getting excited about the fairly unique optical properties of the synchrotron radiation (experiment No. 13). From their point of view the synchrotron is considered to be an intense light source, which emits a continuous spectrum of almost completely polarized light in the vacuum ultraviolet and x-ray region. At 6 GeV the intensity spectrum reaches its maximum at a wave length of about 0.6 Å.

II. PRELIMINARY EXPERIMENTAL RESULTS

In this part of the lecture some preliminary experimental results obtained at DESY will be discussed. The discussion will be limited to one experiment out of each subgroup of Table 1.

1. Bubble-chamber work

The hydrogen bubble chamber (sensitive volume $80 \times 40 \times 40$ cm^3 immersed in a 22.5 kgauss magnetic field, three cameras, repetition rate 1 or 2 cycles per second, power consumption approximately 3 megawatts) was exposed during the Spring of this year to a well collimated, low intensity, hardened 5.5 GeV bremsstrahlung beam. The beam intensity was adjusted to yield an average number of 80 equivalent quanta per burst. This corresponds to a production of between 10 and 15 e^-e^+ pairs per picture in the chamber. During these first runs a total of 400,000 pictures were taken. Thus far, 150,000 pictures were analysed. On the average, one photoproduction event is found in every 30 pictures. The quality of the pictures is determined to a large extent by the thickness and material of the windows through which the bremsstrahlung beam has to pass before reaching the liquid hydrogen. The outer window consists of a 0.5 mm thick stainless steel foil. The inner window is a 3 mm thick aluminium plate. This adds up to a total of about $^1/_{15}$ radiation length of window material, which is just as much as the $^1/_{15}$ radiation length of liquid hydrogen along the beam direction. Figure 1 shows a typical bubble-chamber picture with a three-prong event. For cross-section evaluation, the flux and energy spectrum of the photons were determined by measuring electron-positron pairs in the chamber.

Table 2 is a catalogue of the events which have been analysed so far by the Aachen-Berlin-Bonn-Hamburg-Heidelberg-Munich Bubble-Chamber Collaboration[6]. I shall not discuss the one-prong events without strange-particle production, because of the fact that analysis of these one-prong events is possible only if an assumption is made about the missing mass of the unobserved neutral particle or particles. These events are so-called zero-constraint (OC) events. Especially at high-photon energies are the two samples of one-prong events contaminated with events in which one or more additional neutral mesons are produced.

Table 2

Reaction types and number of events
(based on 150,000 pictures)

Reaction	Threshold (GeV)	Number of events
$\gamma + p \to p + \pi^0 (\pi^0 \ldots)$	0.15	1,465*)
$\gamma + p \to n + \pi^+ (\pi^0 \ldots)$	0.15	1,120*)
$\gamma + p \to p + \pi^+ + \pi^-$	0.31	1,556
$\gamma + p \to p + \pi^+ + \pi^- + \pi^0 (\pi^0 \ldots)$	0.51	691
$\gamma + p \to n + \pi^+ + \pi^+ + \pi^- (\pi^0 \ldots)$	0.51	280
$\gamma + p \to p + 2\pi^+ + 2\pi^-$	0.73	28
$\gamma + p \to p + 2\pi^+ + 2\pi^- + \pi^0 (\pi^0 \ldots)$	0.96	47
$\gamma + p \to n + 3\pi^+ + 2\pi^- (\pi^0 \ldots)$	0.96	28
$\gamma + p \to$ strange particles	0.91	124
	Total	5,339

*) Only 45% of film used.

The only three-prong reaction giving a three-constraint (3C) fit is $\gamma p \to p\pi^+\pi^-$, because all kinematical variables can be measured here, except the energy of the photon. Since an abundant sample of 1556 such three-prong events is available, I shall mainly consider this particular reaction. However, a few remarks about the three-pion production reaction $\gamma p \to p\pi^+\pi^-\pi^0$ will be included in this lecture.

Analysis of these two samples of events revealed **that with great likelihood** two or more of the final particles in these reactions are produced in a resonant state. Some of the possible quasi-two-body states which could lead to two or three-pion final states are listed in Table 3. In the reaction $\gamma p \to p\pi^+\pi^-$, copious production of the 1238 MeV isobar, via channel (a) and of the ρ^0 meson via channel (b), is observed. In the sample of three-pion production events ($\gamma p \to p\pi^+\pi^-\pi^0$) production of the ω meson via channel (b) is the most noticeable feature, but ω production never accounts for more than about 20 or 25% of the total cross-section. This is obvious from Fig. 2 where $p\pi^+$, $\pi^0\pi^-$ and $\pi^+\pi^0\pi^-$ invariant-mass plots of three-pion production events ($\gamma p \to p\pi^+\pi^0\pi^-$) are shown for the photon energy range between 1.8 and 5.5 GeV. At photon energies above 2 GeV some $(N^*)^{++}\rho^-$, according to channel (d), is observed. Between threshold and 0.9 GeV there is some η production according to channel (a).

In Fig. 3 the measured total cross-section for the reactions $\gamma p \to p\pi^+\pi^-$, $\gamma p \to p\pi^+\pi^0\pi^-$ and $\gamma p \to n\pi^+\pi^+\pi^-$ are plotted versus the photon energy. Invariant mass plots for the two-pion production events revealed that the reaction $\gamma p \to p\pi^+\pi^-$ is dominated up to photon energies of about 1.2 GeV by the $(N^*)^{++}\pi^-$ final state. As can be seen in Fig. 4 the cross-section for this channel rises rapidly above threshold to a maximum at about 0.75 GeV, and then falls off steeply again. Above photon energies of 1.4 GeV the reaction is dominated by the $p\rho^0$ channel. It is remarkable that these two channels account for almost all of the total cross-section of the reaction $\gamma p \to p\pi^+\pi^-$. The total cross-section for

Table 3

Photoproduction of quasi-two-body states

A. Two-pion final state (3C)		Threshold (GeV)
(a)	$\gamma p \to N^* \pi$	0.54
(b)	$\to p \rho^0$	1.05
(c)	$\to p f^0$	2.09

B. Three-pion final state (0C)		Threshold (GeV)
(a)	$\gamma p \to p \eta^0$	0.71
(b)	$\to p \omega^0$	1.11
(c)	$\to N^* \pi$	1.00
(d)	$\to N^* \rho$	1.66

ω production in the reaction $\gamma p \to p\pi^+\pi^0\pi^-$ is also given in Fig. 4. It turns out that the ω production cross-section is significantly smaller than the ρ^0 production cross-section at all photon energies investigated so far. It must also be pointed out that the ω peak in the $\pi^+\pi^0\pi^-$-invariant mass plot centres at the correct ω-mass value, whereas the ρ mass in the two-pion production events is found to be (729 ± 5) MeV. This value is considerably lower than the commonly accepted ρ-mass value, but the same tendency towards lower ρ-mass values has been observed in other ρ-photoproduction experiments at CEA and DESY.

We now have to ask ourselves what can be learned from these bubble chamber studies about the mechanism by which the three reactions $\gamma p \to (N^*)^{++}\pi^-$, $\gamma p \to p\rho^0$ and $\gamma p \to p\omega$ proceed.

a) $\gamma p \to (N^*)^{++}\pi^-$

It is observed that the negative pions are strongly peaked forward in the γp centre-of-mass system. The small momentum-transfer events contribute most to the $(N^*)^{++}$ production cross-section. This strongly indicates a peripheral mechanism for the production of the $(N^*)^{++}$ isobar, see graph (1)

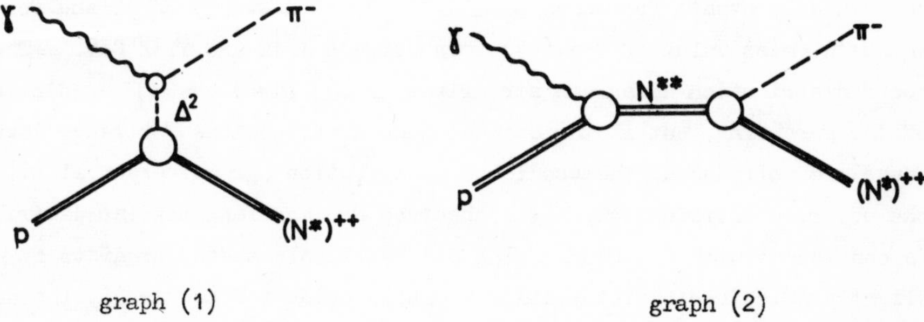

graph (1) graph (2)

On the other hand, the energy dependence and the position of the maximum of the $\gamma p \to (N^*)^{++}\pi^-$ total cross-section may suggest a non-peripheral production mechanism according to which the $(N^*)^{++}_{1238}$ isobar is produced via formation of the 1485, 1512 or 1688 MeV nucleon isobar which must then decay to an appreciable extent via the channel $(N^{**})^+ \rightarrow (N^*_{1238})^{++}\pi^-$ [see graph (2)].

It is interesting to compare the observed cross-section for $(N^*)^{++}\pi^-$ production with the prediction of the Drell one-pion exchange model[7] (OPEM). It is found that at low photon energies the observed cross-section for $\Delta^2 \leq 0.3$ $(GeV/c)^2$ exceeds the pure OPEM cross-section by almost a factor of four. At higher photon energies the discrepancy becomes smaller but remains substantial. Stichel and Scholz[8] have modified the pure OPEM to include corrections for gauge invariance. Their gauge invariant extension of the OPEM amplitude for the process $\gamma p \rightarrow (N^*)^{++}\pi^-$ leads to better agreement with the experimentally observed total cross-sections.

As a further check of the OPEM, the distributions of the $(N^*)^{++}$ decay angles ϑ and φ have been studied. The angle φ is the familiar Treiman-Yang angle between the production plane and the decay plane of the $(N^*)^{++}$ in the $(N^*)^{++}$ rest frame, and ϑ the angle between incident and outgoing proton in the $(N^*)^{++}$ rest frame. The corresponding decay angular distributions are plotted in Fig. 5. The two diagrams contain only events for which $E_\gamma \leq 1.1$ GeV, $\Delta^2 \leq 0.5$ $(GeV/c)^2$ and for which the mass value of the $(N^*)^{++}$ is between 1.12 and 1.32 GeV. The $\cos\vartheta$ distribution disagrees strongly with the $(1 + 3\cos^2\vartheta)$ prediction of the pure OPEM, but if the Stichel-Scholz corrections for gauge invariance are introduced, the predicted distribution agrees very well with the observed distribution. The observed Treiman-Yang angular distribution is consistent with isotropy. The Stichel-Scholz theory predicts a slightly anisotropic distribution. As is evident from Fig. 5, the data are definitely in excellent agreement with these predictions.

b) $\gamma p \to p\rho^0$

The angular distributions in the γp centre-of-mass system clearly show a backward peaking of the protons and a forward peaking of the ρ^0, which become more pronounced at higher photon energies. This suggests a peripheral production mechanism of the ρ^0 mesons. It is interesting to compare the observed four-momentum distributions of events in the region of small squared four-momentum transfers $[\Delta^2 \leq 0.4\ (GeV/c)^2]$ with the predictions of various models. In Fig. 6, $d\sigma/d\Delta^2$ is plotted versus Δ^2 and compared with theoretical predictions of the pure OPEM, a modified OPEM which includes absorption in the final state, and with the prediction of the diffraction approximation of Berman and Drell[9]. According to the judgment of the experts, whose daily business it is to compare experimental data of fairly poor statistical accuracy with theoretical predictions based on even poorer grounds, Fig. 6 favours the diffraction model of Berman and Drell.

The energy dependence of the differential cross-section for ρ^0 production at 0° can also serve as a check of the various models. The pure OPEM predicts $(d\sigma/d\Omega)_{0^\circ} \sim 1/E_\gamma^2$ whereas, according to the diffraction model of Berman and Drell, $(d\sigma/d\Omega)_{0^\circ}$ should rise proportional to E_γ^2. According to what is known so far the differential cross-section for ρ^0 production at 0° seems to increase with increasing photon energy, thus again favouring the diffraction model.

The measured total cross-section at small squared four-momentum transfers, say $\Delta^2 < 0.5\ (GeV/c)^2$, can serve as another check to the predictions of various models. According to the pure OPEM, the total cross-section σ_t is proportional to $f^2_{\pi NN}/4\pi \cdot \Gamma_{\rho\pi\gamma}$ whereas, according to the diffraction model of Berman and Drell, the total cross-section for ρ^0 production in the region of small squared four-momentum transfers should be proportional to $f^2_{\omega\rho\pi}/4\pi \cdot \Gamma_{\omega\pi\gamma}$. With reasonable assumptions for the coupling constants and partial decay widths ($f^2 \pi NN/4\pi \simeq 14$; $f^2 \omega\rho\pi/4\pi \simeq 12$; $\Gamma\omega\pi\gamma \simeq 0.9$ MeV; $\Gamma\rho\pi\gamma \simeq 0.1$ MeV) *), the diffraction model of Berman and Drell gives the correct order of magnitude for the total cross-section, whereas the OPEM would lead to much too small ρ^0 production cross-sections. This is evident from Fig. 7.

The decay angular distribution of the π^- or π^+ in the ρ^0 rest frame, relative to the incident photon direction, represents another check on the predictions of various models. For peripherally produced ρ^0 mesons, the pure OPEM predicts an isotropic distribution of the Treiman-Yang angle and a $\sin^2 \vartheta$ distribution for ϑ, which is the angle between the incident photon and the decay pion in the ρ^0 rest frame. The observed ϑ distribution is isotropic, which is clearly inconsistent with the pure OPEM. I do not know what kind of decay angular distribution is predicted by the diffraction model of Berman and Drell.

c) $\gamma p \to p\omega$

It is observed that the angular distribution of the ω mesons is peaked forward in the γp centre-of-mass system. As can be seen on Fig. 7 one cannot distinguish between the diffraction model and the OPEM on the basis of the measured total cross-section for ω production. It is interesting to notice in Fig. 7 that the total cross-section for ρ^0 production is considerably larger than the corresponding cross-section for ω production at all photon energies. According to the OPEM this ratio should be given by $\Gamma\rho\pi\gamma : \Gamma\omega\pi\gamma$, whereas in the diffraction model of Berman and Drell $\sigma_t(\rho^0) : \sigma_t(\omega) = \Gamma\omega\pi\gamma : \Gamma\rho\pi\gamma$. If one is willing to accept the SU(6) prediction according to which $\Gamma\omega\pi\gamma : \Gamma\rho\pi\gamma = 9:1$, then the observed ratio of $\sigma_t(\rho^0) : \sigma_t(\omega)$ favours the diffraction model.

*) On the basis of the SU(6) prediction according to which $\Gamma\omega\pi\gamma/\Gamma\rho\pi\gamma = 9:1$ the partial width $\Gamma(\rho \to \pi\gamma)$ is assumed to be 0.1 MeV.

Summarizing, we can say: the OPEM seems to be a bad model for the photoproduction of ρ and ω mesons. The OPEM gives a reasonable account of the reaction $\gamma p \to (N^*)^{++}\pi^-$ provided the Stichel-Scholz corrections for gauge invariance are included. For ρ^o and ω photoproduction, the diffraction model of Berman and Drell works better than the OPEM which may be connected with the fact that ρ^o and ω can be coupled directly to a photon.

Finally, at the end of this discussion I would like to mention that all these results are in excellent agreement with the results obtained earlier at CEA with the 12 in. hydrogen bubble chamber[10)].

2. Search for the heavy electron in the mass range between 0.5 and 1.0 GeV

If a heavy electron exists, it should be possible to produce this excited state by inelastic e-p scattering according to the reaction

$$e + p \to p + e^*$$
$$\hookrightarrow e + \gamma \, .$$

One way of searching for the heavy electron would be to scan the recoil proton-momentum distribution. Since the final state is a quasi-two-body state, e* excitation would lead to a peak in the recoil proton-momentum distribution, just as elastic e-p scattering leads to the elastic peak. Other competing reactions, such as electro or photoproduction of one or several pions, cannot give rise to peaks. They lead to a continuous proton-momentum distribution on top of which the e* excitation peak might be observable. If one specifies the incident electron energy, the recoil proton angle and a mass value for the heavy electron, then all other kinematical variables, including the proton momentum, are determined.

A whole range of heavy electron mass values can be scanned by variation of the recoil proton momentum. In this method the sensivity for detecting the heavy electron is rather poor because of the fact that the smallest detectable e* excitation cross-section is determined, to a large extent, by the number of counts in the e* excitation relative to the contribution from underlying background caused by the competing processes. This background can be reduced appreciably by detection of the decay electron in coincidence with the recoil proton. Behrend et al.[11]) used this method at DESY in their search for the heavy electron, to increase the e*-detection sensitivity. They exposed a liquid hydrogen target in a straight section of the synchrotron to the internal electron beam. Two quadrupole spectrometers with appropriate detectors served to detect the recoil proton and decay electron in coincidence. In most of the measurements, the incident electron energy was 4 GeV. For any given mass value, M_e* of the heavy electron, the proton angle φ_p was chosen such that the heavy electron would emerge at an angle of $\vartheta_{e^*} = 10°$, with respect to the incident beam in the laboratory system. All other kinematical variables (for example, proton momentum, momentum of the decay electron) are then uniquely determined. The kinematics are shown in Fig. 8. A heavy electron-mass scale and the corresponding proton angles are given on the abscissa. The proton momentum can be read from the curve marked p_p, and the proton momentum scale on the right. For purely technical reasons, the spectrometer for detection of the decay electron was adjusted to 32° and not to 10°, although at $\vartheta_e = \vartheta_{e*} = 10°$ the probability for acceptance of the decay electron would have been higher. The momentum of the decay electron emitted at an angle of 32°, with respect to the incident beam, can be read from the curve marked $e + p + \gamma$ and the vertical momentum scale on the left. Electrons from electroproduction of pions detected in coincidence with recoil protons have lower momenta. For single pion production events the secondary electron momentum is given by the dotted curve marked $e + p + \pi$ in Fig. 8. If the accepted momentum band of scattered electrons is limited to about 5% (as indicated by the bars in Fig. 8), then electrons from single or multiple production of pions are not accepted, i.e. cannot produce e-p coincidences within the acceptance of the two spectrometers. In the experiment of Behrend et al.[11]), the mass region between 0.5 and 1 GeV was covered by variation of the proton angle between 62° and 54°. The primary energy E_0 and the electron angle ϑ_e were kept constant at 4 GeV and 32°. The recoil proton, electron momenta and angles were chosen according to Fig. 8. As was pointed out before this implies that the e*-production angle was kept constant at 10°. In order to reach an e*-mass value at 1 GeV, the energy E_0 had to be raised to 5.5 GeV and the e* production angle ϑ_{e*} lowered to 7.5°.

If one specifies the mass value of the heavy electron and the e* - e - γ coupling constant λ, one can calculate the differential cross-section $d\sigma/d\Omega_{e*}$ for e* production. For the experiment of Behrend et al. the theoretical cross-sections are given in Table 4 for various heavy electron-mass values between 0.5 and 1.0 GeV and an e* -e - γ coupling constant $\lambda = 1$. For $\lambda \neq 1$, the cross-sections must be multiplied by λ^2. The theoretical cross-sections in Table 4 are based on the one-photon exchange approximation, and on an assumed heavy electron-spin value $I_{e*} = 1/2$. Furthermore, the heavy electron is assumed to have no structure[12].

In the experiment of Behrend et al.[11], no e-p coincidences due to the production of heavy electrons were detected. The experiment yields upper limits for the differential cross-sections for production of a heavy electron of mass M_{e*}. These cross-sections can be expressed in terms of upper limits for λ^2 (the squared e* - e - γ coupling constant) for heavy electron mass values between 0.5 and 1.0 GeV. These are listed in Table 4. It is worthwhile to remember that the decay width $\Gamma_{e^* e\gamma}$ is related to the e* - e -γ coupling constant λ by the equation $\Gamma_{e^*\to e\gamma} = 2\lambda^2 M_{e*}\alpha$, where α is the fine structure constant.

Column 2 of Table 5 gives the order of magnitude λ would have to have if the Pipkin result is to be explained by the existence of a heavy electron of mass M_{e*}[12]. The required order of magnitude is well above the upper limits of λ derived from the e-p scattering experiments at

Orsay[13)] and Hamburg. Therefore, an explanation of the Pipkin results on photoproduction of wide angle electron-positron pairs, based on the assumption of the existence of a heavy electron with a mass between 0.1 and 1 GeV, is ruled out.

Table 4

Results of the search for a heavy electron

$$\left(\frac{d\sigma}{d\Omega_{e*}}\right)_{exp} = \lambda^2 \left(\frac{d\sigma}{d\Omega_{e*}}\right)_{th(\lambda=1)}$$

	p_o = 4 GeV/c ϑ_{e*} = 10°					p_o = 5.5 GeV/c ϑ_{e*} = 7.5°
m_{e*} (MeV)	500	600	700	800	900	1000
$\left(\frac{d\sigma}{d\Omega_{e*}}\right)_{th}$ $\left(\frac{10^{-30}\ cm^2}{sterad}\right)$	7.44	5.32	3.52	2.48	1.84	2.60
$\left(\frac{d\sigma}{d\Omega_{e*}}\right)_{exp}$ $\left(\frac{10^{-33}\ cm^2}{sterad}\right)$	≤ 1.53	≤ 1.07	≤ 0.80	≤ 0.63	≤ 0.52	≤ 1.59
$1/\lambda^2$	≥ 4840	≥ 4960	≥ 4400	≥ 3960	≥ 3560	≥ 1640

Table 5

e^*-e-γ coupling constant as required by the experimental results on photoproduction of wide angle e^-e^+ pairs and e-p scattering

m (MeV)	λ e^--e^+ pair production	λ e-p scattering
120	$0.1 < \lambda < 0.15$	≤ 0.04 *)
200	$0.1 < \lambda < 0.3$	≤ 0.04 *)
400	$0.3 < \lambda < 0.6$	≤ 0.10 *)
500	$0.5 < \lambda < 0.8$	≤ 0.015
700	$0.8 < \lambda < 1.2$	≤ 0.015
900	$1.2 < \lambda < 1.6$	≤ 0.017
1,000	$1.4 < \lambda < 1.9$	≤ 0.025
1,100	$1.6 < \lambda < 2.3$	-
1,500	$2.5 < \lambda < 3.4$	-

*) Orsay measurements, Betourne et al.[13]

3. Electroproduction of pions and photoproduction of dipions

3.1 Electroproduction of pions

Electroproduction of pions and photoproduction of dipions is investigated by Blechschmidt et al.[14] with the apparatus shown in Fig. 9. A 5.2 GeV electron beam with an energy spread of ± 50 MeV and an intensity of about 10^5 electrons/sec hits a liquid hydrogen target of 0.02 radiation length. The production angle of one charged pion and the scattering angle of the electron are measured in two thin-plate spark chambers, SC1 and SC2. The momenta of the two particles are determined by measuring the deflection of the particles in a magnetic field in the two thin-plate spark chambers, SC3 and SC4, behind the magnet. After having passed through holes in the first two spark chambers, SC1 and SC2, the primary electron beam is finally buried in a beam catcher. In a fifth thick-plate spark chamber, electrons are distinguished from heavier particles through shower development.

For triggering the spark chamber, one electron and at least one pion are required. The events arise from single peripheral pion production,

$$e^- p \to e^- \pi^+ n \,, \qquad (1)$$

from peripheral production of a single pion with excitation of a nucleon isobar

$$e^- p \to e^- \pi^\pm (N^*)^{+^0_+} \,, \qquad (2)$$

and from all kinds of multiple pion-production processes. One simple multiple pion-production process would be

$$e^- p \to e^- \pi^+ \pi^- N^+ \,,$$

where the N^+ can be either a proton or an excited positively charged nucleon isobar.

In the experiment, the momenta of the scattered electron and of one charged pion are measured. Therefore, single pion-production processes are kinematically fully determined. They can be analysed completely, whereas multiple pion-production events cannot be analysed completely as long as only one pion is detected.

Thus far, 46 $e^-\pi^+$ and 21 $e^-\pi^-$ events have been observed.

We shall assume that the electron interacts via single photon exchange, as shown in the following graph:

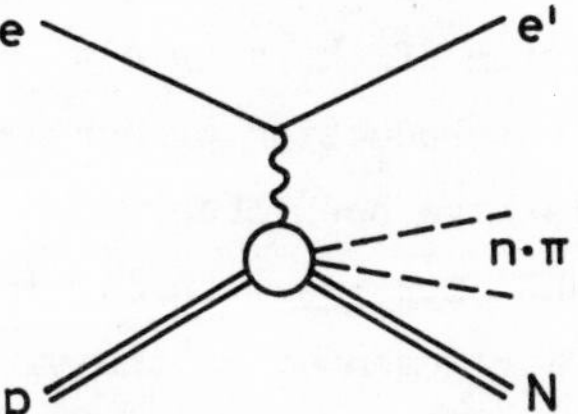

Assuming that the observed events result either from production of a single pion with excitation of a nucleon isobar, $e^-p \rightarrow e^-\pi^{\pm}(N^*)^{+^{0}_{+}}$, see graph (a)

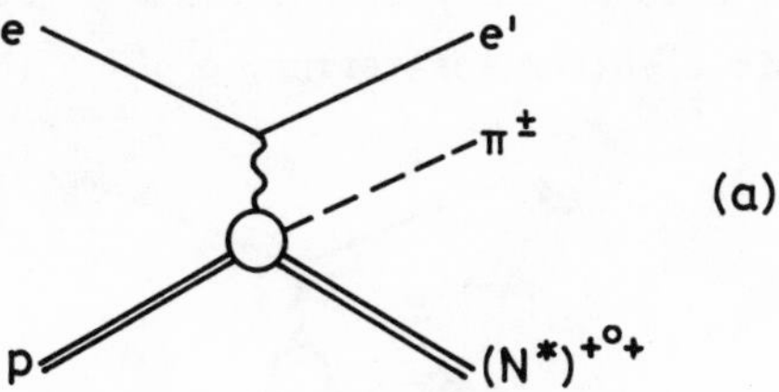

or from single π^+ production, $e^-p \rightarrow e^-\pi^+n$, as shown in graph (b),

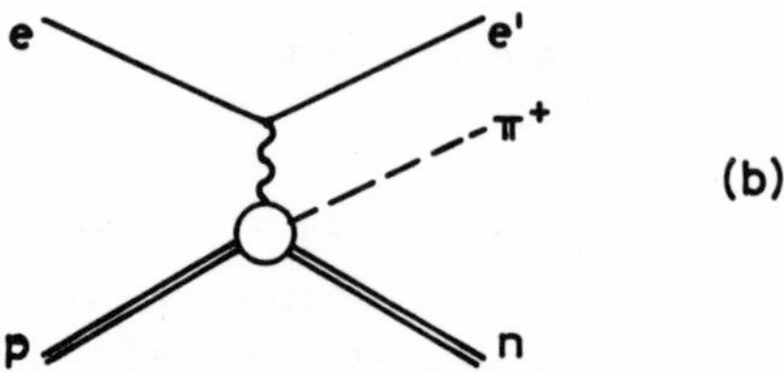

we can calculate the missing mass of the unobserved nucleon or nucleon resonance. The missing-mass resolution in this experiment is ± 60 MeV. Figures 10a and 10b show the resulting missing-mass spectra for the $e^-\pi^+$ and $e^-\pi^-$ production events, respectively. In Fig. 10a the neutron mass is clearly separated from the higher masses. This means that the experiment allows a clear separation of the single π^+ production events, $e^-p \rightarrow e^-\pi^+n$, from other pion production processes. So far, ten such single π^+ production events have been observed. It is desirable to investigate the dependence of the cross-section for these single pion production processes on the squared four momentum of the virtual photon. However, ten events are not sufficient to draw any definite conclusions.

Neither Fig. 10a nor Fig. 10b shows clear evidence for production of a single peripheral pion with excitation of a nucleon isobar. Neither do the data available so far give any evidence for nucleon isobar excitation by inelastic electron scattering, $e^-p \rightarrow e^-(N^*)^+$ [see graph (c)]

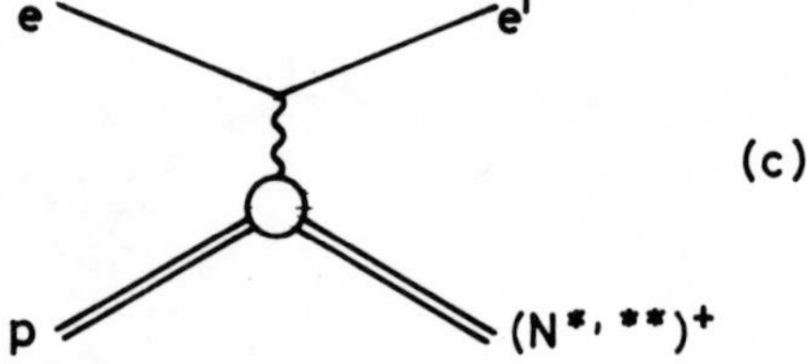

It should also be noted that some of the events, with missing masses of N above the bare nucleon mass, may arise from electroproduction of ρ mesons, as shown in graph (d).

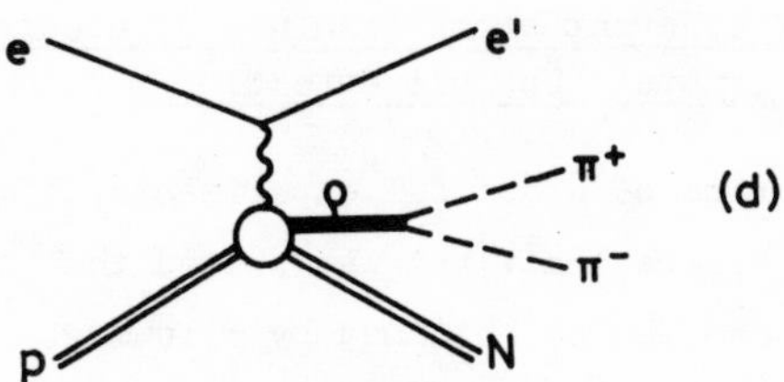

Due to the lack of kinematical information in this experiment, the multiple pion production events cannot be analysed completely. In principle, one cannot distinguish between graph (a) and (d). By improving the statistical accuracy, it is hoped that more can be learned about single and multiple production of pions in e-p collision.

3.2 Photoproduction of dipions

By placing a converter in front of the hydrogen target, photoproduction of dipions can be investigated. The apparatus is shown in Fig. 9. Dipion production events can be analysed completely if the energy of the primary photon is determined. A measurement of the photon energy can be accomplished by the well-known tagging method.

Even without a converter in front of the target, real bremsstrahlung quanta are produced by the electrons within the target itself. Since the apparatus allows the measurement of the momenta of the two charged pions and the opening angle between the two pions, one can calculate, without even knowing the primary photon energy, the effective mass of the $\pi^-\pi^+$ system. Figure 11 shows the effective mass plot for all measured dipion events. The ρ-mass peak is clearly visible around a dipion mass of 720 MeV, with a total width of about 100 MeV. On the basis of these measurements, the total cross-section for photoproduction of ρ^0 mesons in the region between 3 and 5.2 GeV is estimated to be 12 μb. This cross-section value is in good agreement with the cross-section for ρ^0 production derived from the bubble chamber work at CEA and DESY.

4. Production of coherent bremsstrahlung by electrons in a diamond crystal (Überall effect)

In a series of beautiful experiments at the Frascati Synchrotron, Barbiellini et al.[5] were the first who showed that an intense and highly polarized γ-ray beam can be obtained by allowing a high-energy electron beam to strike a suitably orientied single crystal. The shape of the bremsstrahlung intensity spectrum is characterized by typical discontinuities in the form of fairly pronounced peaks. In the Frascati experiments, silicon and diamond single crystal were used as radiators. Similar experiments are being carried out by Timm and collaborators[15] at DESY. They use a diamond single crystal as the radiator. The polarization has not been measured so far, but pronounced spikes have been observed in the bremsstrahlung spectrum when electrons strike the diamond crystal at a small angle with respect to the (110) crystal axis. The diamond single crystal which was used in these experiments had the shape of a $7 \times 4 \times 1$ $(mm)^3$ parallelepiped. The thickness of the crystal along the beam direction is 1 mm, which corresponds to 0.008 radiation length. As can be seen on Fig. 12, the widest face of the crystal has the Miller indices 110. The crystal is mounted in a remotely controlled goniometer placed in a straight section of the synchrotron. Rotation is possible around the horizontal and the vertical axis, both perpendicular to the beam. The crystal axis $(1\bar{1}0)$ is placed parallel to the vertical axis, whereas the (001) axis is placed parallel to the horizontal axis of rotation. The angles of the crystal axis, with respect to the electron beam, can be determined with a precision better than 0.1 mrad. In these measurements, the angle ϑ, between the incident electron beam and the (110) axis, was never greater than 50 mrad; generally, it was much smaller than this. In most measurements the incident beam was either in the (110)(001) or in the $(110)(1\bar{1}0)$ plane. This means that the azimuthal angle, α, between the two planes (001), (110) and $\vec{p}_e(110)$ is either 0 or 90°. However, a few bremsspectra were measured with ϑ relatively large ($\simeq$ 50 mrad), and an azimuthal angle $\alpha = 1.55°$ or $\alpha = 89°$. The advantage of introducing a small deviation of the azimithal angle from 0° or 90° lies in the fact that a particularly high polarization is obtained in the spectral region of the discontinuity which dominates the spectrum. According to numerical calculations by Timm et al., the polarization can be as high as 75% under these conditions, whereas for $\alpha = 0°$ or 90° the theoretical polarization in the first peak is of the order of 50% or lower.

The intensity distributions in the bremsspectra, which were measured by Timm et al. at incident electron energies up to 4.8 GeV, were found to agree well with the theoretical predictions. In these experiments the bremsstrahlung beam was collimated by a collimator having an angle of acceptance of about 0.2 mrad. Since the total natural emission angle of the bremsstrahlung for 4.8 GeV electrons is $2 \times 0.51 \times 10^{-3}/4.8 \simeq 2.13 \times 10^{-4}$, the bremsstrahlung beam is integrated over the emission angle of the photons with respect to the incident electrons. After collimation, the γ-ray beam passes a sweeping magnet, and then strikes the converter of a pair spectrometer in which the resulting e^-e^+ pairs are analysed. A Wilson-type quantameter serves the purpose of monitoring the γ-ray beam intensity.

Figure 13 shows a spectrum as measured by the pair spectrometer. In this particular case, the incident electron energy was 3.8 GeV. The angle ϑ was 5.1 mrad and $\alpha = 0°$. In the first peak, the photons at $k/E_0 \simeq 0.34$ ($E_\gamma \simeq 1.3$ GeV) should have a polarization of 42%. If the angle ϑ is increased from 5.1 to 10.7 mrad, the first peak moves to 2.0 GeV, and is reduced in intensity. The polarization in the peak is then reduced to 33%. The solid and dashed curves in Fig. 13 are theoretical spectra calculated with an exponentially screened Coulomb potential (solid curve) or a Hartree-type potential. The theoretical spectra take into account the divergence in the primary electron beam ($\lesssim 0.1$ mrad), the multiple scattering in the 1 mm thick diamond radiator ($\simeq 0.5$ mrad), the collimation angle of about 0.2 mrad total, and the 2% energy resolution of the pair spectrometer. The energy spread of the incident electron beam was smaller than 0.5% and can therefore be neglected.

REFERENCES

1) F.E. Low, Phys.Rev. Letters 14, 238 (1965).

2) R.B. Blumenthal, D.C. Ehn, W.L. Faissler, P.M. Joseph, L.J. Lanzerotti, F.M. Pipkin and D.G. Stairs, Phys.Rev. Letters 14, 660 (1965).

3) G. Bellettini et al., Physics Letters 18, 333 (1965);
G. Bellettini et al., to be published in Nuovo Cimento;
G. Bellettini et al., Phys.Rev. Letters 3, 170 (1963);
G. Bellettini, Proceedings of the International Conference on Photon Interactions in the BeV-Energy Range, Cambridge, Massachusetts (1963).

4) R.B. Blumenthal, D.C. Ehn, W.L. Faissler, P.M. Joseph, L.J. Lanzerotti, F.M. Pipkin and D.G. Stairs, Phys.Rev. Letters 14, 660 (1965);
R.B. Blumenthal, Thesis, Harvard University (1965).

5) G. Barbiellini et al., Phys.Rev. Letters 8, 112 (1962);
G. Barbiellini, G. Bologna, G. Diambrini and G.P. Murtas, 8, 454 (1962).

6) Aachen-Berlin-Bonn-Hamburg-Heidelberg-Munich Collaboration, DESY report 65/11 (1965).

7) S.D. Drell, Phys.Rev. Letters 5, 278 (1960);
S.D. Drell, Phys.Rev. Letters 5, 342 (1960).

8) P. Stichel and M. Scholz, Nuovo Cimento 34, 1381 (1964).

9) S.M. Berman and S.D. Drell, Phs. Review 133, B 791 (1964).

10) H.R. Crouch et al., Phys.Rev. Letters 13, 636 (1964);
H.R. Crouch et al., Phys.Rev. Letters 13, 640 (1964);
H.R. Crouch et al., Proceedings of the International Symposium on Electron and Photon Interactions at High Energies, Hamburg (1965), to be published.

11) H.J. Behrend et al., DESY report 65/9 (1965).

12) F. Gutbrod and D. Schildknecht, to be published.

13) C. Betourne et al., Physics Letters 17, 70 (1965).

14) H. Blechschmidt et al., DESY report 65/7 (1965).

15) G. Bologna et al., Proceedings of the International Symposium on Electron and Photon Interactions at High Energies, Hamburg (1965), to be published;
U. Timm, DESY reports 64/9 and 65/8;
U. Timm, private communication.

FIGURE CAPTIONS

Fig. 1 : Three-prong event, $\gamma p \to p\pi^+\pi^-$, observed in the 80 cm hydrogen bubble chamber at DESY.

Fig. 2 : Effective mass distributions of (a) $p\pi^+$ and (b) $\pi^-\pi^0$ for $E_\gamma > 1.8$ GeV, and of (c) $\pi^+\pi^-\pi^0$ for $E_\gamma > 1.1$ GeV for the reaction $\gamma p \to p\pi^+\pi^-\pi^0$. The curve in (a) is the phase-space distribution, the curve in (b) is a superposition of phase space and the reflection from ω decay.

Fig. 3 : Cross-sections for (a) $\gamma p \to p\pi^+\pi^-$ and (b) $\gamma p \to p\pi^+\pi^-\pi^0$ and $\gamma p \to n\pi^+\pi^+\pi^-$ as functions of the photon energy.

Fig. 4 : Cross-sections for (a) $\gamma p \to (N^*)^{++}\pi^-$, (b) $\gamma p \to p\rho^0$ and (c) $\gamma p \to p\omega$ as functions of the photon energy.

Fig. 5 : Distributions of the $(N^*)^{++}_{3,3}$ decay angles ϑ and φ for $E_\gamma > 1.1$ GeV $\Delta^2 < 0.5$ $(GeV/c)^2$ and 1.12 GeV $< M_{p\pi^+} < 1.32$ GeV. The curves give the predictions of the OPEM and of the OPEM with the corrections of Stichel and Scholz[8].

Fig. 6 : $d\sigma/d\Delta^2$ for the reaction $\gamma p \to p\rho^0$ for $E_\gamma < 5.5$ GeV. The dashed and dotted curves are the predictions of the OPEM. The solid curve is the prediction of the diffraction model of Berman and Drell[9]. Details are given in the text and in Ref. 6.

Fig. 7 : Cross-sections for (a) $\gamma p \to p\rho^0$ and (b) $\gamma p \to p\omega$ as functions of photon energy for $\Delta^2 < 0.5$ $(GeV/c)^2$. The solid and dashed curves show the predictions of the diffraction model[9] and of the OPEM. For further details see text and Ref. 6.

Fig. 8 : Kinematics for the reaction $e + p \to p + e^* \to p + e + \gamma$, and for various electroproduction processes. Further details are given in the text.

Fig. 9 : Experimental arrangement for the investigation of electroproduction of pions in e-p collisions.

Fig. 10 : Missing-mass distributions M_N for (a) the $e^-\pi^+$ and (b) the $e^-\pi^-$ events derived under the assumption that the events (a) correspond to the reaction $e^-p \rightarrow e^-\pi^+N^o$, and events (b) to the reaction $e^-p \rightarrow e^-\pi^-N^{++}$.

Fig. 11 : Effective mass plot of all measured dipion events.

Fig. 12 : The diamond crystal lattice and diamond single crystal.

Fig. 13 : Photon spectrum for a diamond single crystal oriented at an angle of $\vartheta = 5.1$ mrad and $\alpha = 0$ with respect to the incident electron beam of $E_o = 3.8$ GeV.

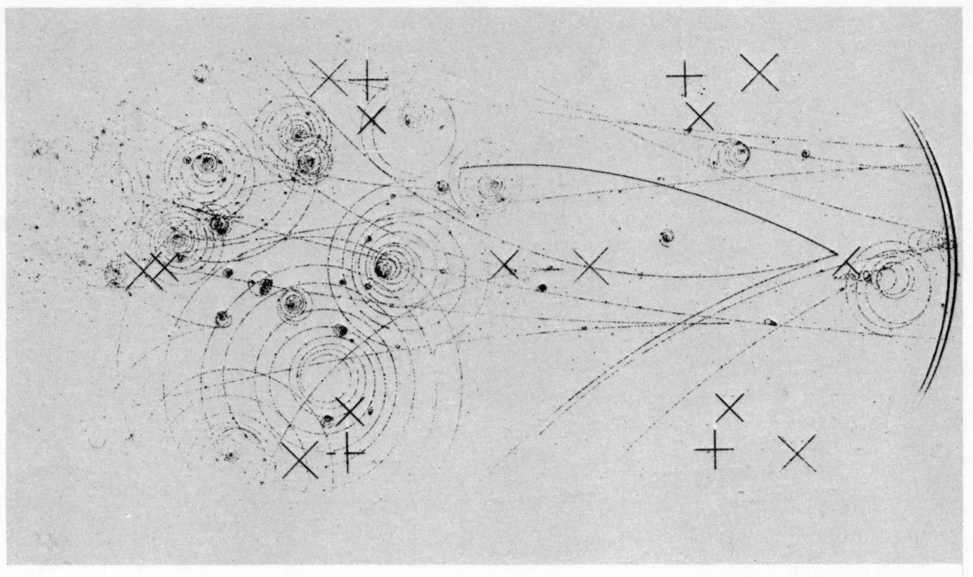

Fig.1

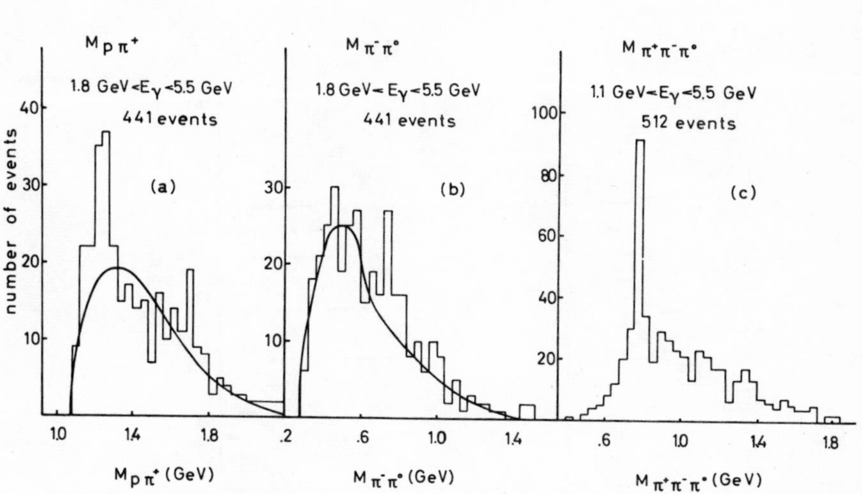
γ p ⟶ p π⁺π⁻π°
M pπ⁺
1.8 GeV<Eγ<5.5 GeV
441 events
(a)
number of events
Mpπ⁺ (GeV)
M π⁻π°
1.8 GeV<Eγ<5.5 GeV
441 events
(b)
M π⁻π° (GeV)
M π⁺π⁻π°
1.1 GeV<Eγ<5.5 GeV
512 events
(c)
M π⁺π⁻π° (GeV)

Fig.2

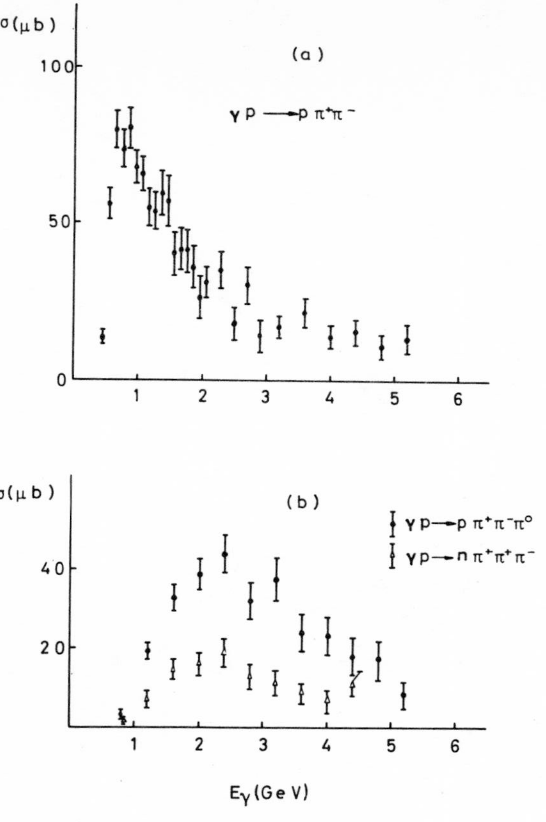
σ(μb)
(a)
γp ⟶ pπ⁺π⁻
100
50
0
1
2
3
4
5
6
σ(μb)
(b)
γp ⟶ pπ⁺π⁻π°
γp ⟶ nπ⁺π⁺π⁻
40
20
1
2
3
4
5
6
Eγ(GeV)

Fig. 3

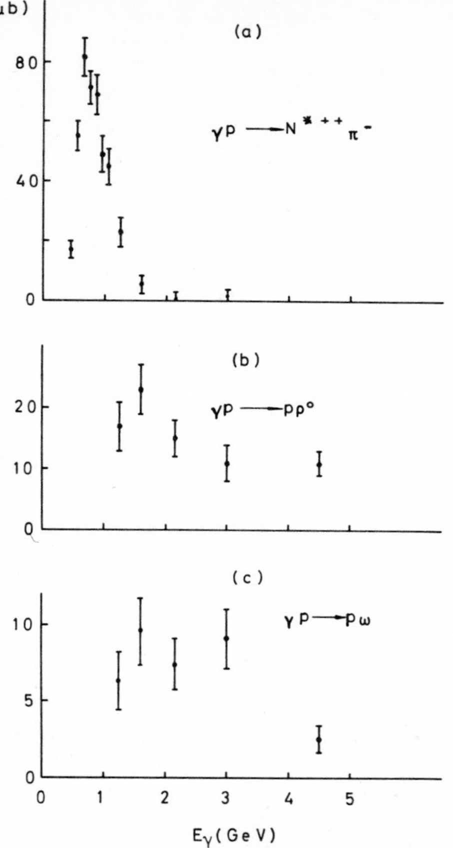
σ(μb)
(a)
γp ⟶ N*⁺⁺π⁻
80
40
0
(b)
γp ⟶ pρ°
20
10
0
(c)
γp ⟶ pω
10
5
0
0
1
2
3
4
5
Eγ(GeV)

Fig. 4

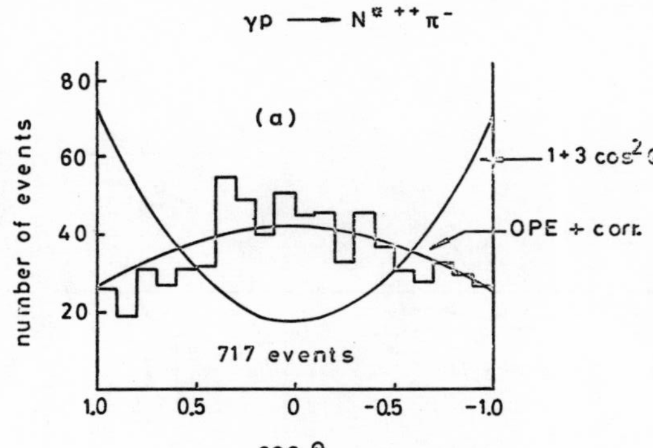
γp ⟶ N*++π⁻
(a)
1+3 cos²Θ
OPE + corr.
717 events
number of events
cos Θ
80
60
40
20
1.0
0.5
0
-0.5
-1.0

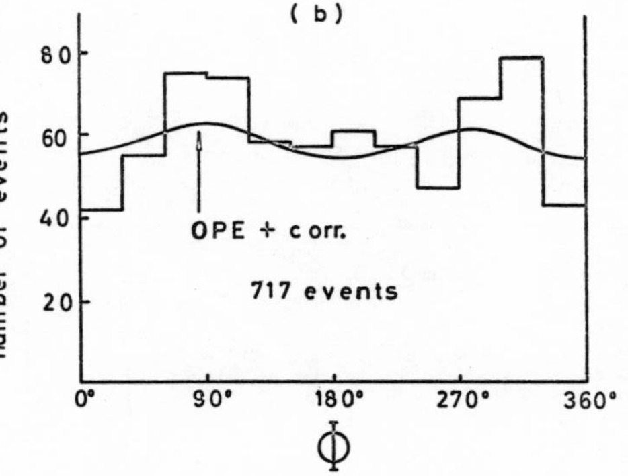
(b)
OPE + corr.
717 events
number of events
Φ
80
60
40
20
0°
90°
180°
270°
360°

Fig.5

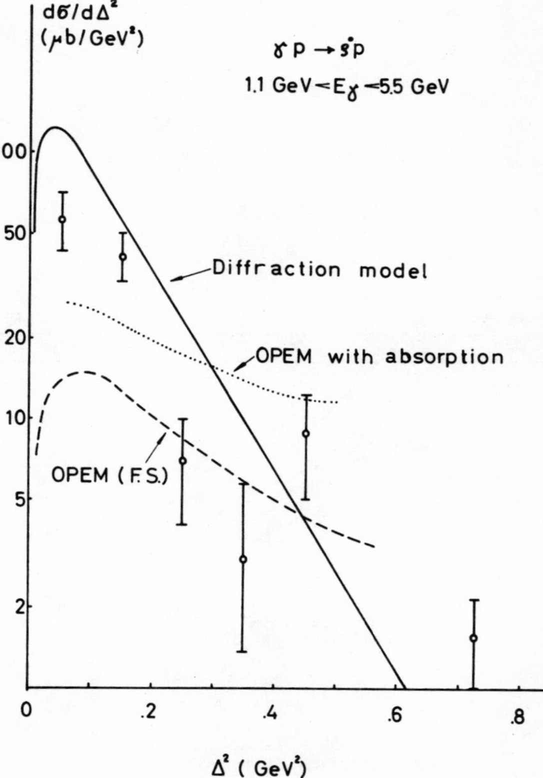
dσ/dΔ²
(μb/GeV²)
γp → ρ°p
1.1 GeV < Eγ < 5.5 GeV
Diffraction model
OPEM with absorption
OPEM (F.S.)
Δ² (GeV²)
100
50
20
10
5
2
0
.2
.4
.6
.8

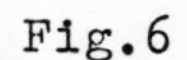
Fig.6

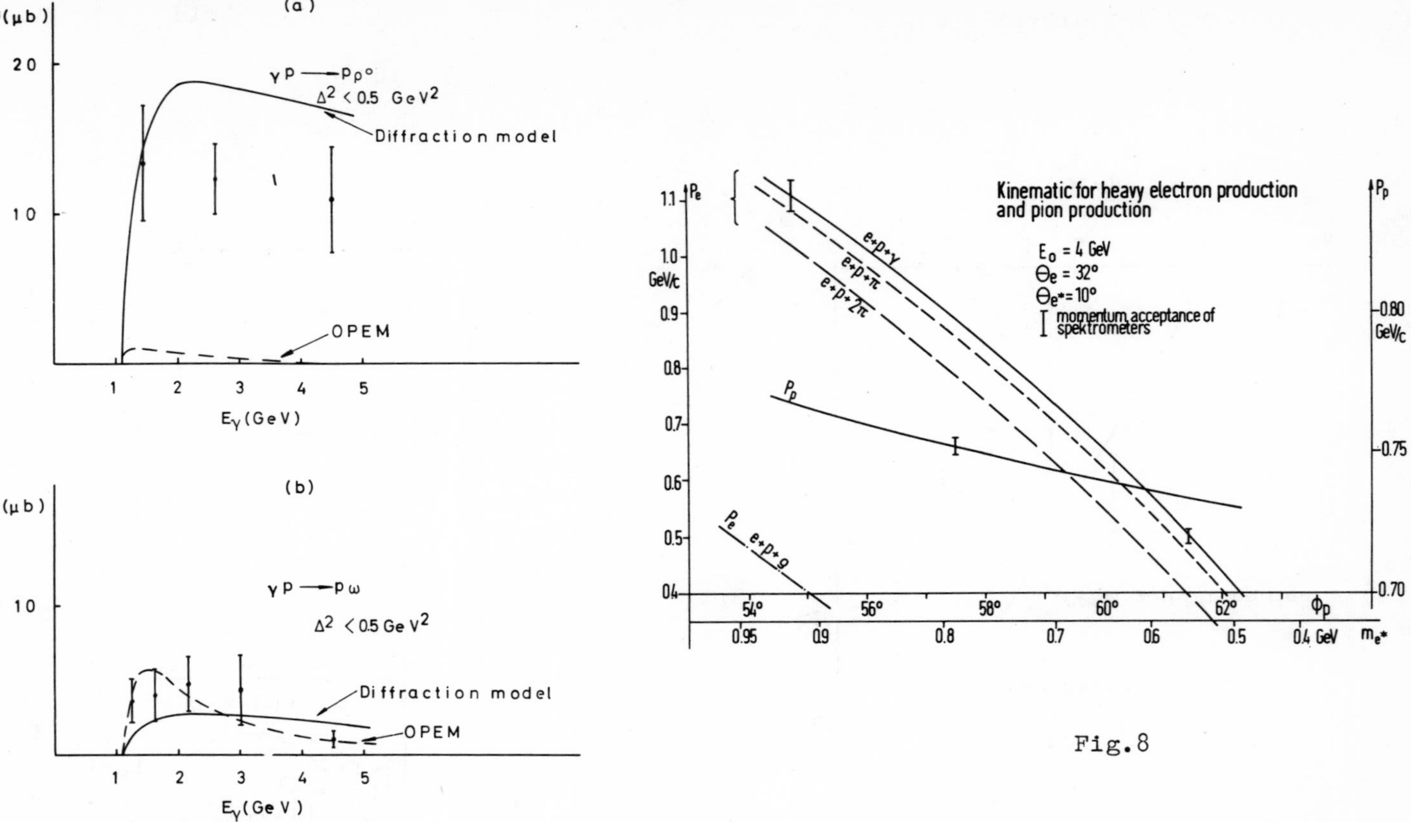
(a)
σ(μb)
20
10
γp ⟶ pρ°
Δ² < 0.5 GeV²
Diffraction model
OPEM
1 2 3 4 5
Eγ(GeV)
(b)
σ(μb)
10
γp ⟶ pω
Δ² < 0.5 GeV²
Diffraction model
OPEM
1 2 3 4 5
Eγ(GeV)
Kinematic for heavy electron production
and pion production
E0 = 4 GeV
Θe = 32°
Θe* = 10°
momentum acceptance of spektrometers
Pe
1.1
1.0 GeV/c
0.9
0.8
0.7
0.6
0.5
0.4
Pp
0.80 GeV/c
0.75
0.70
e+p+γ
e+p+π
e+p+2π
Pp
Pe e+p+g
54° 56° 58° 60° 62° Φp
0.95 0.9 0.8 0.7 0.6 0.5 0.4 GeV me*

Fig.7

Fig.8

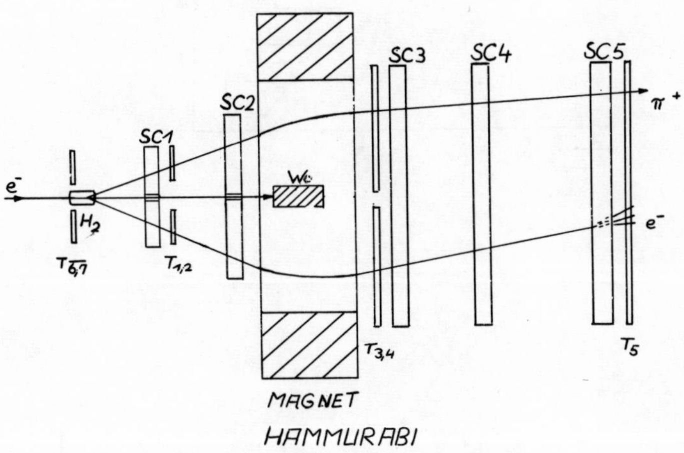
Spark-chamber setup
SC1
SC2
SC3
SC4
SC5
e⁻
H₂
W
π⁺
e⁻
T6,7
T1,2
T3,4
T5
MAGNET
HAMMURABI

Fig.9

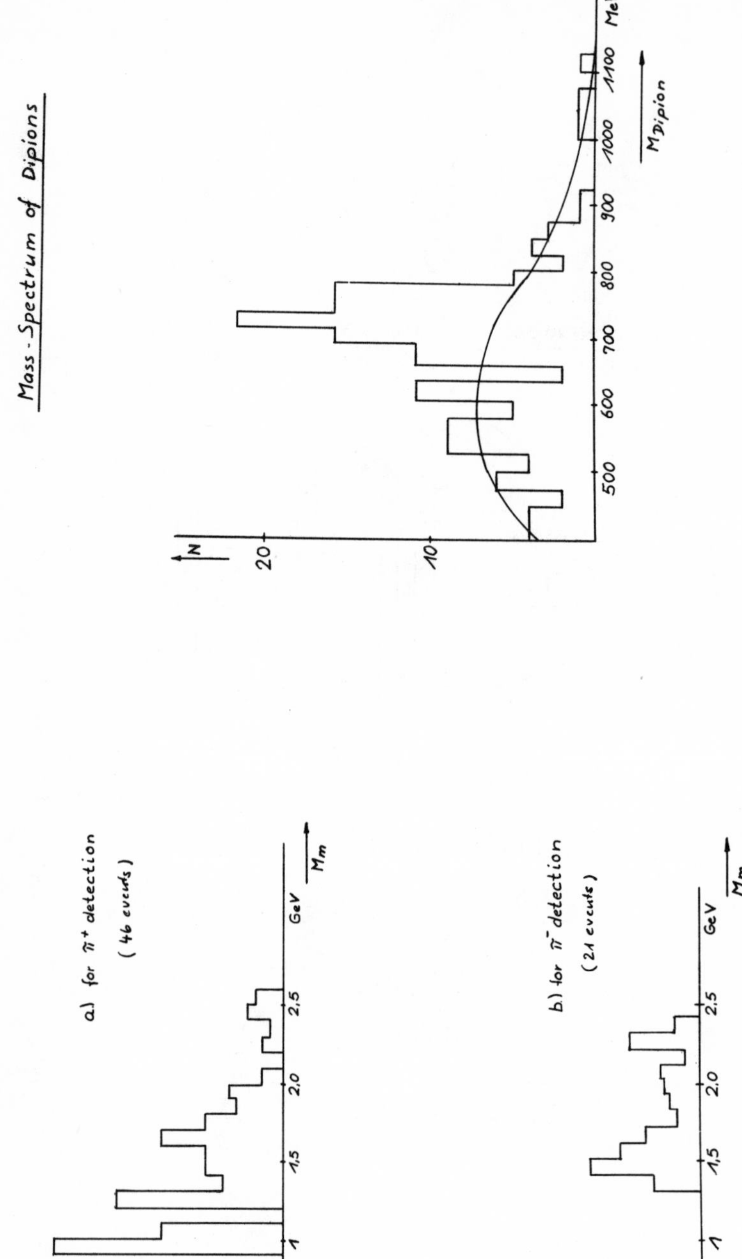

Fig. 10a und 10b

Fig.11

Lage des Targets
im Diamantgitter

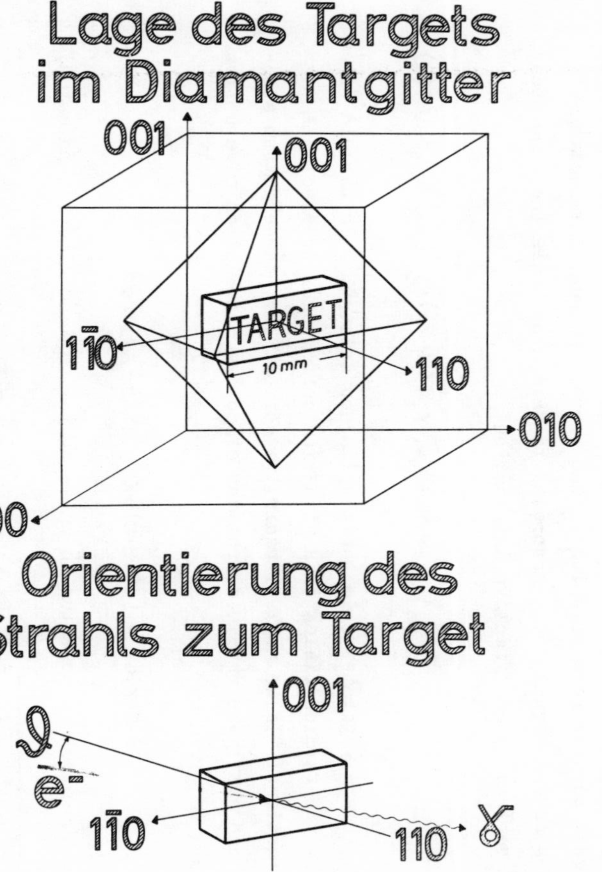
001
001
TARGET
10 mm
1$\bar{1}$0
110
010
100
Orientierung des
Strahls zum Target
001
ϑ
e^-
1$\bar{1}$0
110
γ

Fig.12

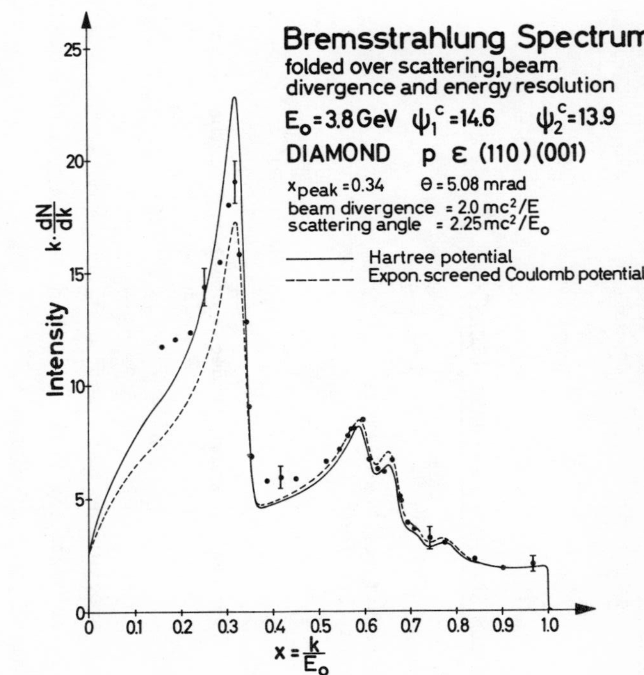
Bremsstrahlung Spectrum
folded over scattering, beam divergence and energy resolution
E_o=3.8 GeV ψ_1^c=14.6 ψ_2^c=13.9
DIAMOND p ε (110) (001)
x_{peak}=0.34 Θ = 5.08 mrad
beam divergence = 2.0 mc^2/E
scattering angle = 2.25 mc^2/E_o
Hartree potential
Expon. screened Coulomb potential
Intensity $k\cdot\frac{dN}{dk}$
$x=\frac{k}{E_o}$
0
0.1
0.2
0.3
0.4
0.5
0.6
0.7
0.8
0.9
1.0
5
10
15
20
25

Fig.13

THE ELECTRON SPECTRUM FROM MUON DECAY*)

J. Lee-Franzini**),
State University of New York at
Stony Brook, New York.

The muon is known to decay according to the reaction

$$\mu^+ \to e^+ + \nu_e + \bar{\nu}_\mu \ . \qquad (1)$$

The energy spectrum of the electron in such a decay has been the subject of many experiments, as well as of many theoretical calculations after the classical paper of Michel[1].

It is our present belief that low-energy weak processes like the muon decay can be described in terms of a local effective interaction; the most popular form of which is the current-current interaction of Gell-Mann and Feynman[2]. According to this hypothesis the muon decay is described by an effective interaction

$$H = \frac{G}{\sqrt{2}} \bar{\psi}_e \gamma_\alpha (1+\gamma_5)\psi_\nu \bar{\psi}_\nu \gamma_\alpha (1+\gamma_5)\psi_\mu \ , \qquad (2)$$

where the only parameter is the coupling constant G. In the old ordering of the fermion fields this situation corresponds to exact V-A interaction. If we neglect radiative corrections, the electron spectrum can easily be obtained from the effective interaction of equation (2) and is given by

$$N(x)dx = [3(1-x) + 2\rho \ (4x/3-1)]x^2 dx \qquad (3)$$

(x is the electron momentum in units of the maximum electron momentum)

---

*) Work supported in part by the United States Office of Naval Research under Contract No. Nonr-266(72).

**) The author gratefully acknowledges the United States National Science foundation for a travel grant, and the New York State Research Foundation for a grant-in-aid.

where ρ is the celebrated Michel parameter and is equal to 0.75 in the current-current local interaction theory.

Small deviations from the spectrum of Eq. (3) can be generated by

i) the interaction is not strictly local;

ii) the interaction is not strictly V-A.

One example of case i) theories, is the assumption that the weak current

$$j_\alpha = \bar{\psi}_\ell \gamma_\alpha (1 + \gamma_5) \psi_\nu \qquad \ell = \mu, e \tag{4}$$

is coupled to a vector-boson field W, whose inverse mass is a measure of the non-locality of the interaction. In this case Eq. (3) is still valid to a very good approximation (~ one part in a thousand) except for the value of ρ which is now given by[3)]

$$\rho = 0.75 + [m_\mu/m_W]^2/3 , \tag{5}$$

where m_μ is the muon mass and m_W is the mass of the vector boson, very often referred to as the intermediate boson W.

Possibility ii) which would destroy the formal beauty of the current-current interaction scheme needs to be rewritten as:

$$\begin{aligned} H = {} & G_V(\bar{\psi}_e \gamma_\alpha \psi_\mu) \left(\bar{\psi}_\nu \gamma_\alpha (1 + \gamma_5) \psi_\nu\right) - \\ & - G_A(\bar{\psi}_e \gamma_\alpha \gamma_5 \psi_\mu) \left(\bar{\psi}_\nu \gamma_\alpha (1 + \gamma_5) \psi_\nu\right) . \end{aligned} \tag{6}$$

If $G_A \neq G_V$, the spectrum of Eq. (3) is modified by the addition of a small term

$$\left[\frac{|G_A|^2 - |G_V|^2}{|G_A|^2 + |G_V|^2} \frac{m_e}{m_\mu} \frac{1 - x}{x}\right] x^2 \, dx . \tag{7}$$

Since the current-current scheme has much supporting evidence, we will ignore this term in the following.

About three years ago there was no information available on the possible existence of a vector boson coupled to the weak current, except that its mass had to be greater than the K-meson mass. For the case $m_W \simeq 500$ MeV we have $\rho = 0.763$ and for $m_W = 1000$ MeV $\rho = 0.753$. The best experimental values for the ρ parameter at that time were

$$\rho = 0.78 \pm 0.025^{4)}$$
$$\rho = 0.751 \pm 0.034^{5)} \qquad (8)$$

giving some indication that a more precise measurement could be relevant to the question of the W meson. It is clear, however, that in order to be sensitive to a mass in the range 0.5 to 1 GeV, an accuracy of a few parts in a thousand was required. With this aim in mind, an experiment was designed and carried out at the Nevis Cyclotron Laboratories by M. Bardon, P. Norton, J. Peoples, A. Sachs and myself[6)].

Traditionally, two methods have been employed to measure the electron spectrum in muon decay; a) The spectrometer method, and b) The visual method.

The spectrometer method

In this method the momentum of the electron is measured by defining the electron orbit by a series of slits in a magnetic field. Counting rate versus magnetic field yields the momentum spectrum. This method has very high momentum resolution and the possibility of collecting a large number of events. There are difficulties, however, connected with scattering at the slits, possible dependence of the effective solid angle on the field setting, and finally the system usually has a very small luminosity.

The visual method

In this type of experiment the muons stop in the sensitive volume of a visual detector (for example, a bubble chamber). The decay electron spirals in the chamber's magnetic field and from measurements of the track the momentum is determined. It is possible in this case to apply rigorous geometrical criteria on the events, thus obtaining a sample for which the detection efficiency versus momentum is well known. The problems connected with this method are:

i) wide momentum resolution (6%) due to multiple scattering in the liquid traversed by the electron;

ii) it is very difficult to collect more than a few tens of thousands of events because of the slow measuring process.

In both methods, however, one has to take into account the amount of matter encountered by the electron in the system. The effect of the matter is threefold:

i) it reduces the electron momentum by ionization, and introduces a spread about the average energy loss[7] (usually referred to as the Landau distribution). Although the average energy loss has been experimentally verified to a good degree of accuracy (1%), the tail of this distribution has not been experimentally tested extensively (~ 10%)[8].

ii) It introduces a further spread in energy by the bremsstrahlung process. The Bethe-Heitler formulation of this effect is probably accurate to 10% only.

iii) It introduces fluctuations on the trajectory through multiple scattering, thus necessitating a detailed knowledge of the resolution function in order to precisely measure the momentum of the electron.

In conclusion, to improve past measurements by an order of magnitude, we need a system which combines the virtues of both methods, i.e. large solid angle, precise momentum determination, geometrical event selection, fast data gathering and, finally, with as little matter in the electron path as possible, and that little amount of matter being strategically placed such that to first order it does not affect the momentum resolution. The resulting experimental set-up is shown in Fig. 1. It consists of four single-gap thin sonic spark chambers placed inside a 1 m diameter magnet. The field is ~ 6.6 kgauss, and shimmed with current coils so that it is uniform to 1/2000 on the average, and at worst to 1/1000. A π^+ beam is incident along the field and is stopped in the 3 mm thick target counter. The system is triggered by a stopping π followed by an emerging positron curving through the four sonic spark chambers, and giving a count in the pair of counters following chamber IV. The momentum is essentially given by the separation of the sparks along the line through chambers I and II, corrected by the cosine of the angle at which the track crosses the line. Note, there is vacuum between chambers I and II, and that most of the material is at I or II, where to first order it does not affect the momentum measurement. Chamber III is used to measure the angle, chamber IV excludes tracks which have scattered. The pulse height from the target counter is recorded to determine the energy lost in the counter by the emerging positron. The sonic spark-chamber system, which is capable of accepting up to 20 events per second, has been described before in detail in the literature[9]. The accuracy with which spark positions are determined is indicated by the measured r.m.s. deviations from a straight line, which was ± 0.3 mm. The momentum bite of this modified 180° spectrometer is approximately 35% of the maximum accepted momentum, so that with two field settings we can examine half of the electron spectrum with ample overlap of the two regions. In the experiment we have recorded a total of 1.3×10^7 events at six different field settings. The absolute accuracy and the resolution of the momentum measurement can be determined from the sharp fall-off of the spectrum near the end point. For events associated with a given range of energy loss in the target counter (i.e. pulse height), the fall-off is consistent with an integrated gaussian with standard deviation of 0.14 MeV. This effective resolution of the system is caused primarily by the statistical uncertainty in the pulse height, and by multiple scattering in the thin windows and foils.

Considering the spectrum at 6.62 kgauss for each region of pulse height, the mid-point of the fall-off can be plotted versus pulse height (Fig. 2). The straight line through such points can be extrapolated to zero pulse height, indicating the measured momentum of 52.62 0.02 MeV/c for a positron starting at the upper edge of the target counter. Adding the correction of 0.19 MeV/c for momentum lost in material after the target and for the effect of the shape of the spectrum, a value of 52.81 ± 0.02 MeV/c is obtained. This should be compared with the maximum momentum of 52.83 MeV/c for the theoretical spectrum, assuming zero rest mass for the neutrinos. Alternatively, one can use the 0.02 ± 0.02 MeV/c difference to set a preliminary "upper limit" of five electron masses for the muon neutrino.

The data from two of the runs were analysed and are shown in Fig. 3. The smooth curve is the theoretical spectrum (with $\rho = 0.75$) of Eq. (3) plus the radiative corrections as given by Kinoshita and Sirlin[10]. In addition, the radiation loss distribution (corresponding to the 0.005 of radiation length traversed by the positron) and the Landau distribution of the ionization loss after the target counter (0.19 MeV) have been folded in. A χ^2 fit to the data gives a best value

$$\rho = 0.747 \pm 0.005 . \qquad (9)$$

This result is a preliminary one based on 0.8×10^6 events which correspond to a statistical error of 0.002. The quoted error reflects the uncertainty in the estimate of the remaining corrections.

A similar experiment using wire chambers has been performed by the Telegdi group at Chicago, with more limited statistics, and they have reported a value of 0.746 ± 0.01[11].

REFERENCES

1) L. Michel, Proc.Phys.Soc(London) A63, 514 (1950).

2) R.P. Feynman and M. Gell-Mann, Phys.Rev. 109, 193 (1958).

3) T.D. Lee and C.N. Yang, Phys.Rev. 105, 1671 (1957);
T.D. Lee and C.N. Yang, Phys.Rev. 108, 1611 (1957);
T.D. Lee and C.N. Yang, Phys.Rev. 119, 1410 (1960).

4) R.J. Plano, Phys.Rev. 119, 1400 (1960).

5) M. Block, E. Fiorini, E. Kikuchi, G. Giacomelli and S. Ratti, Nuovo Cimento 23, 1114 (1962).

6) M. Bardon, P. Norton, J. Peoples, A.M. Sachs and J. Lee-Franzini, Phys.Rev. Letters 14, 449 (1965).

7) L. Landau, Nuclear Phys. 3, 127 (1957).

8) E.L. Goldwasser, F.E. Mills and A.O. Hanson, Phys.Rev. 88, 1137 (1952).

9) M. Bardon, J. Lee, P. Norton, J. Peoples and A.M. Sachs, Informal Meeting on Film-less Spark Chamber Techniques and Associated Computer Use, CERN, Geneva (1964) 64- 30, p. 41 (to be published).

10) T. Kinoshita and A. Sirlin, Phys.Rev. 113, 1652 (1959).

11) V. Telegdi, Argonne Weak Interaction Conference, 1965 (to be published).

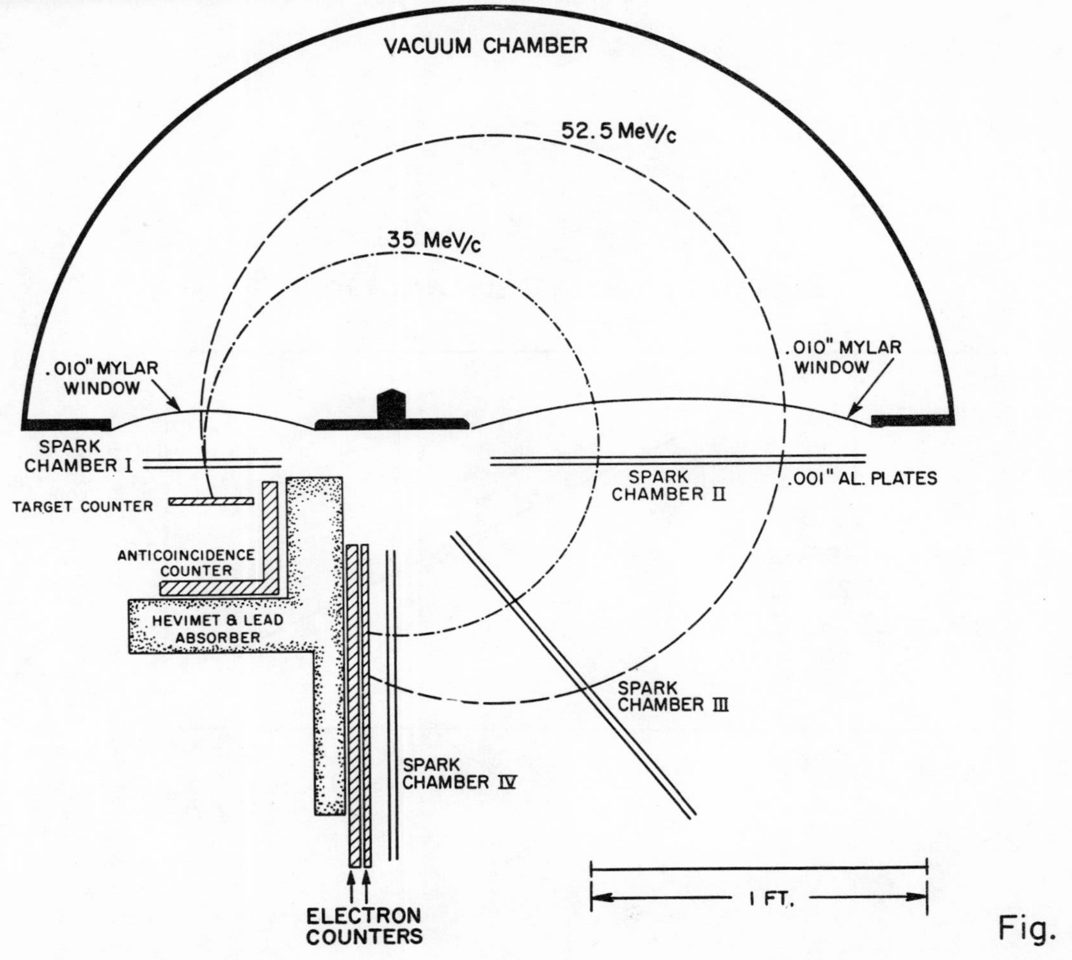
VACUUM CHAMBER
52.5 MeV/c
35 MeV/c
.010" MYLAR WINDOW
.010" MYLAR WINDOW
SPARK CHAMBER I
SPARK CHAMBER II
.001" AL. PLATES
TARGET COUNTER
ANTICOINCIDENCE COUNTER
HEVIMET & LEAD ABSORBER
SPARK CHAMBER III
SPARK CHAMBER IV
ELECTRON COUNTERS
1 FT.

Fig. 1

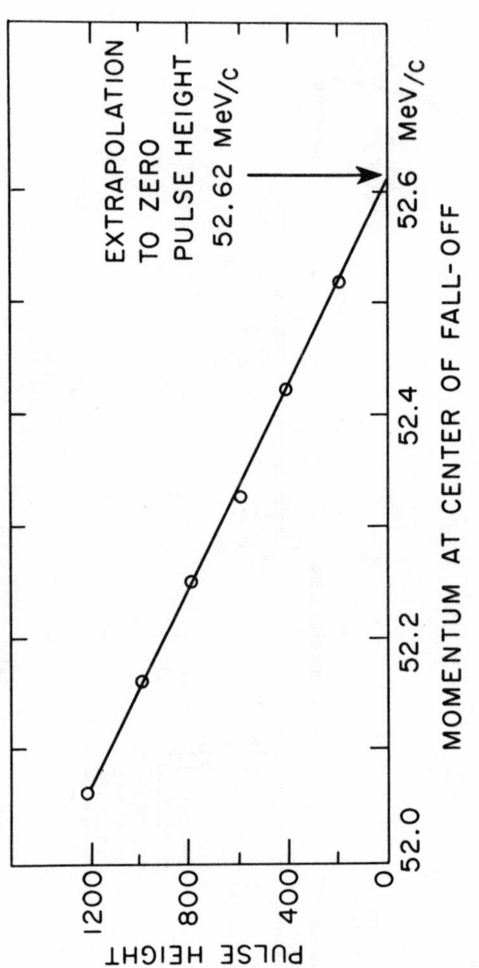
EXTRAPOLATION
TO ZERO
PULSE HEIGHT
52.62 MeV/c
PULSE HEIGHT
1200
800
400
0
52.0
52.2
52.4
52.6
MeV/c
MOMENTUM AT CENTER OF FALL-OFF

Fig. 2

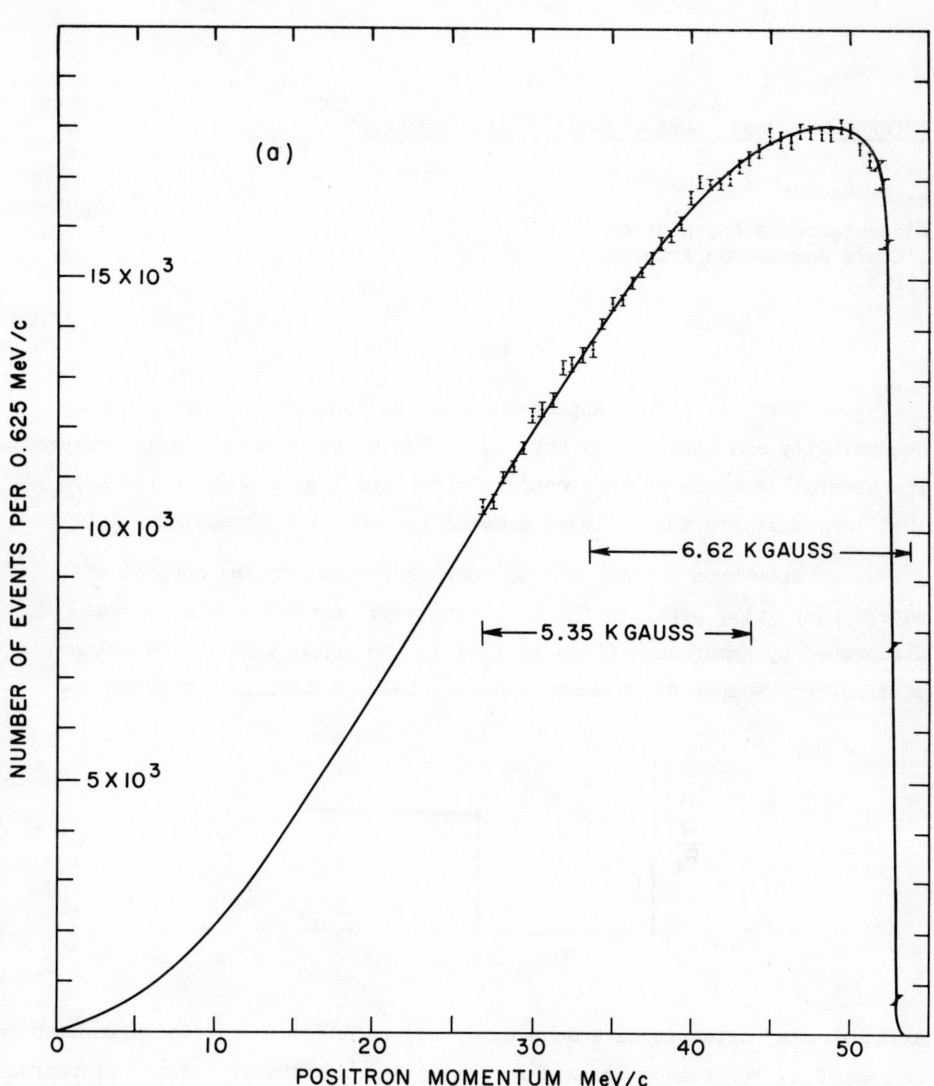
(a)
15 X 10^3
10 X 10^3
5 X 10^3
NUMBER OF EVENTS PER 0.625 MeV/c
6.62 K GAUSS
5.35 K GAUSS
0
10
20
30
40
50
POSITRON MOMENTUM MeV/c

FIG. 3

SUPERCONDUCTORS: SUPERCONDUCTING AND OTHERWISE*)

L. N. Cooper**),

Laboratoire de Physique de
l'Ecole Normale Supérieure,
Paris.

I. SUPERCONDUCTING

There is little apparent technical connection between superconductivity and high-energy physics. The price of my admission however, was several lectures on superconductivity, and I give them in the hope that they will provide at least a moral for particle physicists.

Once upon a time, not so long ago, talks on the subject of superconductivity began in the following way: superconductivity was discovered by Kamerlingh Onnes in 1911[1]; he found that the resistance of a mercury sample disappeared suddenly below a critical temperature.

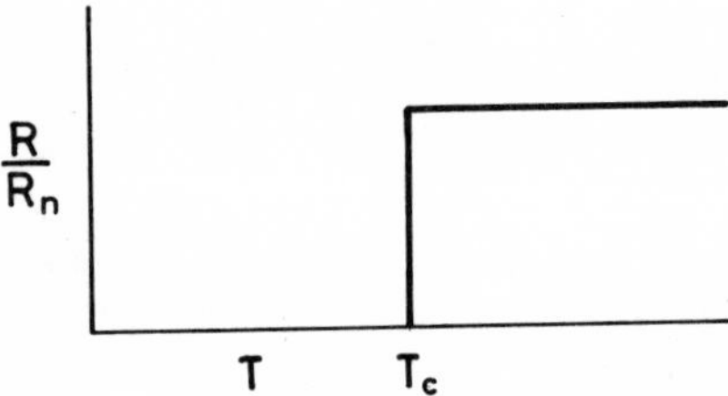

Although many attempts have been made, no one has even succeeded in measuring any d.c. resistance below this critical temperature. Soon afterwards Onnes discovered that relatively small magnetic fields destroy superconductivity. This critical field is a function of temperature, as below.

---

*) This work was supported in part by the United States National Science Foundation and Atomic Energy Commission.

**) Alfred P. Sloan Fellow; permanent address, Brown University, Providen[ce]
Rhode Island.

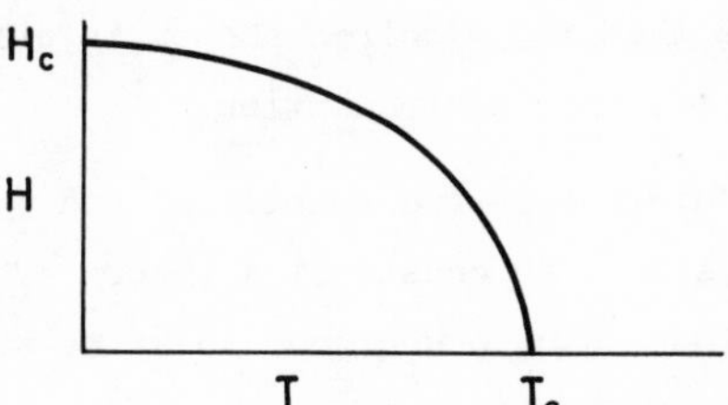

In 1933 Meissner and Ochsenfeld[2)] discovered what has come to be known as the Meissner effect: below the critical temperature, the magnetic field is expelled from the interior of the superconductor.

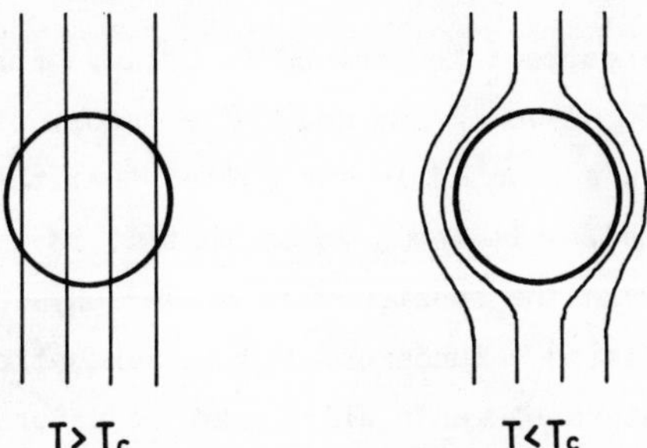

In addition, the electronic specific heat increases discontinuously at T_c and then vanishes exponentially near $T = 0$. This and other evidence indicates the existence of an energy gap in the single particle excitation spectrum.

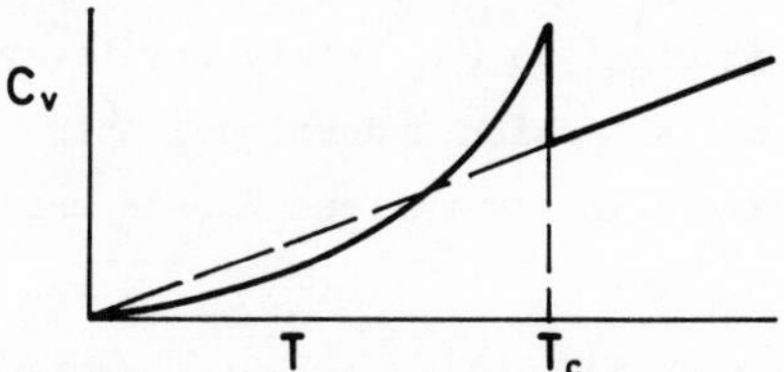

And it has recently been discovered[3)] that the transition temperature varies with the mass of the ionic lattice as

$$\sqrt{M}\, T_c = \text{constant}. \qquad (I.1)$$

This discovery indicates that the electron-phonon interaction is somehow responsible for the superconducting transition.

All of these things are true sometimes, and by focusing our attention on them we were able to construct a theory of superconductivity. However, in discussing these same phenomenon today we would be obligated to say that superconductors show no d.c. resistance except in certain instances when they do; type I superconductors show a Meissner effect, while type II superconductors do not; some superconductors show a specific heat curve, like the one above, while others do not; some superconductors seem to have no energy gap and others show no isotope effect.

Was it then incorrect to pretend to believe what we heard and what we ourselves said? I would say no. The theory of superconductivity which we shall discuss was created in the belief that these were the phenomena to be explained. It seems to me that it is that belief which is of significance, for in the construction of any physical theory there is first involved a decision - fundamentally an aesthetic decision - as to what are the qualitative elements which must be understood. Nature is complicated enough so that it is usually impossible to explain everything at once. Therefore, in attacking any problem, and I think this is true from the problem of the planets to the problem of the atom, one attempts to isolate first the essential qualitative features. These must be simple enough so that the mind can encompass them (and there is a risk, a risk that what has been isolated might be completely irrelevant). If the right choice has been made and the situation understood, then the structure created may support that marvellously various facade which is the face of reality.

In what follows I would like to describe how such a decision came about in constructing a theory of superconductivity, and later discuss some of the recent developments in this field. First, I have to say a few things about electrons in a metal. In general, the situation is very complicated. However, it is possible if one is not overwhelmed to construct a relatively simple and successful single particle model.

In the Bloch theory[4] of the normal metal, the conduction electrons described by single particle wave functions are independent of one another. Bloch's theorem states that in the periodic potential produced by the fixed lattice and the conduction electrons themselves, the single electron wave functions will be modulated plane waves,

$$\varphi_\kappa(\underline{r}) = u_\kappa(\underline{r}) e^{i\underline{k}\cdot\underline{r}} , \qquad (I.2)$$

where $u_\kappa(r)$ is a two-component spinor with the lattice periodicity. We use κ to designate simultaneously the wave vector $\underline{k}$, and the spin state σ: $\kappa \equiv \underline{k},\uparrow$; $-\kappa \equiv -k\downarrow$. The single particle Bloch functions satisfy a Schrodinger equation

$$\left[-\frac{\hbar^2}{2m}\nabla^2 + V_0(r)\right]\varphi_\kappa = \mathcal{E}_k\varphi_\kappa , \qquad (I.3)$$

where $V_0(r)$ is the periodic potential and, in general, might be a linear operator to include exchange terms. The single particle energies, $\mathcal{E}_k$ display discontinuities at band edges, but for simplicity we can consider the conduction band to be only partly filled so that the Fermi surface is far from any band edge. Although it is not essential we often use the 'free electron' or plane-wave approximation with an effective mass m*:

$$\begin{aligned} \varphi_\kappa &= u_\kappa e^{i\underline{k}\cdot\underline{r}} \\ \mathcal{E}_k &= \frac{\hbar^2 k^2}{2m^*} \quad . \end{aligned} \qquad (I.4)$$

In essence this amounts to replacing $V_0(r)$ by a momentum dependent potential which is then approximated by an effective mass.

The Pauli exclusion principle requires that the many electron wave function be antisymmetric in all of its co-ordinates. This means that no two electrons can be in the same Bloch state, or that the many electron wave function can be written:

$$\Phi = \frac{1}{\sqrt{N!}} \sum_{\substack{\text{permutations of} \\ \kappa_1 \,\ldots\, \kappa_N}} (-1)^P \varphi_{\kappa_1}(r_1)\varphi_{\kappa_2}(r_2) \,\ldots\, \varphi_{\kappa_N}(r_N) \;. \quad (I.5)$$

The energy of the entire system is then

$$W = \sum_{i=1}^{N} \mathcal{E}_i \;, \quad (I.6)$$

where $\mathcal{E}_i$ is the Bloch energy of the ith single electron state. The ground state for the system is obtained when the lowest N Bloch states are "filled" by single electrons, which can be pictured in momentum space as the filling in of a Fermi sphere. In the ground-state wave function there is no correlation between electrons of opposite spin, and only a statistical correlation of electrons of the same spin. (The only way that electrons are correlated with others is through the general anti-symmetry requirement on the total wave function.)

The single-particle excitations are given by wave functions identical to the ground state except that a one-electron state $k_i < k_f$ is replaced by another $k_j > k_f$. This may be pictured in momentum space as opening a vacancy below the Fermi surface and placing an excited electron above. The energy difference between the ground state Φ_0 and the excited state $\Phi_{i,j}$ is

$$\mathcal{E}_j - \mathcal{E}_i = \epsilon_j - \epsilon_i = |\epsilon_j| + |\epsilon_i| \;, \quad (I.7)$$

where for later convenience we define ϵ as the energy measured relative to the Fermi energy, $\epsilon_i = \mathcal{E}_i - \mathcal{E}_F$.

This seemingly simple theory already represents a great advance in answering such questions as why ordinary metals conduct at all. In the earliest free-electron theories, the electron was visualized as drifting along, let us say accelerated by an electric field, till it hit something. How far should it drift before it hit? One would guess approximately, 10^{-8} cm, since this is the distance between ions. The mean-free-path of electrons is orders of magnitude larger however; this was not understood until the problem could be handled with the quantum theory and the electron treated as a wave.

The Bloch wave propagates through a periodic lattice with an infinite mean-free-path. To understand resistance, which disappears in a superconductor, one then considers deviations from the perfectly periodic lattice. The electrons, for example, scatter from impurities, and one can calculate scattering matrix elements and find that the resistivity is proportional to the impurity content. The electron may also collide with a vibrating ion and scatter into a different momentum state, and this also produces resistance. It requires an electric field to keep the electrons moving in one direction, so it became fixed in peoples' minds that the fact that the ions were vibrating, and the fact that the electrons interact with the ions, caused resistance. Further, the electrons interact with each other because of the Coulomb force, which gives an average energy of 1 eV/atom. The average energy in the superconducting transition may be estimated from T_c as 10^{-8} eV/atom. Thus, the qualitative changes which occur in passing the transition temperature involve an energy change orders of magnitude smaller than the Coulomb energy. The Coulomb interaction is present in every metal; some of them appear to become superconductors, and some do not, so one is led to guess that this interaction, even with its large energy, is somehow irrelevant.

If one attempts to construct the simplest possible model of metal, what is left? One has the system of N Fermions in a container of volume Ω. They must satisfy Fermi statistics and so fill up a Fermi sphere.

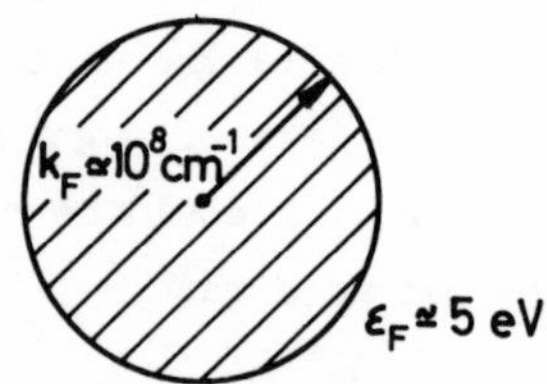

This is a system of free electrons which will be normal, certainly not superconducting, so this is not enough. We need something else, perhaps the interactions of the electrons with each other. There are many such interactions, some of them very complicated. This was why the isotope effect was so significant. We see that as the mass of the ion becomes infinite, the transition temperature goes to zero and we have no superconductor. Since the transition temperature varies with the ionic mass, this suggests that the dynamics of the ions is somehow involved in the transition. The transition into the superconducting state is somehow related to the electron-phonon interaction, the interaction which produces resistance.

In 1950 Frölich[5)] and Bardeen[6)] pointed out that since electrons could interact with lattice vibrations to produce resistance, the electrons would also interact with virtual lattice vibrations to produce an electron self-energy. This they observed would be proportional to an average phonon energy squared giving the isotope effect. However, the self-energy did not produce a phase which had the properties of a superconductor.

It is apparently true that it is the interaction of the electrons with the lattice vibrations which produces superconductivity. But it is an interaction between two electrons via the exchange of a phonon which does this. When an electron collides with a phonon it may be scattered and this produces resistance. At T = 0 when there are no lattice vibrations and the phonon part of the resistivity goes to zero, it is still possible for the electron to excite a virtual phonon which, interacting with another electron, produces an electron-electron interaction.

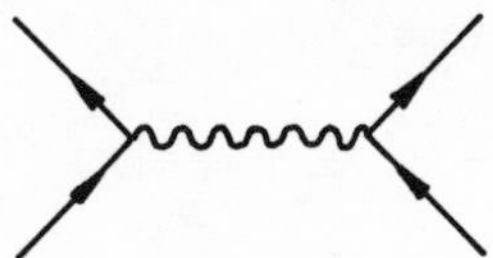

This came as a shock. People still ask how it is possible that there be an interaction of this sort without phonons actually being present. I have a list of classical examples to illustrate this point, with which I will not bore you, since I presume everybody here knows how boson exchange can produce a force between two fermions. The interaction via phonon exchange turns out to depend on the relative phases of the electrons. Most of the interesting things which go on in metal occur near the Fermi surface, and it turns out that the phase of the interaction is such that the interaction is attractive, roughly within a region $\hbar\omega$ of the Fermi surface, and so you can eliminate the phonon field and replace it by an effective potential. This potential is velocity dependent but the details turn out not to be terribly relevant. All that is important is that you can get an attractive effective potential due to phonon exchange.

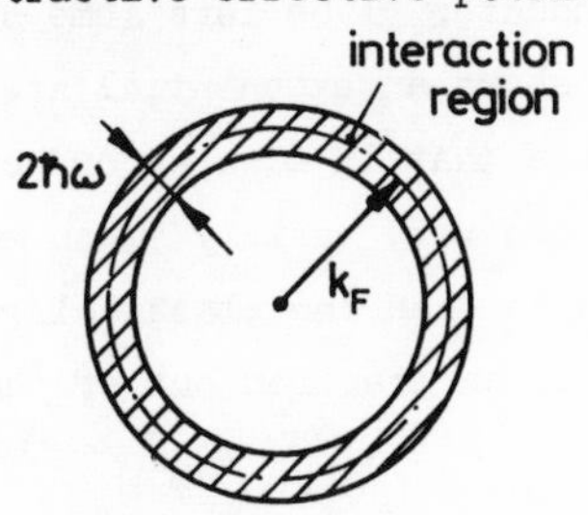

The interesting thing is that this effective potential is attractive as opposed to the Coulomb force which is repulsive. Thus, it becomes possibl that two electrons might actually attract one another.

Let us return for a moment to a consideration of the specific heat. The specific heat tells us the number of degrees of freedom that are available to absorb a given amount of energy. The principle of microscopic egalitarianism guarantees that each will get an equal share which is $\frac{1}{2}k_B T$. For the classical gas of N particles there are 3N degrees of freedom. Therefore,

$$E = \tfrac{3}{2}\, Nk_B T \tag{I.8}$$

and

$$C_v = \left(\frac{\partial E}{\partial T}\right)_v = \tfrac{3}{2}\, Nk_B \; . \tag{I.9}$$

The ideal Fermi gas at low temperatures has many fewer degrees of freedom. The electrons at the bottom of the Fermi sphere require too much energy for excitation (at low temperatures) and, thus, are unavailable. Approximately the number of degrees of freedom for an ideal Fermi gas is about $3 \cdot k_B T \cdot 2N(0)$, where $N(0)$ is the density of states per unit energy of one spin at the Fermi surface. The specific heat crudely is then

$$C_v \simeq 6N(0)\, k_B^2\, T \; . \tag{I.10}$$

(The correct value is $C_v = \frac{2\pi^2}{3} N(0) k_B^2 T$.)

The ideal superconductor (I believe some still exist) which was mentioned at the beginning, shows an exponential specific heat at low temperatures. This indicates that in a superconductor the number of degrees of freedom goes to zero very rapidly, much more rapidly than in a Fermi gas, already more rapid than the classical gas. It suggests that it is difficult to excite an electron out of the Fermi Sea. This, among other things, led to the notion of an energy gap. Somehow, in some way not understood, near the Fermi surface a small but volume independent amount of energy is required to get an excitation. Thus, some suggested that what produced the phenomenon of superconductivity was an energy gap for the excitation of an electron. Others suggested otherwise; this resulted in a certain amount of not too dispassionate discussion at that time when it was not at all clear what was relevant.

If this is reminiscent of some of the things you go through, be reassured; it is when matters are not clear that everyone is most insistent about what the right road is. In one of his parables[7] Kafka tells us that in the day of Alexander the imperial sword pointed the way to India; the armies possibly might never reach India, but the direction was clear. Nowadays, he says sadly, there are many swords and they point in all directions. Not only is the passage impossibly difficult, but the way to begin is completely obscure.

Suppose we believe that the essence of the matter is that somehow the electrons in a metal interact relatively weakly due to the exchange of phonons, and somehow this interaction produces an energy gap. How can this come about? How does one calculate the ground state of such a system? You might use perturbation theory, but you find that the ground state calculated looks nothing like the superconducting ground state. It seems clear that the ground state must be qualitatively different from the free particle ground state. Still, if one wants non-perturbation solutions it is not easy to see which direction to take.

To some of us the heart of the matter seemed to be that the density of states, N(0), is very large.

$$\frac{N(0)}{\Omega} = \frac{\text{number of states available to electrons at the Fermi surface}}{\text{cm}^3} \simeq$$

$$\simeq \frac{10^{22}\ \text{levels}}{\text{eV cm}^3}$$

for an average metal. This, as it would normally present itself, is a very degenerate problem, and because of this degeneracy the system behaves in what seems at first to be a very strange manner.

To see what can occur in such a degenerate situation consider a system of N levels, each of which is connected with all others by, let us say, a constant matrix element -V. Assume for the moment that all of the diagonal elements are 0. The Hamiltonian matrix then appears as below:

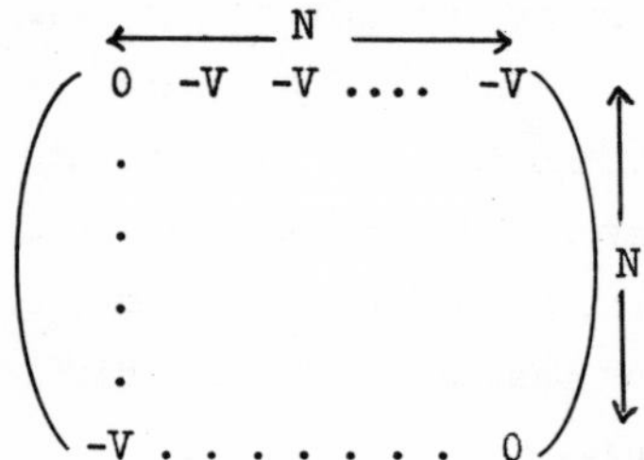

When this is diagonalized the N levels originally at E = 0 split so that one level is depressed to $E = -(N-1)V$, while the $(N-1)$ levels are each raised by the amount +V

N levels; E = 0

N- 1 levels; E = +V

1 level E = -(N-1) V

The level which is depressed is a linear combination of the original N levels, which is qualitatively a different state. In addition, since N is usually proportional to the volume of the system, while V is usually proportional to Ω^{-1}, the product NV is volume independent.

Degenerate systems seem quite easily to give results of this kind. This suggests that there is a common phenomenon which occurs in such degenerate systems - not terribly dependent on the details of the interaction - which leads to states qualitatively different from the original states, and which are separated from the original states by a volume independent energy. If one believed that an energy gap was somehow relevant to an understanding of superconductivity, this would be a very suggestive result.

Let us consider again the many electron system. Is there anything about it which is like the system we have analysed above? In this system (for two-body interactions), the electrons scatter in pairs and, since we assume total momentum is conserved, the momentum of the initial and final configurations is equal. There are a vast number of many electron configurations but most of these are not connected to each other (since, for example, they might differ by more than two single particle states.) In the matrix of these configurations there are submatrices of states all of which are connected to one another. These are the configurations which differ by a single electron pair. Or more simply, if we consider an electron pair of some momentum, this will be able to scatter into all other pair states of the same total momentum, which are not already occupied by other electrons. We are led to attempt to diagonalize first such a submatrix.

Every such singlet spin zero-momentum pair state can scatter to every other.

Suppose we consider then such an electron pair of zero total momentum and with singlet spin (for S-state interactions the exchange term cancels the direct term for triplet spin pairs.) We can put in the statistical effect of the other electrons by limiting the region of interaction to that above the Fermi surface. The Schrödinger equation for the electron pair, ψ, is

$$(H_0 + H_1)\psi = E\psi \,, \tag{I.11}$$

where H_0 is the kinetic energy operator of the pair and H_1 is the two-body interaction. If we expand ψ in plane wave pair states

$$\psi = \Sigma\, a_i \varphi_i \,, \tag{I.12}$$

where $H_o \varphi_i = 2\mathcal{E}_i \varphi_i$, we obtain

$$(2\mathcal{E}_j - E)a_j + \sum_i \langle j|H_1|i\rangle a_i = 0 \,. \tag{I.13}$$

To make things as simple as possible write the matrix element as

$$\begin{aligned} \langle j|H_1|i\rangle &= -V \qquad \mathcal{E}_F \le \mathcal{E} \le \mathcal{E}_F + \hbar\omega \\ &= 0 \qquad \text{elsewhere} \end{aligned} \tag{I.14}$$

which amounts to introducing a factorizable potential in a small interaction region. We then obtain

$$a_j = \frac{VC}{2\mathcal{E}_j - E} \,, \tag{I.15}$$

where

$$C = \sum_i a_i \Rightarrow N(0) \int_{\mathcal{E}_F}^{\mathcal{E}_F + \hbar\omega} a(\mathcal{E})\, d\mathcal{E} \,, \tag{I.16}$$

and N(0) is the density of electron states at the Fermi surface. The consistency condition yields an eigenvalue equation:

$$1 = N(0)V \int_{\mathcal{E}_F}^{\mathcal{E}_F + \hbar\omega} \frac{d\mathcal{E}}{2\mathcal{E} - E} . \tag{I.17}$$

For E between $2\mathcal{E}_F$ and $2(\mathcal{E}_F + \hbar\omega)$,

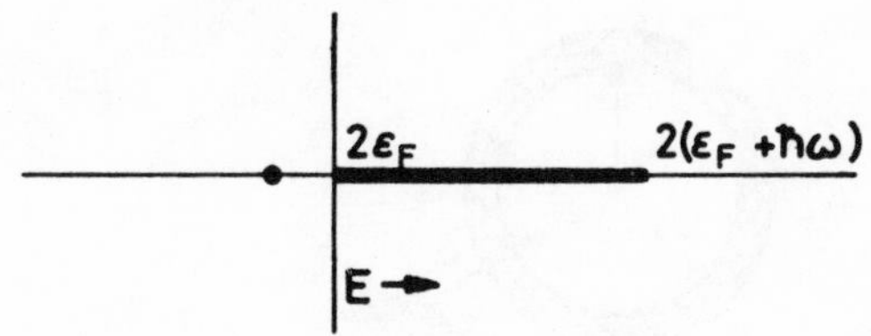

the solutions obtained are very like ordinary plane waves. For $E < 2\mathcal{E}_F$ however, a new type of solution appears whose energy is

$$E - 2\mathcal{E}_F = - \frac{2\hbar\omega}{e^{1/N(0)V} - 1} \simeq 2\hbar\omega \, e^{-1/N(0)V} , \tag{I.18}$$

and whose character is quite different. It is a localized state with an extension (for the magnitudes of the parameters that are usual) of about 10^{-4} cm. The normal Fermi gas is thus unstable to the formation of such electron-electron correlations.

We now attempt to remove at least this instability by diagonalizing first that part of the total Hamiltonian matrix which corresponds to the correlation of such zero momentum singlet pairs. It is not trivial to do this because the pairs interfere with each other statistically, i.e one electron pair can scatter everywhere except where other electrons are, so it is necessary to solve even this simplified problem in some self-consistent (and often approximate) way.

To gain an understanding of the solution that results, imagine that we open a space below the Fermi surface so that instead of having complete occupation below and nothing above, as in the normal metal, the occupation probability is smeared over a region of the order of the amount of energy gained in the transition, Δ.

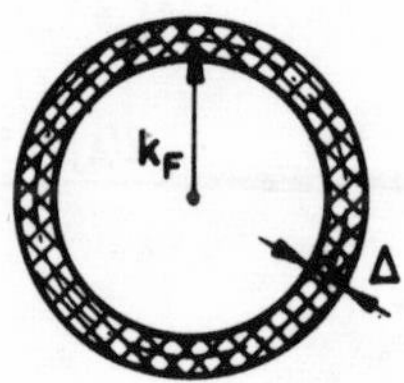

The number of pairs involved is then of the order of

$$\text{Number of pairs} \sim N(0)\hbar\omega\, e^{-1/N(0)V} ,$$

and the condensation energy is estimated to be the number of pairs involved multiplied by the energy per pair, which in the weak coupling limit is:

$$W_n - W_s = W_{normal} - W_{superconducting} \sim N(0)(\hbar\omega)^2\, e^{-2/N(0)V} . \qquad (I.19)$$

This is to be compared to the condensation energy in the weak coupling limit in the B.C.S. theory[8] which is just

$$W_n - W_s = 2N(0)(\hbar\omega)^2\, e^{-2/N(0)V} . \qquad (I.20)$$

The factor $(\hbar\omega)^2$ gives the isotope effect, while the exponential makes the binding energy extremely sensitive to the magnitude of the interaction; it suggests why the condensation energy can be so small.

An elegant way[9] of obtaining these and other relevant results is through the use of the equations of motion for the field operators ψ^* and ψ, which create and annihilate electrons. For a two-body interaction this can be written

$$\left(i\hbar \frac{\partial}{\partial t} + \frac{\hbar^2}{2m} \nabla^2\right) \psi_\alpha(\underline{r}, t) = \sum_\beta \int v(\underline{r}\underline{r}')\, \psi_\beta^+(\underline{r}'\, t) \times$$

$$\psi_\beta(\underline{r}'\, t)\, \psi_\alpha(\underline{r}\, t)\, d\underline{r}' \; . \tag{I.21}$$

From this, taking time-ordered vacuum expectation values of products of ψ and ψ^*, an equation of motion for the Green's function

$$G_{\alpha\beta}(xx') = -i \left\langle T\left(\psi_\alpha(x)\psi_\beta^*(x')\right)\right\rangle \tag{I.22}$$

can be constructed. On the right-hand side occur the vacuum expectation values of products of four-fermion operators of the form $\langle T(\psi^+ \psi \psi \psi^+) \rangle$. The equations are rigorous but cannot be solved unless they are approximated in some way. The essential approximation made in constructing a theory of superconductivity from these equations is that if the expectation value is taken for the N particle ground state then, on inserting a set of intermediate states in a typical matrix element

$$\sum_\alpha \langle N|\psi\psi|N + 2, \alpha\rangle \langle \alpha, N + 2|\psi^+\psi^+|N\rangle \quad ,$$

a kind of 'minimal leakage' occurs. That is the contribution from all intermediate states, other than the N+ 2 particle ground state, are small enough to be neglected. The overwhelming contribution comes from the matrix element between the ground state of the N and N+ 2 particle ground states, when one creates or annihilates a single correlated electron pair. This is a drastic approximation (it is equivalent to the assumption that it is the electron pair interactions which are important) but when made it yields Gorkov's equations[9] which give the theory of superconductivity in a gauge invariant and rather elegant fashion.

The picture that results can be summarized as follows. For the normal metal at T = 0 all of the single particle states are occupied up to k_F, and empty above. For the superconductor there is occupation of states on both sides of the Fermi surface so that if the states $\underline{k}\uparrow$ is occupied, then so also is the state $-\underline{k}\downarrow$, or if the state $\underline{k}\uparrow$ is unoccupied then so also is the state $-\underline{k}\downarrow$. The actual wave function is a linear combination of many pair states around the Fermi surface.

The single particle excitation spectrum of the superconductor can be put into a one-to-one correspondence with the single particle excitations of the normal metal. These excitations in the superconductor obey Fermi statistics but their energy of excitation is

$$E = \sqrt{\epsilon^2 + \Delta^2} \ .$$

Thus, as opposed to the single particle excitations of the normal metal, the lowest excitation energy is $|\Delta|$, which gives the desired energy gap.

In the ground state of the superconductor all the electrons are in singlet pair correlated states of zero total momentum. In an n electron excited state n electrons are in "quasi-particle" states. These are very similar to the normal excitations; they are not strongly correlated with any of the other electrons; they exist against a background of all the other electrons which are still correlated - as they were in the ground state; they can be easily scattered or further excited. On the other hand, the background electrons (those which remain correlated) still behave like a superfluid and are very hard to scatter or to excite. Thus, one can identify two almost independent fluids. The correlated portion of the wave function has the properties of the superfluid: the resistance to change, the very small specific heat; whereas, the excitations behave very much like normal electrons, and have an almost normal specific heat and resistance. When a steady electric field is applied to the metal the superfluid electrons short out the normal ones, but with higher frequency fields, the resistive properties of the excited electrons can be observed.

II. OTHERWISE

I would like now to indicate briefly how the alteration in the ground state of the system produces those curious properties characteristic of superconductors. The fact that the specific heat, C_V, is exponential at very low T, comes about directly from the energy gap because, as we have said, C_V measures the number of degrees of freedom, and at very low T where $k_B T$ is smaller than the gap, there remain very few degrees of freedom.

To explain the loss of conductivity imagine that we have a normal metal at T = 0; then the only resistance is due to impurities. Let us assume the impurities are heavy so that the electrons scatter elastically. How does normal resistance come about? If we switch on an electric field some electrons will transfer from one side of the Fermi surface to the other, and there will be a current. On switching off the electric field the electrons scatter elastically from the impurities and they become randomized over the Fermi surface and, after a while, there is no current, because after they scatter, so to speak, they do not know which direction they are scattering into; the electron system is left at a higher temperature, which is transferred eventually to the lattice.

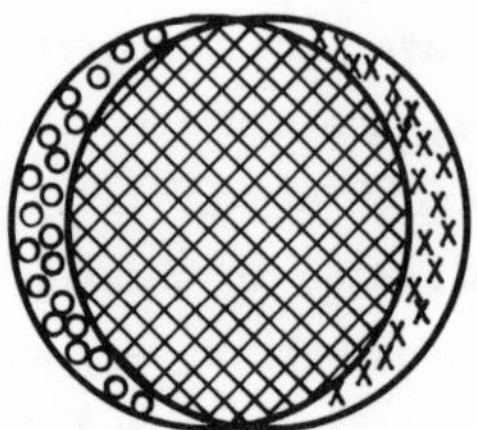

For the superconductor we have constructed, it is just this process which cannot occur because if all electrons are paired in the ground state, you cannot have elastic scattering of a single electron. In the normal case, all you need to get an excitation is an infinitesimal amount of energy, while in the superconductor you need a finite amount of energy - the gap.

One possibility for a current carrying state is obtained by giving each pair a small velocity (this state is almost as good as before - it has a little more kinetic energy) so that each pair can still scatter into any other pair. You still have the coherence. You might ask why this state does not relax into the old one. It could, but a large number of pairs would have to scatter at once, a process which is intrinsically different from the single particle processes which produce resistance in a normal metal.

The Meissner effect can, perhaps, be thought of in the following way. The pair wave functions in an ideal superconductor occupy a spatial (coherence) distance of the order $\xi_0 \sim 10^{-4}$ cm, which is large compared to the distance between electrons in a metal, $\sim 10^{-8}$ cm. This makes them behave like a large molecule. Large molecules are diamagnetic in the ground state, and one can think of the diamagnetism as being associated with these molecules.

One does not expect the magnetic field to go to zero discontinuously as it crosses the boundary. It will be screened out over some distance λ, the penetration depth, the field going to zero exponentially. In the ideal type I superconductor, the penetration depth is of the order of 10^{-5} cm.

These are some of the phenomena of classical superconductivity. I would like to discuss now some of the things that have turned up recently, and which especially illuminate the nature of the superconducting state.

The existence of pairing in the ground state seems to be particularly well illustrated by the phenomenon of flux quantization. If a superconducting cylinder with a hole in it is placed in a magnetic field parallel to the axis of the cylinder, the material of the cylinder expels the field (Meissner effect) but the field can penetrate the hole;

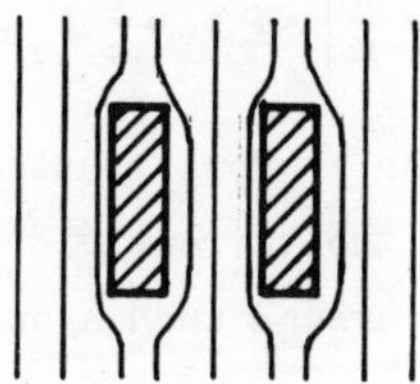

Supposed to be the x section of an s.c. cylinder in a uniform magnetic field.

the flux through the hole is:

$$\Phi = \int H \cdot dA \ . \qquad \text{(II.1)}$$

For a normal metal the flux in the hole depends linearly on the external field, but for a superconductor the following striking behaviour is observed[10].

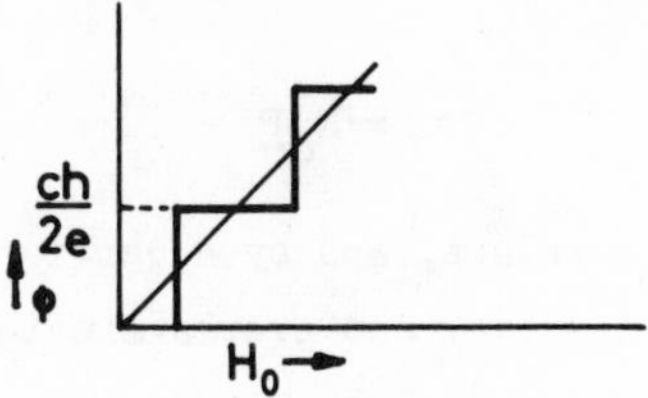

Below the transition temperature the flux permitted through the hole does not depend linearly on the external field but rather comes in steps of ch/2e. This can be understood with the simple argument which follows.

The current operator is

$$\underline{J} = -\frac{e}{m}\left[\underline{P}+\frac{e}{c}\mathcal{N}\underline{A}\right], \tag{II.2}$$

where $\underline{P}$ is the momentum operator, $\mathcal{N}$ the number operator for electrons and $\underline{A}$ the vector potential. Taking the expectation value for some current-carrying state $|I>$ gives

$$<\underline{J}>_I = -\frac{e}{m}\left[<\underline{P}>_I+\frac{e}{c}N_e\underline{A}\right], \tag{II.3}$$

N_e being the number of electrons. Now, consider a curve, c deep inside the superconducting cylinder where the current flow is zero because of the Meissner effect. Integrating around such a curve gives:

$$\int_c <\underline{J}(\underline{r})>_I \cdot d\underline{\ell} = 0 = -\frac{e}{m}\int_c <\underline{P}>_I \cdot d\ell - \frac{e^2N_e}{mc}\int \underline{A}\cdot d\underline{\ell} . \tag{II.4}$$

The integral in the last term is just the enclosed flux Φ. Thus, the flux through the hole is

$$\Phi = -\frac{c}{eN_e}\int_c <\underline{P}>_I \cdot d\underline{\ell} \tag{II.5}$$

Onsager[11]) characterized a superfluid as a fluid, each of whose carriers had a common momentum, i.e. for a superfluid

$$<\underline{P}>_I = N_c\underline{P} , \tag{II.6}$$

where N_c is the number of carriers, and by a carrier we mean some grouping of the constituents, repeated over and over again throughout the system.

Therefore, Bohr's quantization condition yields:

$$\int \langle \underline{P} \rangle_{\ell} \cdot d\underline{\ell} = N_c \int \underline{P} \cdot d\underline{\ell} = N_c \, \nu h \, , \qquad \text{(II.7)}$$

where ν is some integer. From this it follows that

$$\Phi = - \frac{ch}{e} \frac{N_c}{N_e} \, \nu \, . \qquad \text{(II.8)}$$

The observed flux units are ch/2e, and so the picture is right if

$$N_c = \frac{1}{2} N_e \, . \qquad \text{(II.9)}$$

This suggests that two electrons form a carrier, each pair having a common momentum. The argument may be made more formal by imposing an appropriate condition on the wave function. This is very direct evidence for the existence of pairs with a common momentum, almost without any assumptions. If one thinks in terms of carriers with a common momentum then the argument shows that the carriers have charge 2e.

It has been suggested that flux quantization is a logical consequence of superconductivity. I do not believe that this is so. The connection requires a symmetry property, in the case discussed above the rotational invariance of the electron-electron interaction which produces superconductivity. If one could construct in some practical way a sample which did not possess the necessary rotational invariance, it might be possible to obtain a superconductor for which the flux inside the hole was not quantized.

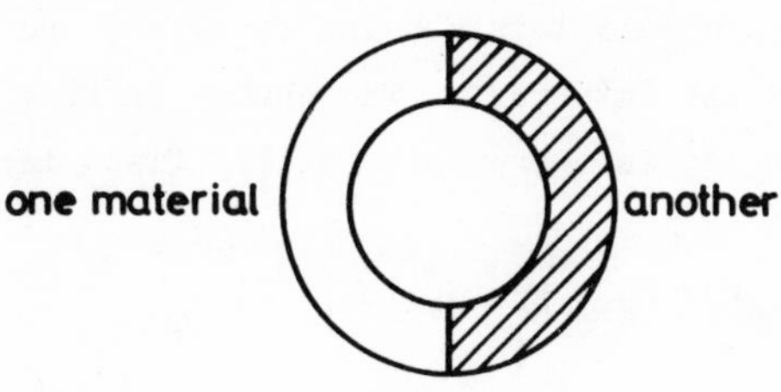

The spatial extension of the pairs in the superconductor is exhibited by phenomena involving superimposed films. If a thin superconductor is superimposed on a thin normal material as below,

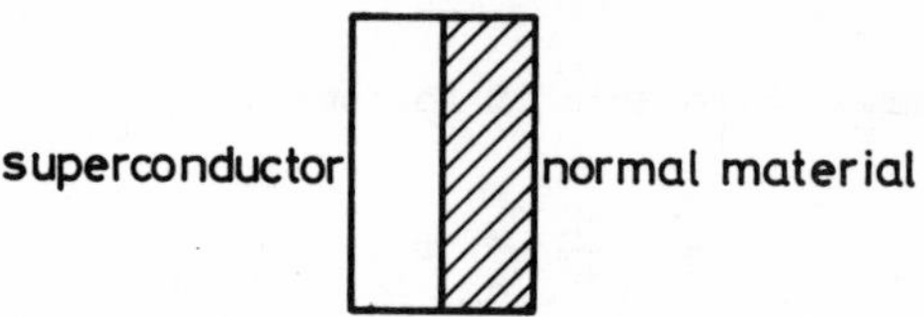

and if the barrier between the two metals is low enough so that the electrons can pass freely from one to the other, then the transition temperature of the superconductor will be depressed, and for thin enough specimens the entire sample can become a superconductor.

To understand this we observe that the electron-electron interaction is different in the two materials. An interaction sufficient to produce a superconductor (at a given temperature) exists only in the material normally superconducting. However, the pair wave function can extend beyond the superconductor and penetrate the normal material. The situation is not too different from that of the deuterium nucleus in which a short-range potential holds together a wave function that extends far beyond the boundaries of the potential, and for samples smaller than the spread of the pair wave function ($\xi_0 \simeq 10^{-4}$ cm) the effective interaction is reduced by the ratio of the thickness of the superconductor to the thickness of the entire sample.

Another phenomenon possibly observed long ago, but only recently appreciated is that of type II superconductors. A type I superconductor is one which displays a Meissner effect and thus expels the magnetic field from its interior. The change in free energy per unit volume due to this complete expulsion of the field is:

$$F_n - F_s = \frac{1}{8\pi} (H_c^2 - H^2) , \qquad \text{(II.10)}$$

where H_c is the critical magnetic field. It might seem preferable (and this was considered by Landau a long time ago) that the specimen breaks up into many superconducting and normal regions. This would lose a little condensation energy but would gain much of the magnetic energy lost due to the field expulsion.

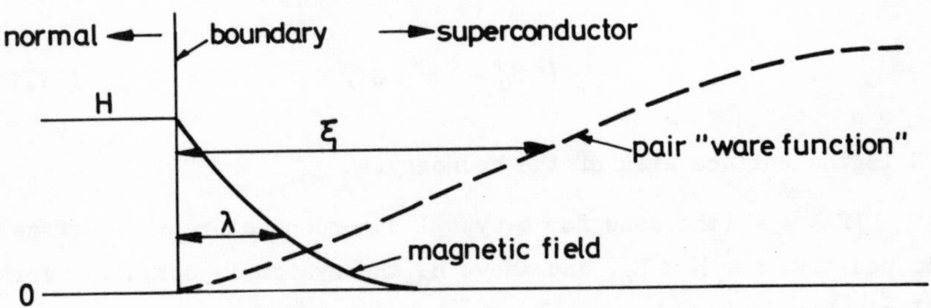

Whether or not this field penetration occurs depends upon the surface energy at the boundary. At the boundary of a superconductor the magnetic field drops from its maximum value to zero in the distance, λ. This gives a small correction to the free energy since the field is not completely expelled, and within this region the specimen goes from normal to superconducting.

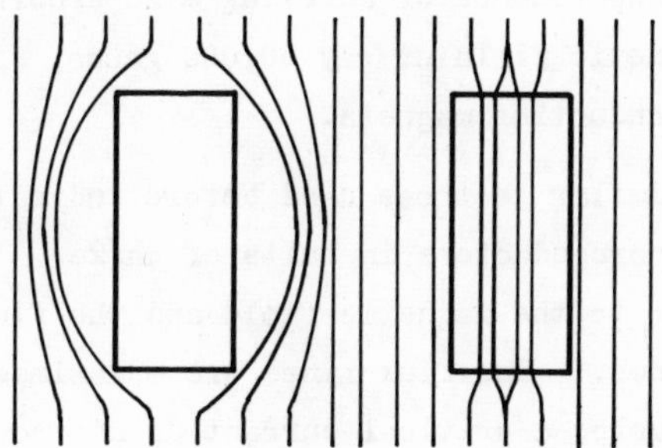

Due to the field penetration the increase in free energy, because of the magnetic field expulsion, is made smaller but, at the same time, since part of the material is nearer to normal there is a loss of condensation energy. (This loss may be thought of as resulting from an increased kinetic energy of the pairs.) As a consequence, there is added to the free energy of the superconductor a surface term

$$+ \frac{1}{8\pi} (\xi H_c^2 - \lambda H^2) S , \qquad (II.11)$$

where S is the surface area of the boundary.

If $\lambda < \xi$ (the case for a type I superconductor) the surface term is positive for $H < H_c$, and above H_c the system is normal. For type II superconductors however $\lambda > \xi$; more precisely,

$$\frac{\lambda}{\xi} \geq \frac{1}{\sqrt{2}} \qquad (II.12)$$

and the surface energy can become negative. In this case H can exceed H_c penetrating the specimen as above. From the point of view of technology the nice thing about these superconductors is that if the magnetic field goes right through them, they do not have such low critical fields. Thus, one can have a superconductor carrying a supercurrent and at the same time supporting a magnetic field of say 80,000 gauss; it thus becomes possible to build superconducting magnets.

Arguments similar to those used before indicate that the flux penetrates type II superconductors in units of ch/2e. In the presence of currents perpendicular to the magnetic field and the flux lines, there are forces on the flux lines. The flux lines are sometimes pinned (perhaps to dislocations) and below a critical current do not move. Above this current they begin to move as through a viscous medium, and the current working to move them encounters resistance. Thus, one can have d.c. resistance in a superconductor.

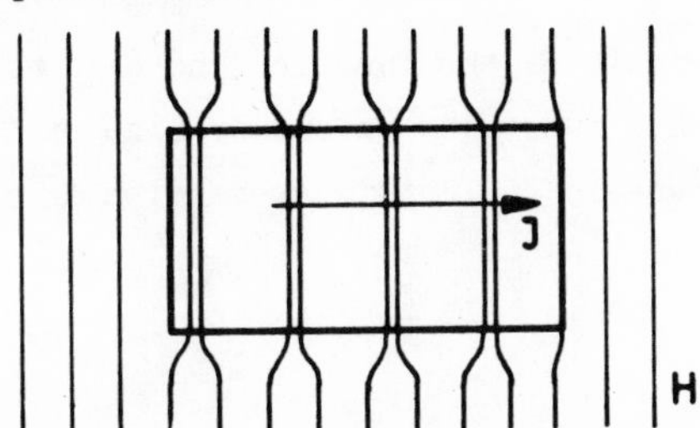

The enormous range of coherence of the ground state superconducting wave function leads to quantum effects on a macroscopic scale. These are illustrated in a striking way in some recent tunnelling experiments. If two superconductors are superimposed so that between them there is a very thin barrier, depending on the height and width of this barrier,

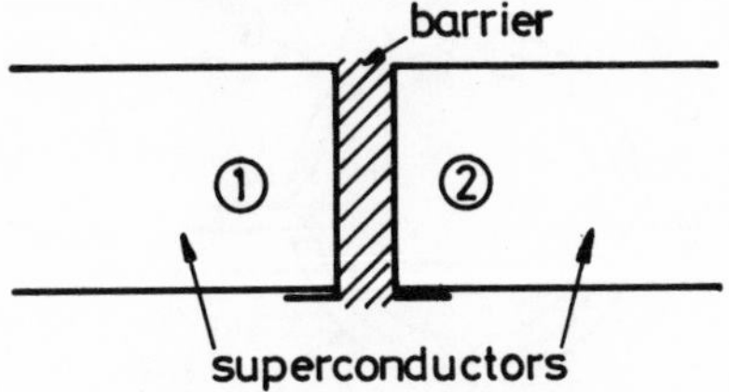

electrons may tunnel from one side to the other.

It was realized by Josephson[12)] that in addition to tunnelling by quasi-particle excitations, it was also possible that pairs themselves might tunnel through. This leads to the possibility that a super current might flow (that is a current for zero external potential) in spite of the barrier. This is very reasonable in hindsight, since we would expect that as the barrier goes to zero (as the two superconductors become one) some current should go continuously into the usual super current.

For zero external potential the Josephson current is given by

$$J = J_0 \sin \delta_0 , \tag{II.13}$$

where J_0 is a characteristic of the function and δ_0 is the phase difference between the pair wave functions on either side of the junction. In the presence of a magnetic field this expression is modified and becomes:

$$J = J_0 \sin\left(\delta_0 + \frac{2e}{\hbar c}\int_c A \cdot d\ell\right) , \tag{II.14}$$

where the integral is taken over the path between the first and second superconductor. For a broad junction the presence of the vector-potential term produces phase differences for various paths of current flow, which produce oscillations in the current as the magnetic field is varied.

If one had two identical junctions as below

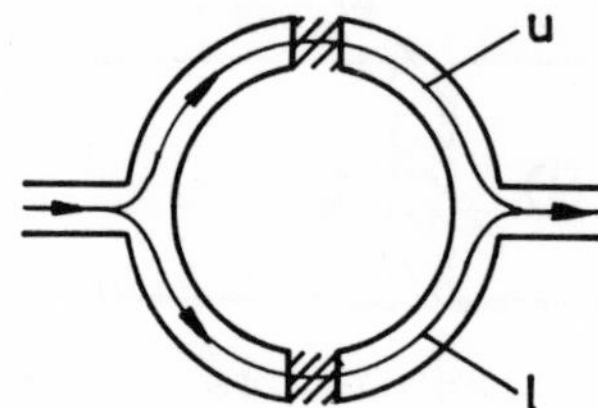

the total current that flows, the sum of the currents through the upper and the lower paths, is:

$$J = J_0 \sin\left[\left(\delta_0 + \int_u A \cdot d\ell\right) + \sin\left(\delta_0 + \int_l A \cdot d\ell\right)\right] . \tag{II.15}$$

The path integrals can be related to the enclosed flux Φ to give

$$J = J_0 \left[\sin (\delta_0 - \eta\pi) + \sin (\delta_0 + \eta\pi)\right] , \tag{II.16}$$

where η is the ratio of the enclosed flux to the standard flux unit ch/2e

$$\eta = \Phi \Big/ \frac{ch}{2e} \ . \tag{II.17}$$

This yields

$$J = J_0 \sin \delta_0 \cos \eta\pi \ , \tag{II.18}$$

which varies with the external magnetic field and has maximum absolute value when η is an integer or when Φ is an integral multiple of ch/2e. Since the interference depends only on the enclosed flux, it is possible to arrange things so that the magnetic field is confined to the hole, making possible an observation of the effect of the vector potential[13)].

I have not had the time to do more than sketch some of the phenomena and theory associated with superconductivity, and to suggest their variety. However, it is probably clear that if we were to attempt now to describe the qualitative properties of superconductors (without preconceptions), it might be difficult to make ourselves believe those simple and essential things, zero resistance, Meissner effect, energy gap and so on, with which our story began; and without this belief it might not be so easy to construct a theory.

REFERENCES

1) H.K. Onnes Commun.Phys.Lab.Univ. Leiden No. 124 C (1911).

2) W. Meissner and R. Ochsenfeld, Naturwissenschaften 21, 787 (1933).

3) E. Maxwell, Phys.Rev. 78, 477 (1950);
C.A. Reynolds, B. Serin, W.H. Wright and L.B. Nesbitt, Phys.Rev. 78, 487 (1950).

4) F. Bloch, Z. Physik 52, 555 (1928).

5) H. Frölich, Phys.Rev. 79, 845 (1950).

6) J. Bardeen, Phys.Rev. 79, 167 (1950).

7) F. Kafka, Parables (Knopf).

8) J. Bardeen, L.N. Cooper and J.R. Schrieffer, Phys.Rev. 108, 1175 (1957).

9) L.P. Gorkov, J.Exptl.Theoret.Phys. (USSR) 34, 735 (1958), translated: Soviet Phys. JETP 7, 505 (1957);
L.P. Kadanoff and P.C. Martin, Phys.Rev. 124, 670 (1961).

10) B.S. Deaver and W.M. Fairbanks, Phys.Rev. Letters 7, 43 (1961);
R. Doll and M. Näbauer, Phys.Rev. Letters 7, 51 (1961).

11) L. Onsager, Phys.Rev. Letters 7, 50 (1961).

12) B.D. Josephson, Physics Letters 1, 251 (1962).

13) R.C. Jaklevic, J. Lambe, A.H. Silver and J.E. Mercereau, Phys.Rev. Letters 12, 159 (1964);
R.C. Jaklevic, J. Lambe, A.H. Silver and J.E. Mercereau, Phys.Rev. Letters 12, 274 (1964).

DISCUSSION 1

CHAIRMAN : Professor A. Pais

Secretaries : C. Noack and K. Lassila

Noack : Could you explain again, in detail, your concepts of "approximate kinematical" versus "approximate dynamical" symmetry? In particular, what is the conceptual difference between the isospin symmetry and that in LS coupling of atomic spectroscopy?

Pais : I think this is a very good point to belabour because many of the controversial questions that have arisen in regard to SU(6) will perhaps become more clear if we bear in mind, very precisely, what the difference is between these two types of approximate symmetries.

In the case of isospin we like to argue as follows. There is an interaction Hamiltonian of the form

$$H_{int} = H_{strong} + e \cdot H_{el.magn.} \cdot \qquad (1)$$

The first part preserves isospin, while the second violates it. This division of the interaction into two parts is characterized by a parameter e which is independent of the state of the system under consideration. In the limit $e \to 0$ we have a strict symmetry which is isospin. I call this a kinematical symmetry because this limit condition does not involve any dynamical variables; in other words, the limit condition is itself covariantly defined.

Turning to the case of Russel-Saunders coupling, the Hamiltonian of an atom is

$$H = \sum \frac{p_i^2}{2m} - \sum_i \frac{z e^2}{r_i} + \sum_{i<j} \frac{e^2}{r_{ij}} + H' \quad ,$$

where H' contains the relativistic corrections, such as spin-orbit coupling. In situations where H' is negligible, L^2 and S^2 are good quantum numbers separately (Russel-Saunders coupling). Now, to characterize this approximation we should not use $e \to 0$ since then the Coulomb terms also vanish; there is no atom, in other words. Rather one sees that the spin-orbit term (as well as the rest of H') depends on v/c (through $\underset{\sim}{\ell} = \underset{\sim}{r} \times \underset{\sim}{p}$ for example). Thus, the approximation of LS coupling is essentially one in which the explicit velocity dependence of the Hamiltonian is neglected. Whether this is justified, i.e. whether the matrix elements of $\underset{\sim}{p}$ are small depends on where you are in the spectrum. In heavy atoms, it is not true for the L shell; in fact, spin-orbit coupling is very important in the L shell (Roentgen spectra). When you come to the outer shells (but still the bound states), Russel-Saunders coupling becomes much better. Then, for the excited states with very elliptic orbits, the matrix elements of $\underset{\sim}{p}$ become important. So you see, the general idea is that it is not just a parameter that goes to zero, but that the expectation values of certain dynamical operators like $\underset{\sim}{p}$ are small in certain physical situations. That is what I call an approximate dynamical symmetry. Here the limit in which the symmetry is valid is not covariantly defined.

<u>Question</u> : However, since the spin-orbit term is proportional to the magnetic moment μ, if one lets $\mu \to 0$, then this term also vanishes.

Pais : Sure, but $\mu = e\hbar/2mc$, and you do not want to let e go to zero because then there would be no atom at all. The validity of Russell-Sanders coupling amounts to saying that v/c is small, and this depends on the dynamical situation.

Hertel : If you neglect $e \cdot H_{el.magn.}$ in Eq. (1), this may, however, also be formulated by stating that the dynamical conditions must be such that H_{strong} is much bigger than $H_{el.magn.}$. Then, should one draw the conclusion that isospin is also a dynamical symmetry?

Pais : It is, of course, quite true that in high-energy scattering, for example, the Coulomb interference is appreciable only in the forward direction. Nevertheless, it is possible (at least so is the current thinking) to state that isospin is a good symmetry if you neglect all electromagnetic effects. A group is a mathematically sharp concept, and in the case of isospin you can define it by $e \to 0$. After having said that, I agree with you that there are certain regions where this neglect is more important and others where it is less important.

Altarelli : I should like to know if there is a simple procedure, starting from a Young diagram, to guess the form of the tensor in terms of upper and lower indices.

Pais : There is a systematic way to do this, but it is rather complicated. Instead let me give you examples. Consider the following representations in SU(6):

i) $\underline{405}$. It has the Young diagram

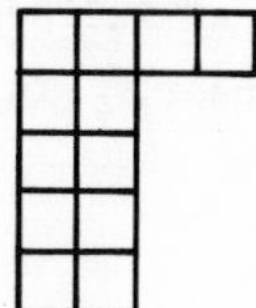

The tensor of this representation is a traceless tensor $T^{\widehat{\alpha\beta}}_{\widehat{\gamma\delta}}$ ($\frown$ denotes a symmetric pair of indices). How do I check that this is the correct tensor?
There are $\frac{6.7}{2} = 21$ possibilities for choosing the indices α and β, and 21 ways of choosing γ and δ. The tracelessness of the tensor gives 36 conditions so that the tensor finally has $21 \cdot 21 - 36 = 405$ different components.

ii) <u>280</u> has the Young diagram

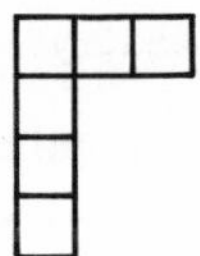

and its tensor is $T^{\widehat{\alpha\beta}}_{\widehat{\gamma\delta}}$, again traceless ($\wedge$ denotes an anti-symmetric pair). Here, the dimension is $21 \cdot 15 - 35 = 280$. Note that there is one less trace condition because of the identity $T^{\widehat{\alpha\beta}}_{\widehat{\beta\alpha}} \equiv 0$. In cases of practical interest it is usually easy to find the tensor in this way.

<u>Nagy</u> : If one has a tensor of the form $T^{a_1 \cdots \alpha_n}_{\beta_1 \cdots \beta_m}$ and it is traceless in the sense that the upper indices have a symmetry corresponding to a Young diagram, and the lower indices also have a symmetry corresponding to another Young diagram, then it is an irreducible tensor. If you raise or lower the indices with the aid of the Levi-Civita symbols, do you get an irreducible tensor?

<u>Pais</u> : Yes. This is true for the unitary <u>unimodular</u> case. For the more general case of U(N) it can also be applied, but with a little more care. In the general linear group you need four kinds of indices, and you need four kinds of Young diagrams per tensor. In the case of the unitary unimodular

group, by raising all indices with the Levi-Civita symbols, you can have one and only one kind of index, namely all upper. Then by looking at the symmetries of that set of indices you can find all irreducible representations.

Question : As far as I understand the Young tableau technique can be used only for the group SU(n), but not for other compact groups?

Pais : Young diagrams can be used for all Lie groups. They can be used for finite dimensional representations of non-compact groups, as you will see later on in my lectures. They can also be used for the exceptional groups, orthogonal and symplectic groups, for anything (cf. Diu, Nuovo Cimento).

Question : You have developed representation theory of unitary groups in the tensor basis. What is the relation of this to the physical particles?

Pais : I am going to talk about the physics of SU(6) in detail in the coming lectures, but let me just give the physical background of the relation $\underline{6} = (\underline{3},\underline{2})$ now. In SU(3), the defining representation is a triplet. Let me call it

members	p	n	λ
with charges	q	q - 1	q - 1
hypercharges	y	y	y - 1
and baryon number	b	b	b .

(in a pure quark model one would have $q = 2/3$, $y = 1/3$, $b = 1/3$). In SU(2), the defining representation is a doublet t^1, t^2 where t^1 stands for a wave function with spin up ($\uparrow$), t^2 for spin down ($\downarrow$).

In SU(6), the defining representation is a sextet so we have six states $S^a(a = 1,2,\ldots,6)$. Now the relation $\underline{6} = (\underline{3},\underline{2})$ tells us that I can write these states as

$$t^1p,\ t^1n,\ t^1\lambda,\ t^2p,\ t^2n,\ t^2\lambda\ ,$$

or

$$p\uparrow,\ n\uparrow,\ \lambda\uparrow,\ p\downarrow,\ n\downarrow,\ \lambda\downarrow.$$

This is the mathematical basis I was talking about:

$$S^a = t^\alpha T^A\ ,$$

where t^α refers to spin and T^A to SU(3).

Lipschutz : The representation 56, which contain the baryons, is fully symmetric. How can one make a bound-state model of the baryon without getting involved in parastatistics for the quarks?

Pais : This is a very broad question and an important one, so let me first state the problem. If I look upon the baryon as a bound state of three sextets (quarks with spin), then the Pauli principle says that its spatial wave function must be totally antisymmetric since there is total symmetry in the internal variables (spin and unitary spin). In simple thinking about non-exchange potentials, however, we are used to saying that the ground state of such a system has no nodes, which means that it is fully symmetric in the space co-ordinates. If we think of the baryon as a bound state of three quarks (which I do not guarantee), we have a bit of a puzzle.

This is a very important physical question to which, to this day, nobody knows the answer for sure. Let me, however, remark that I can talk about SU(6) and its consequences without having necessarily to answer this question. When I write down the tensor of the 56, it means that this transforms like a state built up of three sextets, not that it is such a state. This means that I am not forced to answer your question. That is very important. In fact, all of the results obtained from SU(6) so far are completely independent of the realistic existence of quarks. They have been obtained solely from the tensor analysis of a group. I want to stress that these two things are to be kept apart. That does not mean that I think your question is uninteresting; on the contrary, I think it is one of the most exciting questions in particle physics. I think that the study of models in which one tries to make a realistic picture of this bound three-particle system is very interesting; however, I do not think that any such study made so far is conclusive.

Question : Do you mean to say you think that quarks do not exist?

Pais : I mean to say that I am uncommitted. I can only say one thing: I do not believe that the nucleon is something very fundamental. There must be something more profound. You cannot live for ever with these 56 objects.

Question : Do I understand properly that in addition to having 1/3 integral values of charge, hypercharge and baryon number, the quarks should obey Fermi's statistics?

Pais : The simple picture which I made involves this assumption but this assumption can be challenged. There is another kind of statistics, now called parastatistics, which has an intrinsic triality. I will only say: provided you do it properly, I allow you to challenge any property of a quark as long as the three-quark system shall be orthodox; i.e. the baryon shall obey Fermi statistics. But if, in making this baryon, you want to use quarks and you want to be a bit unorthodox, that is part of the game of the coming years.

Question : Can one remove the difficulty brought up by Lipschutz by introducing three different sets of triplets, as in the theory of Nambu?

Pais : Yes. Let me sketch the idea. Assume that you have three sets of quarks, each an SU(6) sextet but distinguished from one another by some attribute I do not know, say red, white and blue. If the baryon is made up of a red, a white and a blue quark, they are all different fermions and you thus get rid of the forced antisymmetry of the spatial wave function. Thus, by introducing another quantum number and making the wave function antisymmetric in that quantum number, you reconcile Fermi statistics, symmetry of the spatial wave function and the three-quark structure of the baryon. That is another way out.

Franzini : Could you elaborate on the necessity of extending the total antisymmetry of a system of fermions to all internal variables, especially if these refer to approximate symmetries?

Pais : That is an old problem, mentioned long ago by O. Klein. Let me take a simple example, protons and neutrons. Suppose that you consider them as different particles and, therefore, construct a wave function which is antisymmetric in the protons and separately antisymmetric in the neutrons.

Then you can make a mapping to another language in which you have a two-valued variable which says that '1' is proton and '2' is neutron. (This does not mean that they have to be degenerate.) Now, you can show mathematically that antisymmetry in all co-ordinates including the new variable is completely equivalent to the conventional Pauli principle.

Franzini : Could you get out of the difficulty with the symmetry of the 56 by saying that the Pauli principle does not apply to all the internal variables?

Pais : It is not logically excluded. Let me try to summarize the possibilities:

i) quarks need not exist (i.e. the baryon need not be a bound state of three quarks;

ii) quarks need not be fermions, but they may obey some form of parastatistics;

iii) quarks may have some other, as yet unknown, internal degrees of freedom, and then the wave function of the bound state baryon may be spatially symmetric;

iv) the ground state may be antisymmetric (with exchange forces or other more complicated forces);

v) SU(6) will not survive.

Personally, there is something unattractive to me in every one of these possibilities.

Snow : If there is no non-relativistic quark model of the baryon so that the quarks are moving relativistically and SU(6) breaks down, does the question of symmetry or antisymmetry of the baryon wave function remain?

Pais : That is really what I mean by the fourth possibility. We may perhaps take a three-particle model with non-relativistic forces. Because the particles are very, very heavy you need a lot of binding. Yet, this does not exclude a non-relativist picture. It is that naive way of thinking that leads us to believe there is a puzzle. Now, please distinguish two kind of relativistic questions. One is the relativistic scattering of supermultiplets as a whole on each other. There a relativistic treatment is essential. Yet the nucleon may be a non-relativistic bound system. If it is not, that may wel remove the question of symmetry.

Wong : We know that the symmetries of strong interactions are not exact symmetries. This implies that there is mixing of supermultiplets. Is this consistent with the group-theoreti approach to the problem of strong interactions?

Pais : What you ask is essentially if configuration mixing is important. The simplest thought is, of course, that configuratio mixing should always be strong because the interactions are strong. It is a miracle I do not understand that configuration mixing seems to remain under control, so that you can recognize the supermultiplets. The mystery of the higher symmetries lies in the successes. We have surely got hold of part of what the theory of the future is to contain, but we do not know what the theory is to be like.

DISCUSSION 2

CHAIRMAN Professor A. Pais
Secretary : C. Noack

Hamprecht : Could you please explain a bit further the philosophy underlying the calculation of the magnetic-moment ratio, and make clear the assumptions that go into it?

Pais : First of all, the assumptions I make deal only with nucleon densities taken at zero momentum, i.e. in the extreme non-relativistic limit. Before I begin stating the assumptions, let me quote a simple but important mathematical lemma. Consider the direct product of a totally symmetric representation (n) of SU(N) and its complex conjugate (n^{N-1}). Then this direct product is simply reducible (which means that it contains no representation more than once), and it contains the adjoint representation once. Now, in SU(3) it is an old assumption that the electromagnetic form factors should transform like an octet, i.e. like the adjoint representation. The natural analogue for SU(6) is then that the form factors should again transform like the adjoint representation, which in this case is the $\underline{35}$. My first assumption is, therefore, that the baryon density transforms as a $\underline{35}$. This means that it is given by the tensor $B^*_{abc}\, B^{abd}$. Note that because of the lemma stated there is one and only one way of building a tensor like this. An immediate consequence of this assumption is that there is no undetermined parameter in the ratio of the magnetic moments; you know immediately that this ratio is fixed in SU(6). In SU(3), the situation is different: $\underline{8} \otimes \underline{8}$ contains $\underline{8}$ twice, and one parameter is therefore left undetermined, which is the neutron-proton magnetic-moment ratio. With the same argument one sees that SU(3) predicts the magnetic moment of every member of the decuplet to be proportional to its charge. The same result is true for quarks. All this follows immediately from the lemma stated and the assumption that the electromagnetic form factors transform like an $\underline{8}$.

Now, let me consider the charge operator. We know what it looks like in SU(3): it is a member of an octet, in fact, it is

$$Q = \begin{pmatrix} 2/3 & & \\ & -1/3 & \\ & & -1/3 \end{pmatrix}$$

Similarly, in SU(6) the charge is supposed to be a member of a $\underline{35}$. We maintain that it belongs to an octet, and since it carries no spin, it transforms like $(\underline{8},\underline{1})$.

For the magnetic moment I again maintain that the octet is dominant, but since the magnetic-moment operator carries spin it must transform like $(\underline{8},\underline{3})$. These are the assumptions that go into the calculation of the magnetic-moment ratio.

Snow : How compelling is the assumption that the form factors should transform like a $\underline{35}$?

Pais : One always starts with the simplest assumption possible. There is no reason to go beyond that until you are forced to. Let me give you an example. In the strict SU(3) symmetry there is the well-known relation $\mu_\Lambda/\mu_N = 1/2$. In the derivation of such relations one assumes that the magnetic moments of the two particles are expressed in a common magneton, which means that one assumes they are mass-degenerate. But SU(3) is broken and the masses of the Λ and the N are not equal. Beg and I have made a very simple correction for this, expressing the magnetic moment of each particle in its own magneton, using the physical mass. The result is $\mu_\Lambda/\mu_N = 0.78$, while the experimental world-wide average is 0.77 ± 0.2.

You see, I cannot prove from first principles that electromagnetism must transform like an 8, but I know that if it does not, this μ_Λ/μ_N ratio becomes arbitrary. Since the agreement is so good, the assumption is apparently a very good starting point. You see, that is what particle physics is: we make the simplest assumption which has predictive power, then you guys become very happy and say "now we can chop somebody's head off", and just once in a while our head stays on!

Zichichi : We know that the magnetic moment has a very large anomalous part, but what amazes me is that in spite of the enormous renormalization the relation between the magnetic moments is preserved. What does that mean?

Pais : Let me assume, for the moment, a world with strict SU(3) symmetry, which is conceptually possible. In that world I can do electrodynamics, and I can renormalize. If in that world the electrocurrent transforms effectively as an 8, it is no mystery that this relation exists. You can check it through from diagram to diagram: for every renormalization diagram for the neutron you have one for the lambda; you look at every vertex and at every coupling constant and you find this relation preserved. That is the strength of group theory.

Wong : We know that SU(3) and, therefore, also SU(6) is not an exact symmetry but actually badly broken. Is it not really embarrassing that in spite of this strong violation the magnetic-moment ratio comes out so well?

Pais : That is what I called the master problem of higher symmetrie We see phenomenologically that some things fall into place, but we really do not understand dynamically why this is so. In the case of the magnetic-moment ratio, look again at the relation $\mu_\Lambda/\mu_N = \frac{1}{2}$ predicted by SU(3). We have seen that one can incorporate a very elementary type of symmetry break ing, namely the mass corrections, and get this value right. Since the neutron-proton mass difference is negligible in th context, the magnetic moment ratio is not affected by this type of symmetry breaking.

Question : We have seen that the D/F ratio, undetermined in SU(3), is fixed in SU(6). Is this true also for the higher resonances?

Pais : Not in general. An indeterminacy may reappear. For example, look at the 70 representation. Some people have speculated (mostly on grounds of simplicity) that the 70 should be the next representation to be occupied by baryon resonances. If you would like to know how these interact with the meson field, you have to ask about $70^* \otimes 70$, and that contains 35 twice.

Lipschutz : Do you have any comments on the work of McGlinn, Coleman and others who say that a relativistic generalization of SU(6) cannot exist?

Pais : In my first lecture I stressed the difference between a kinematical and a dynamical symmetry, and I have been so insistent about this difference precisely because of all these theorems on "trouble with SU(6)". Now, I can simply say: these theorems show that there is trouble with a <u>kinematical</u> interpretation of SU(6). I can look at these theorems and at the same time have a certain optimism about SU(6), because I look upon SU(6) as an indication of where the dynamics is going.

The important content of the McGlinn theorems is to show that certain lines of development are closed. They show (if anyone doubled it) that SU(6) certainly is a dynamical symmetry, but they do not say that SU(6) is impossible.

<u>Daniels</u> : Can you comment on the interpretation of SU(3)? Does it reflect a symmetry in space or is it "accidental"? I am drawing on the analogy with the three-dimensional oscillator which has a so-called "accidental" symmetry larger than just rotational invariance. This symmetry is, however, a symmetry of phase space.

<u>Pais</u> : I think we really do not know. At this moment it is a harmless attitude to think that we have a direct product of space-time symmetries on the one hand and a kinematical SU(3) on the other. It is, however, not inconceivable that SU(3) may be due to some such "accidental" structure of the over-all spectrum of all elementary particles. But I do not think we know.

<u>Altarelli</u> : What can be said about the assignment of quantum numbers (charge, hypercharge, etc.) to quarks?

<u>Pais</u> : The reason why we are so interested to know if quarks exist and if they are fractionally charged or not is that we like to have this picture of a baryon being a bound state of three quarks of one kind, and nothing else. If this picture is correct then we have to have $q = 2/3$ and $y = 1/3$.

If, however, one assumes an additional particle, a kind of supernucleus, which is a spinless SU(3) singlet but carries charge, then $q = 2/3$ is no longer necessary. The same is true if there is more than one set of quarks.

Now, I have said before that the predictions of SU(6) are, in general, independent of q and y. There is one notable exception, however. Let us assume that the baryon is built up of three quarks and a supernucleus, and that in a static approximation we can write the magnetic-moment operator as a sum of the magnetic-moment operators of the three quarks:

$$\vec{\mu} = \vec{\mu}^{(p)} + \vec{\mu}^{(n)} + \vec{\mu}^{(\lambda)} \propto q \cdot \vec{\sigma}^{(p)} + (q-1)\vec{\sigma}^{(n)} + (q-1)\vec{\sigma}^{(\lambda)} .$$

This is very similar to what is usually done in nuclear physics. We know the factors q, $q-1$, $q-1$ because the simple-reducibility lemma I quoted before tells us that the magnetic moments of the quarks are proportional to their charges. We can then calculate the matrix elements of this magnetic-moment operator between proton states and neutron states. One finds

$$\frac{\mu(N)}{\mu(P)} = \frac{3q-4}{3q+1} ,$$

which is equal to $-2/3$ for $q = 2/3$ only.

As I said before, I am non-committed about the existence of quarks, but this result makes me feel that if there exists only one kind of quarks and if the internal symmetry is not largerthan SU(3) then they should be fractionally charged.

DISCUSSION 3

CHAIRMAN : Dr. D.H. Sharp
Secretary : K. Lassila

Lipshutz : How do Dashen and Frautschi handle the infrared divergence problem in the relativistic mass-shift problem, for example, the neutron-proton mass difference?

Sharp : Let me formulate the question in a little more detail. This point is extremely important for practical calculations of electromagnetic effects. When you are doing calculations with electromagnetism, the forces are due to photon exchange and are of long range so you get the infrared divergence. The presence of this infrared divergence will make it necessary to modify the formalism introduced in the lectures. The question then amounts to asking for the form this modification takes, and for some justification of it. I had not intended to discuss this topic here, because in these lectures I am discussing octet enhancement and, principally, the calculation of the A matrix elements. For these A matrix elements there is no infrared divergence when one works to lowest order violations appear in factors like δM or δg, while the A matrix is determined just by the strong interactions. But if one worked to higher order in the symmetry-breaking, then some infrared divergences could appear in the A matrix. And of course they appear in an important way in the driving terms in electromagnetic problems even in order e^2. In the neutron-proton mass difference problem, one obtains an infrared divergence from Coulomb scattering or when one has a photon connecting an initial and final charged line in πp scattering ("inner bremsstrahlung"). Let me briefly outline how we cope with this problem. For purposes of introduction, let me go back to the non-relativistic case and look at Coulomb scattering. The phase shift is given by the familiar expression consisting of an argument of a gamma function and a logarithmic term

$$\delta_\ell = \arg \Gamma\left(\ell + 1 + \frac{i\delta b}{2q}\right) - \frac{\delta b}{2q}\, \ell n\, 2\, qr$$

[where δb is the small coupling constant (e)] for a given partial wave ℓ. The logarithmic term, we note, is independent of ℓ and of the scattering angle ϑ so that it will appear as a phase factor which multiplies the entire S matrix (i.e. it can be factored out of each partial wave). For the ℓth partial wave

$$S_\ell = e^{2i\,\delta_\ell} \quad .$$

We define a new S-matrix $\hat{S}_\ell$ in which we remove an infrared divergent factor

$$\hat{S}_\ell = \exp\{-i(\delta b/2q)\, \ell n(g(s)/\lambda^2)\}\, S_\ell \quad ,$$

where g(s) is still to be determined. One loses some information in doing this, as time-delay experiments will be affected (such experiments essentially measure $d\eta/ds$, the change in the phase shift with respect to energy (see recent Physical Review papers by Goldberger and Watson). This will not concern us now; the lost information can always be reinserted later.

A new scattering amplitude can be introduced

$$\hat{A} = \frac{e^{2i(\eta + \delta\hat{\eta})} - 1}{2iq} =$$

$$= \frac{e^{2i[\eta + \delta\eta + (\delta b/4q)\, \ell n(g(s)/\lambda^2)]} - 1}{2iq} \quad ,$$

where $\delta\eta$ is the change in the strong interaction phase shift brought about by turning on the electromagnetic interaction. Next we examine the change in the amplitude. In the lecture we had an amplitude

$$\delta A(s) = A_N(s) - A_0(s)$$

given by the difference in the "new" and "old" amplitudes. Since the amplitudes $A_N(s)$ and $A_0(s)$ have been redefined so that a phase has been taken out, $\delta A(s)$ will be correspondingly redefined

$$\delta\hat{A}(s) = \hat{A}_N(s) - \hat{A}_0(s) = \delta A(s) + \frac{\delta b}{4q}\left(\frac{e^{2i\eta}}{8}\right) \ell n \frac{g(s)}{\lambda^2} \quad ,$$

where the exponentials in the A's have been expanded since the coupling constant δb is small. The quantity $\delta\hat{A}(s)$ will remain finite as r goes to infinity (i.e. as $\lambda - 0$) for any value of the arbitrary function g(s), so we have a well-behaved amplitude. Therefore, we can pick g(s) in any convenient way; in particular, we shall pick g(s) in such a way as to minimize the sensitivity of the dispersion integrals to the high-energy behaviour. We have

$$\delta\hat{A}(s) = \frac{e^{2i\eta}}{q}\left[\delta\eta + \frac{\delta b}{4q} \ell n \frac{g(s)}{\lambda^2}\right] ,$$

where we have used $\delta A = (\delta\eta/q)\ e^{2i\eta}$ which follows from $A = e^{2i\eta - 1}/2iq$. We now pick an appropriate energy dependent function for g(s) so that

$$\delta\hat{A}(s) = [\delta\eta - \delta\eta_{Born}]\left(\frac{e^{2i\eta}}{q}\right)$$

is the change in the redefined amplitude. The reason for doing this can be seen from the well-known fact that in potential theory the scattering phase shift tends to its Born approximation value as the energy approaches infinity. Thus, the redefined amplitude is more convergent at infinity than that given by either the Born amplitude itself or by $\delta A(s)$ itself, so as mentioned above, the dispersion integrals are less dependent on the high-energy behaviour.

This is a satisfactory solution to the infrared divergence problems in the non-relativistic case; the infrared problems in the relativistic problem are more difficult. There are effects associated with photon exchange (analogous to that in the Coulomb scattering problem above) and with bremsstrahlung. The amplitude can be redefined as above, but in the relativistic case we do not know that the high-energy behaviour is given by the Born approximation and we do not know the best choice for the function $g(s)$, so choosing it so that we get the Born phase change would be only a guess. Furthermore, the calculation is extremely difficult and the subtraction procedure too complicated to carry out, so further simplifications are made, the validity of which is difficult to evaluate. We need not go into this though, since this problem does not enter into our calculation of the A matrix and octet enhancement.

Hamprecht : At what stage does the infrared difficulty arise in a bootstrap calculation (using the N over D method)?

Sharp : First, it is evident it does not arise in our strong interaction calculation where we calculate A as N/D with N and D given functions determined solely by the strong interactions. Suppose you now want to calculate perturbations around this strong interaction problem - additional shifts in positions of bound-state poles and residues. These would bring in the infrared problems in the case of long range perturbations.

Lassila : Last month's Phys.Rev. Letters contained a paper by Fulco and Wong in which they put particles with the quark properties into the bootstrap equations and obtained predictions which were precisely the same as those following from SU(6). This seems, possibly, to indicate something deep about the bootstrap approach. Perhaps if you are familiar with this work you could comment on its implications, relevance and (maybe) reliability?

Sharp : Unfortunately, I have not read this work on the bootstrapping of quarks. The heart of a bootstrap calculation consists in finding a set of channels and a set of forces which you have some reason to believe are the dominant ones and which form a good approximation. I cannot, offhand, imagine what set of channels would be appropriate for such a calculation so will not comment further.

Blum : Would it be possible to list all the assumptions that went into formalism developed in these lectures, and to what extent they are likely to affect the result?

Sharp : Do you mean which assumptions go into this S-matrix perturbation theory calculation method, or into the whole formalism of octet enhancement as these are two separate questions?

Blum : You are trying to calculate electromagnetic mass shifts which, in the end, will produce a number. How reliable is that number?

Sharp : That is a hard, but fair, question. Some assumptions are made in setting up the general formalism, and some more come in at the stage of doing practical calculations, I will try to answer this by making a list of different assumptions and commenting on each for the example of the neutron-proton mass shift calculation.

i) The nucleon is assumed to be describable as a pion-nucleon bound state in practical calculations. This assumption limits the number of different channels you have to include. The

Chew-Low model and its generalizations make this assumption, and it produces a pretty good description of some features of low energy πN scattering. When you do an N/D calculation, a nucleon bound state is found with a mass that is approximately correct and a coupling constant of the right order of magnitude (within a factor of 2) is found. The bootstrap calculation answers do depend to a certain extent on parameters. Now this assumption (that the nucleon is a πN bound state) seems to produce the right answers, so ad hoc it looks like a reasonable one. But the nucleon couples to a lot of other channels, and no one has really shown that the effect of these other channels is small. Also, rather drastic assumptions have been made as to what the forces producing the bound state are. The forces included are those due to nucleon (N) and N^* exchange; the effect of including higher resonance exchange or vector-meson exchange in other channels is not too certain but estimates which are made indicate that the effect is small. The estimates themselves can be questioned and are at best good to a factor of 2. You see, the trouble is a dispersion theory bootstrap calculation really consists of an infinite set of coupled equations. Most of them we can't even write down because we don't know the singularity structure of the relevant scattering amplitudes. In any case we have to draw a line somewhere because we have an infinite number of equations. So we end up with some drastic approximation to our full set of equations. This in itself isn't necessarily so bad; what makes it so bad is that we can't even begin to estimate the effect of the neglected terms. As things stand, this just doesn't seem like an adequate formulation of strong interaction dynamics to me. Whether methods to cope with these difficulties can be developed within the framework of dispersion theory, or whether a wholly new technique is called for, are challenging questions for the future.

ii) We assume that the partial wave amplitude has the usual analyticity properties implied by the Mandelstam representation. iii) We have assumptions pertaining to S-matrix perturbation theory. Here, we assume certain things about the N and D functions, which are alright if the assumptions about the analytic properties of the partial-wave amplitude are alright.

Question : Then might it not be better to call this assumption 2' since it does not sound very independent?

Sharp : Additional assumptions besides that contained in ii) are made, such as the one about the D function going to a constant. Let us recall what was done. After making the assumptions about the partial-wave amplitudes and about N and D, we went on to calculate the changes in the partial-wave amplitudes. The assumptions made in order to calculate these changes were not all contained in ii); one such assumption, a crucial one, was that we could write an unsubtracted dispersion relation for $D_{new} D_{old} \delta A(s)$. It is true that one can show that the change in the partial wave amplitude goes to zero at ∞, but we still needed an independent assumption about the behaviour of D. This behaviour at infinity was, we saw, tantamount to assuming there were no elementary particles in the theory.
iv) Next, we assumed we could work to first order in small quantities, as e the electric coupling.

Hertel : Yes, I calculate the position and residue of the pole. You are asking how one would tell experimentally whether a particle is a bound state or an elementary particle? With the deuteron it is clear, but with the nucleon not so clear. The problem of formulating a precise theoretical and experimental criterion for distinguishing between bound states and elementary particles is one of the most subtle facing us today. When Reggeism was first proposed we thought we could tell by checking the high-energy region for Regge asymptotic behaviour;

if we found such behaviour we would know that the particle exchanged was composite. This distinction is in question now, because Gell-Mann, Goldberger, Low and Zachariasen have done an elaborate series of calculations beginning with a Lagrangian field theory and then deriving the Regge asymptotic behaviour, so that the Regge pole criterion is now not very clear at all. Another possible way of distinguishing is provided by the Levinson theorem, but we do not know that it is true in relativistic many-channel problems.

Hertel : People have calculated this nucleon as a bound state but if you then look at the phases that come out they are in bad agreement with experiment.

Sharp : There is a calculation by L.A.P. Balazs who used his own techniques. I think his results agree at low energies (up to several masses above the pion threshold) with the experimental π-N phases.

Hertel : I asked this question because there is a paper by Stech and Rothleitner showing that under quite general assumptions this question can be experimentally answered by measuring the sign of the $P_{1/2\,1/2}$ phase shift at the inelastic threshold. If the sign is negative, it is most likely that the nucleon is a bound state, but if it is plus then the nucleon certainly is not. If the phase shift is plus, then it is most likely that the nucleon is produced by many channels. If it is a CDD pole, the situation is so complicated that you can say almost nothing.

Henri : Can you give other examples where this N/D technique can be, or has been, used to derive results that can be compared with experiment besides the neutron-proton case?

Sharp : Yes, let me list some.

i) The pattern of symmetry breaking in B and Δ supermultiple contains many examples. These will be presented in tomorrow's lecture.

ii) The mass differences in other baryon and isospin multiplets have been calculated by Gilman.

iii) The magnetic moments of nucleons have been computed.

iv) Problems connected with the electromagnetic corrections in nucleon-nucleon scattering have been calculated by Goldberg. Coulomb force problems in nuclear physics can probably be similarly treated.

v) Electromagnetic properties of the deuteron can be calculated.

Lipshutz : Have there been any exact SU(3) symmetric bootstraps calculated besides that for the vector-meson octet?

Sharp : There have been a lot of calculations on SU(3) symmetric reciprocal bootstraps for the B and the Δ supermultiplets. It may well be that an exact SU(3) symmetric bootstrap does not exist for these systems. But this is not crucial for our purposes, however, for one can place the difference between the real bootstrap and the SU(3) symmetric bootstrap into the driving term and still incorporate this into the octet enhancement formalism.

DISCUSSION 4

CHAIRMAN : Professor A. Pais

Secretary : C. Noack

Blum : What is changed when passing from SU(6) to $SU(6)_W$? If the two groups are isomorphic the only thing I can think of in physical terms is that you use a new identification of particles.

Pais : That is precisely true. $SU(6)_W$ has a $\underline{35}$ just like SU(6), but the states of that $\underline{35}$ are obtained by a re-shuffling of the states of the usual $\underline{35}$. Also the η' comes in. In principle, the relation of spin to W spin is something very similar to the relation of isospin to U spin in SU(3) (and, in fact, was invented by the same people): looking at isospin you find, for example, that π^+, π^0, π^- form a triplet and K^+, K^0 a doublet, whereas U spin puts π^-, K^- in a doublet and K^0, $\frac{1}{2}(\pi^0 + \sqrt{3}\cdot\eta), \overline{K^0}$ in a triplet.

Blum : This looks, then, as if SU(6) is really something like an accident: you can re-shuffle the particles and still get something.

Pais : No, that is not the way to look at it. Suppose you have the ordinary rotation group. There are subgroups of this: I can turn around the z axis - I get a cylinder. I can also turn around the x axis - I get another cylinder. Now, for convenience in some problems you may want to look at rotations around the axis of the earth; if you send a satellite into orbit another axis may be more appropriate. So you look at different, but isomorphic subgroups of a big group. There is nothing accidental about that.

Question : In $SU(6)_W$ you need four components for a quark spinor, whereas in SU(6) you need only two. How then can the two groups be isomorphic?

Pais : SU(6) is the static group, but even at rest it is useful to use four components if I wish to distinguish quark and antiquark. I can use

$$\begin{pmatrix} \alpha \\ \beta \\ 0 \\ 0 \end{pmatrix} : \text{quark}; \quad \begin{pmatrix} 0 \\ 0 \\ \alpha \\ \beta \end{pmatrix} : \text{antiquark}$$

with $|\alpha|^2 + |\beta|^2 = 1$. Likewise for $SU(6)_W$. You see, I have introduced $SU(6)_W$ as a subgroup of SU(6,6), that is why the four components appeared. I have defined

$$W_1 = \tfrac{1}{2}\gamma_4\,\sigma$$
$$W_2 = \tfrac{1}{2}\gamma_4\,\sigma_2$$
$$W_3 = \tfrac{1}{2}\sigma_3 \quad .$$

These matrices can be written as:

$$W_1 = \frac{1}{2}\begin{pmatrix} \sigma_1 & 0 \\ 0 & -\sigma_1 \end{pmatrix}, \quad W_2 = \frac{1}{2}\begin{pmatrix} \sigma_2 & 0 \\ 0 & -\sigma_2 \end{pmatrix}, \quad W_3 = \frac{1}{2}\begin{pmatrix} \sigma_3 & 0 \\ 0 & \sigma_3 \end{pmatrix},$$

and I can use two-dimensional representations for the quarks and the antiquarks at rest.

Question : But you do it differently for quarks and antiquarks.

Pais : Of course I do. That is the way it is made, because I want to get different physics out of it. But, both for quarks and antiquarks you still have a representation of a group isomorphic to SU(6).

Cooper : Does unitarity break down for $SU(6)_W$?

Pais : Collinearity is a well-defined concept experimentally, but dynamically it is not self-contained. Even if all outer lines of a graph are collinear, the dynamics will involve virtual states with other momenta. You see, forward scattering is connected to non-forward scattering through unitarity, and $SU(6)_W$ says something about forward scattering, but nothing about non-forward scattering. In respect to unitarity breakdown, $SU(6)_W$ does not answer the question.

<u>Cooper</u> : Does this mean that the set of states is not a complete set?

<u>Pais</u> : That is quite true, collinear states are incomplete because not everything goes forward and backward. In regard to unitarity let me say this: SU(6,6) or SL(6,C) is an attempt to formulate a dynamical symmetry which is intrinsically broken by kinetic energy. In some situations, nevertheless, it may be that this breakdown apparently is not very serious, for reasons we do not understand, but in most cases it is not so.

<u>Hamprecht</u> : Why cannot the symmetry breaking of SU(6,6) help us with the unitarity dilemma?

<u>Pais</u> : Perhaps it can. But one would like to do it in a systematic way, and we do not know how to do that.

<u>Question</u> : SU(6) is a group of rank 5, yet we have an obvious interpretation for only 3 of the 5 commuting generators (Y, I_z, S_z). Is there any physical interpretation for the remaining two generators?

<u>Pais</u> : In the defining representation the operators you are referring to are $S_z \otimes Y$ and $S_z \otimes I_z$. Beyond that I know of no physical interpretation.

DISCUSSION 5

CHAIRMAN : Professor N. Cabibbo
Secretaries : A. Saeed and R. Van Royen

Question : Is SU(6,6) well supported by experiments?

Cabibbo : Because of various difficulties, SU(6,6) cannot be enforced as an exact symmetry. The most we can expect is that certain subgroups of it become approximate symmetries, in particular, dynamical situations; for example, $SU(6)_W$ in collinear processes. This has been shown to be sufficient to obtain most of the good results, and to avoid some of the bad ones.

Lassila : Franzini's results on C conservation, just reported, puts stringent upper limits on possible C violation. How bad is this for the theories that allow large C violation? Would you like to comment on this and the relation to your recent Physical Review Letter?

Cabibbo : For the C violation aspect the result is clear! However, the spirit of that work is still worth commenting on. It can be proved*) that in certain physical situations C and/or T symmetry follows from other requirements; for example, the matrix element of the electromagnetic current among physical proton states of momentum $\vec{P}$ and $\vec{P}'$, and can be written as

$$<\vec{P}'|\gamma_\mu|\vec{P}> = \frac{1}{(2\pi)^3}\,\bar{u}(p')\left[F_1\gamma_\mu + \frac{F_2}{2M}\sigma_{\mu\nu}k_\nu + \frac{iF_3}{2M}k_\mu\right]\times u(p),$$

with $k_\mu = p'_\mu - p_\mu$.

*) N. Cabibbo, Phys.Rev. Letters 14, 965 (1965).

T reversal requires F_3 to be zero, but the same follows from current conservation, $k_\mu j_\mu = 0$.

Similar results can be obtained for large classes of vertex functions with two particles on the energy shell. In some cases these results can only be obtained by invoking I-spin and SU_3 invariance. While it is not possible to prove similar theorems for complicated processes which involve many particles, it could be that the dynamics are such that the approximate C invariance of the vertex function is "propagated" into an approximate C invariance of the strong interactions. C invariance would then have a dynamical basis. I have the impression that we do not know enough about the dynamics of strong interactions to obtain convincing conclusions, but this line of thought is worth pursuing. An extensive discussion of C and CP will be given by Professor Prentki.

<u>Question</u> : What is the difference between SU(6,6) and SU(12)?

<u>Cabibbo</u> : The two groups have a very different structure; SU(6,6) being non-compact, while SU(12) is compact. U(12) cannot be used to build Lorentz invariant amplitudes, its basic invariant being $\Psi_\alpha^+(x)\,\Psi_\alpha(x)$, which is not Lorentz invariant. On the other hand, the two groups give rise to the same interesting subgroups $SU(6) \otimes SU(6)$, $SU(6)_W$ and $SU(3) \otimes SU(3)$. In fact, the positive parity part of the algebra of SU(12) (see the lecture on the Gell-Mann approach) is identical with the compact part of SU(6,6).

<u>Remiddi</u> : Are there known reasons for which mesons must have B = 0 and S integer?

<u>Zichichi</u> : There is a relation for B number, spin and lepton number of the form $e^{i\pi(2S+B+L)} = 1$ by Michel. Michel's theorem says that integer spin will have $B+L = 0$.

<u>Cabibbo</u> : What Michel does is to show that the relation quoted by Zichichi allows a minimal extension of the Poincaré group to include baryon and lepton gauge transformations of the kind $e^{i(\alpha B+\beta L)}$. In general, the direct product of the Poincaré group by the baryon and lepton gauge contains three <u>independent</u> superselection rules, due to the absolute conservation of fermion parity $F = e^{i\,2\pi\,J_3}$ (= $e^{i\,2\pi s}$ for particles of spin s, equal to +1 for boson, -1 for fermions) of baryon and lepton charge.

The Michel relation implies that F is a function of B and L, so that the resulting group is a little smaller than the direct product: Poincaré group $\otimes\ e^{i\alpha B} \otimes e^{i\beta L}$

<u>Kotanski</u> : The smallest group which contains the Lorentz group is SL(6,C). Why has very little work been done on it?

<u>Cabibbo</u> : SL(6,C) is a subgroup of SU(6,6) which is defined by having the two invariants: $\bar{\Psi}_\alpha \Psi_\alpha$ [as in SU(6,6)] and also $\bar{\Psi}_\alpha \gamma_5\ \Psi_\alpha$.

The baryon and meson multiplets can be written in exactly the same form as discussed by Pais in his lectures; only you get many more independent couplings of the multiplets among themselves, since each pair of indices α and β can be saturated either directly with δ^α_β or with $(\gamma_5)^\alpha_\beta$. The predictions are, therefore, less restrictive than those of SU(6,6). If we consider what parts of the group commute with the Dirac equation (see lectures by Pais), we find smaller groups than from SU(6,6). For particles at rest: SU(6) generated by

$$\lambda_i , \sigma_K \lambda_i ;$$

for particles of motion along direction 3: SU(3) $\otimes$ SU(3) generated by

$$\lambda_i , \sigma_3 \lambda_i ;$$

for coplanar and more complicated process: SU(3) alone rests.

DISCUSSION 6

CHAIRMAN : Professor J.S. Bell

Secretaries : N. Lipshutz and F.N. Ndili

Question : Would you please explain the meaning of Watson's theorem?

Bell : Watson's theorem concerns the effect of strong interactions on the predictions of time reversal or PCT invariance. For example, PCT implies for transition matrix elements

$$(\bar{F}'|T|\bar{I}') = (I|T|F), \quad (1)$$

where the bar indicates antiparticles replacing particles, and the prime denotes reversal of spins. Because of the interchange of initial and final states this is not very useful. To get something useful one needs to use also unitarity. The unitary condition on the S matrix is

$$S^+S = 1 .$$

This should be true whether or not the weak interactions are switched on. If δS is the increment in S due to weak interactions we have therefore, to first order,

$$(\delta S^+)S + S^+(\delta S) = 0 .$$

The weak interaction transition operator T is conventionally related to δS by

$$\delta S = -i(2\pi)^4 \ \delta^4(P_F - P_I)T ,$$

where the δ function arises from conservation of energy and momentum. Thus

$$T^{+}S - S^{+}T = 0$$

whence

$$(I|S^{+}T|F)^{*} = (F|S^{+}T|I) \ . \tag{2}$$

If we could ignore strong interactions we could set S = 1 here, and then from (1) and (2) we have

$$(F|T|I) = (\bar{F}'|T|\bar{I}')^{*}$$

- particle and antiparticle decay amplitudes are complex conjugates.

More generally, we are concerned with initial states that would be stable in the absence of weak interactions; then even allowing for strong interactions

$$S|I\rangle = |I\rangle \quad \langle I|S^{+} = \langle I| \ .$$

We can choose to work with final states F that diagonalize S:

$$S|F\rangle = e^{2i\delta}|F\rangle$$

$$\langle F|S^{+} = e^{-2i\delta}\langle F| \ .$$

The phase shift δ is just the ordinary scattering phase when F is a system of two bodies below the threshold of inelasticity. Then (2) reads

$$\langle I|T|F\rangle = e^{2i\delta} \langle F|T|I\rangle^*$$

and (1) can be written

$$\langle \bar{F}'|T|\bar{I}'\rangle = e^{2i\delta} \langle F|T|I\rangle^* .$$

Thus particle and antiparticle decay amplitudes fail to be complex conjugates by the phase shift of the final state.

<u>Daniels</u> : You have mentioned that it is impossible so far to find an internal symmetry group which leaves Dirac's equation invariant. Is it not asking too much for this invariance? By analogy, the wave equation of electrons is Dirac's equation and not the Klein-Gordon equation with an internal symmetry added. Should we not instead look for a relativistic wave equation which treats, say, octets a priori and is related to Dirac's equation just as Dirac's equation is related to the Klein-Gordon equation?

<u>Bell</u> : There are two levels to this question. If I want to construct a Lagrangian field theory of spin ½ quarks, then perhaps you are right, since such a Lagrangian may not be the correct starting point. On the other hand, if I forget about formulating a field theory and simply ask what symmetry properties the S matrix might have, I know that each spin ½ particle I consider must be described by the Dirac equation, since the Dirac equation is simply the transcription to a moving frame of the fact that a spin ½ particle has a given transformation property in the rest frame and that Lorentz invariance is a true symmetry of

nature. This statement holds irrespective of the possibility that the fundamental objects of nature may satisfy very complicated equations of motion.

Daniels : Is it really surprising that $SU(6)_W$ works and yet is not the internal symmetry group of the particles?

Bell : I do not object to considering $SU(6)_W$ as a symmetry of the system, but it is difficult to understand it in this way in the context of field theory. For instance, in a given reaction in which the incident and final particles move in a given line, I have no reason to assume that during the interaction they were also moving in the same straight line. It is, therefore, better to abandon attempts to make a field theory, and instead to ask what symmetry operations commute with the minimal set of observables needed to describe the incident and final particles, that is, to describe the S matrix.

Blum : If one interprets SU(6) or SU(6,6) as a dynamical symmetry in the sense of Pais, is there any way to estimate the range of validity of the symmetry, so that one would know which results of the theory are to be trusted?

Bell : The answer to your question is not yet known. One would hope that in trying to formulate an exact theory that one might obtain insight into the conditions under which these symmetries hold, but such insights have not yet been forthcoming.

Løvseth : As far as I understand, it is the successful results of SU(6) which have inspired people to try and make it relativistic. To which extent can these results be obtained from a quark model, and cannot we in that case factor off SU(3).

<u>Bell</u> : Let me first say something in general about the quark model. There is no essential difficulty as long as you stick to a composite system of slow quarks. At one time people thought that large binding energy meant relativistic particles, but this is not true. For the mesons let us assume a potential like the one shown on the figure with a depth of nearly 2M (M = quark mass) and a shape near the bottom like $V \simeq \frac{1}{2} \alpha r^2$. We then have

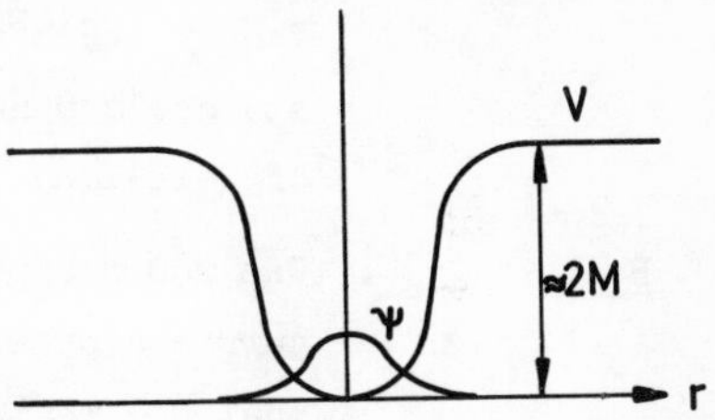

V

≈2M

ψ

r

$$\overline{E_{kin}} = \frac{1}{4} \hbar\omega = \frac{1}{2} M < v^2 > \qquad \omega = \sqrt{\alpha/M}$$

$$< v^2 > = \frac{\hbar}{2} \sqrt{\alpha/M^3} \; .$$

Thus, $v \to 0$ as $M \to \infty$, and we have a non-relativistic system. However, we should also expect excited states of this system. A typical length is $\ell = \hbar/cM_p$ leading to energy level spacings of the order of $\Delta E = \hbar\omega = \hbar v/\ell = vc\, M_p \lesssim$ few 100 MeV. We should, therefore expect low excited states.

Let me also make one point concerning the baryons. Fermi statistics require an antisymmetric space part. People think this is a strange property of the ground state, but this I think is a heritage from nuclear physics where you have a Serber force of type $v(1 + P_r)$, where P_r is the exchange operator. This leads to a preference of symmetric states. Here, the situation might be opposite; the force might be of type $v(1 - P_r)$ which prefers antisymmetry.

What could we obtain from such a system? It is plausible that we could explain the multiplet structure and the static moments. The relations for the form factors and the Johnson-Treiman relations for scattering processes seem more implausible from such a non-relativistic viewpoint.

Question : Why does one assume that the symmetry operators of the theory transform one-particle states only into one-particle states, and not into many-particle states, as in the method of Gell-Mann.

Bell : One would expect that by considering spatial integrals of currents or even of first moments of current distributions that one would be led to states quite similar to the initial state, since such integrals are relatively smooth functional operators. This statement must be modified somewhat because of the unpleasant vacuum properties of such operators. Hence, one should not consider open ended expressions like

$$\int J\, d^3x | \rangle$$

but rather should always close off such expressions by putting definite states on the left, viz.

$$\langle | \int J\, d^3x | \rangle$$

to avoid the possibility of having the infinite extent of the vacuum make difficulties.

Osborn : The Born approximation in non-relativistic potential scattering can be made unitary by dispersion methods. Can this be done with the effective 'Born approximation' given by symmetry principles?

Bell : As to your first question, I have been informed that Professor Salam is attempting just such a programme. The relevant question is: how much of the original symmetry does one expect to survive? In the case of a weak interaction, the unitarizing corrections are small, second order effects, and so one would expect much of the symmetry to survive. On the other hand, in strong interaction physics, one does not believe that the Born approximation is a good approximation, so one expects the modification to be large and since the modifications are brought about by processes which do not respect the symmetry, one has no reason to expect much of the symmetry to be preserved.

Wong : Will you please tell us the present experimental status for the existence or non-existence of quarks?

Bell : There is no evidence for the existence of quarks. All experiments so far indicate only upper limits on the cross-section for quarks below a certain mass.

DISCUSSION 7*)

CHAIRMAN : Professor L.A. Radicati

Secretaries : W. Blum and C. Noack

Radicati : Let me start off by adding a remark I should have made this morning on the hydrogen atom problem. It may have appeared that the guess of the correct group SO(4) was rather arbitrary. This is not quite so. We know not only the degeneracy n^2 of an energy level, we also know from experiment how it splits under the influence of a magnetic field. This already tells us that we have the algebra $\vec{L}$ of angular momentum. Now, we need a step operator to go from one ℓ to another, i.e. an operator with non-vanishing matrix elements $<\ell|M|\ell+1>$. The smallest choice is a vector operator, so we have a second angular momentum-type operator. We are thus led in a natural way to considering SO(4).

Kotanski : What new results do you expect from the introduction of non-compact groups into the field of elementary particles?

Radicati : According to me, none. I think it is a waste of time. I simply wanted to discuss some simple examples to see whether by the knowledge of these I can be wiser than before. If, from looking at the levels of the oscillator, we would have been able to infer its dynamics, then we would have learned something. This remark does not apply to non-compact groups in general as there are quite a few interesting ones.

Hamprecht : We have seen that in the analysis of the hydrogen atom the knowledge of the degeneracy is essential. Suppose that in the case of elementary particles there is a symmetry in

*) These discussions refer to two lectures by Professor Radicati on "Non-Compact Groups". They are not included in the book as the text was not submitted in time for publication.

some internal space and that we have no external means of breaking this symmetry, thus removing the degeneracy. Would this have any relevance to the problem?

Radicati : One does need the full knowledge of the degeneracy otherwise one does not get anywhere. But presumably nature always removes degeneracies.

Question : In the case of the harmonic oscillator you arrived at the conclusion that you need operators $A^{\pm}$ such that $[A^{\pm},H] = \mp A^{\pm}$. From this we could continue by putting $[A^{+},A^{-}] = 1$, thus identifying $A^{\pm}$ with the well-known raising and lowering operators. You then get an algebra with basis A^{+},A^{-}, $A_3(=H)$ and 1. This seems to me an equally reputable algebra.

Radicati : You are probably right.

Cooper : Is not one really just looking for an operator whose eigenvalues give the observed spectrum?

Radicati : Yes. I would like to stress that A^{+} and A^{-} are not at all invariants. The algebra we constructed is not a symmetry of the problem, it is a "spectrum-generating algebra".

Mathews : The n^2 dimensional representations of SO(4) were picked out to describe the degenerate levels of the hydrogen atom by suppressing all representations D(j,k) with $j \neq k$. The argument was used that $J^2 - k^2 = 4\vec{L}\cdot\vec{M}$ is a pseudoscalar, and as there is no pseudoscalar in the problem, this quantity must vanish. However, we have an identical situation in the case of the Lorentz group and still all representations D(m,n) are considered significant. In the light of this fact could you explain the suppression of representations in the present case?

Radicati : One can construct a physically significant pseudoscalar $\underline{W} \cdot \underline{P}$ (the helicity) from the Lubanski operator $W_\mu = \epsilon_{\mu\nu\rho\lambda} J_{\nu\rho} P_\lambda$ and the momentum operator P_μ in the case of the Lorentz group. This is not really relevant, however, since the question concerns the correspondence between SO(4) and the homogeneous Lorentz group, where there are no operators P_μ. Ultimately the justification for taking only D(j, j) must rest on the fact that our system has only states with degeneracies $(2j+1)^2 = n^2$ and that, after all, we cannot think of any pseudoscalar in the problem.

Iwao : Since SO(4) is a group of rank two we need two Casimir operators in order to specify the irreducible representations. The pseudoscalar Casimir operator $\vec{L} \cdot \vec{M}$ is identically zero. Can we still say this is an SO(4) group?

Radicati : Yes, by requiring $\vec{L} \cdot \vec{M} = 0$ we specify a certain class of representation of SO(4), i.e. the ones with $j = k$. Another question is the following: is there a group which has only these representations and not the ones with $j \neq k$?

Prentki : There is none since the direct product $D(j,j) \times D(k,k)$ contains representations with $j \neq k$. The situation is different in the eightfold way. If you multiply representations of triality zero, you obtain only those with triality zero. This group is $SU(3)/Z_3$.

Daniels : In the oscillator problem we found the invariant $-x_1^2 - x_2^2 + x_3^2$ of SO(2,1). Could one identify the first two terms with the negative Hamiltonian which looks quite similar?

<u>Radicati</u> : I do not think the invariant has any physical meaning. The quantities which have a physical meaning are the ones constructed from the generators.

DISCUSSION 8

CHAIRMAN : Professor L.A. Radicati

Secretary : W. Blum

__Ndili__ : What is the physical meaning of the "principal series" and the "complementary series" you derived for the pion-nucleon field?

__Radicati__ : The "complementary series" is a representation which has, as far as I know, no physical realization. The "principal series" is just the representation of the homogeneous Lorentz group.

__Lipshutz__ : Should you really be pessimistic about the fact that you cannot find a classical Hamiltonian that corresponds to a particular non-compact group?

__Radicati__ : I am not, since there is no reason to expect a classical Hamiltonian.

__Lipshutz__ : Is not U(6,6) a natural generalization of SU(6)?

__Radicati__ : U(6,6) is only __one__ possible choice for a non-compact group containing SU(6).

__Lipshutz__ : Since one cannot find dynamics uniquely from a non-compact group, would this mean that group theory is at a dead end?

__Radicati__ : I do not know, we shall see next year.

__Iwao__ : Is there any reason to choose specific representations from the compact $U(6) \otimes U(6) \subset U(6,6)$ (non-compact) as, for example, in SU(6) for mesons and baryons

Mesons in SU(6) : __35__, __405__, ...

Baryons in SU(6) : __56__, __700__, etc.

Radicati : In U(6,6), Gell-Mann suggested the choice of

Mesons : $\underline{143} = (6,6)^-, (35,1)^+, (1,35)^+, (1,1)^+, (\bar{6},6)^-$
Baryons : $\underline{364} = (56,1)^+, (21,6)^-, (6,21)^+, (1,56)^-$.

I do not know the reason for this choice, probably Gell-Mann has chosen these by physical intuition.

Iwao : The deuteron magnetic moment is different from the sum of those of proton and neutron. Do you think SU(6) theory can explain the difference?

Radicati : I have not done calculations on this subject. My personal opinion is that one would not get anything from it, since this is a very detailed nuclear problem depending on tensor forces which are certainly spin and isotopic-spin dependent.

Busza : At the end of this course I got the feeling that unitary symmetry theories are reaching a dead end, and that there is little hope of a revolutionary breakthrough in elementary particle physics coming from these theories.

Bell : I do not share your pessimism, since I do not think that one can expect that there will be a sudden revolutionary breakthrough. We are rather in the middle of a continuing process.

DISCUSSION 9

CHAIRMAN : Professor J. Prentki
Secretaries : C. Noack and W. Blum

Iwao : All the estimates of C or CP violation have been done in a pole approximation, so the result is not reliable. Can one get more reliable results by assuming some symmetry, SU(3) for example, as there are several processes associated with e^+e^- pairs?

Prentki : It is true that such estimations can only give a rough idea on orders of magnitude. In order to apply symmetry, we would have to know the transformation properties of the C-violating current K_μ.

Hertel : In the Sekman theory of C violation, particles may acquire an electric dipole moment. Does this also apply to the neutrino, and what would be the order of magnitude?

Prentki : The electric dipole moment is caused by a term $\mathcal{L}' = \bar{\Psi}\sigma_{\mu\nu}\gamma_5\Psi$ in the Lagrangian. If you have a two-component neutrino theory then the Lagrangian must be invariant under the transformation $\Psi \to \gamma_5\Psi$, which takes $\mathcal{L}' \to -\mathcal{L}'$; so $\mathcal{L}' = 0$ in the case of γ_5 invariance, and no electric dipole moment for a two-component neutrino is allowed by the Sekman theory.

Lipschutz : Would you expect that this new K_μ current would affect things like the Lamb shift?

Prentki : Not to any observable extent, since the heaviness of the lowest hadron-antihadron mass depresses its contribution below observability.

DISCUSSION 10

CHAIRMAN : Professor N. Cabibbo

Secretary : H.C. Dehne

Di Capua : Have you any argument in favour or against the estimate of the quark's mass made by Lee, Gürsey and Nonemberg in their specific quark model? The estimate was of a few GeV and was based on the violation of the mass formula.

Cabibbo : The argument of Lee, Gürsey and Nonemberg is suggestive but not compelling

Hertel : What is the physical meaning of the constant f which appears in the statement

$$< p' | I^+ | n > = f \, \delta^3 (\vec{p}' - \vec{p}_n) \quad ?$$

Cabibbo : The deviation of f from the value 1 is due to the violation of exact isospin symmetry by electromagnetism. Thus, to a proton with momentum p′, in reality one can find an associated γ ray and, hence, the overlap between the states $|p>$ and $I^+|n>$ is not perfect, $I^+|n>$ containing no part corresponding to $p + \gamma$.

Blum : Do you believe "saturation" to be of "kinematical" or "dynamical" origin?

Cabibbo : I do not know, but I think it is probably dynamical. The idea is still so new that as yet no one understands what conditions are necessary for saturation to occur and, hence, it is difficult to speculate in any more detail. For instance, so far we have dealt only with single-particle states, and these are probably insufficient to produce saturation.

Lipshutz : Are there any simple, obvious effects that one would expect to see if there was a breakdown of Lorentz invariance at very high energies, and very small distances?

Bell : It is hard to say how a theory will break down until a new theory which incorporates violations of the old has been developed. Such checks as have been performed (for example, γ decay of ultra-relativistic particles as a check of the independence of c on source velocity) have shown no detectable violations. In general, it is very hard to violate Lorentz invariance without producing drastic consequences.

Comment by Noack : Gürsey and co-workers have considered de Sitter universes and can, in this way, produce arbitrarily small violations of Lorentz invariance.

Rejoinder by Bell : This theory, however, is not concerned with very high energies and very small distances.

Wong : Why is it difficult to give sum rules for the generators corresponding to S, P and $t_{\mu\nu}$?

Cabibbo : It is not difficult to write them down, it is only difficult to interpret them. The only currents which take part in currently observable processes are:

j^3_μ, j^8_μ : Electromagnetism

$\left.\begin{matrix} j^1_\mu + i j^2_\mu \\ g^1_\mu + i g^2_\mu \end{matrix}\right\}$: $\Delta S = 0$ Weak interaction

$\left.\begin{matrix} j^4_\mu + i j^5_\mu \\ g^4_\mu + i g^5_\mu \end{matrix}\right\}$: $\Delta S = 1$ Weak interaction.

Now, from PCAC ($\partial_{\mu} g_{\mu} \propto \varphi_{\pi}$) we can relate the g's to pion omission, but we do not know enough about processes other than pion-proton scattering to apply Adler's technique.

CLOSING CEREMONY

Alla cerimonia ufficiale di chiusura, che ha avuto luogo Sabato, 9 Ottobre, nel salone dell'albergo Jolly in ERICE, hanno partecipato: Monsg. Dr. S. Cassisa, in rappresentanza di S.E. il Vescovo di Trapani, l'On.le Nino Montanti dell'Assemblea Nazionale, S.E. il Prefetto di Trapani, Dr. G. Napoletano, il Sindaco di Erice, Prof. A. Savalli, il Sindaco di Trapani, Prof. A. Calcara, l'Assessore al Turismo della Giunta Provinciale Amministrativa, Prof. S. Giurlanda, il Presidente della Federazione degli Industriali Siciliani, Avv. G. Messina, ed altre Autorità civili e militari.

Il Direttore Generale del Centro Europeo per le Ricerche Nucleari, Chiarissimo Professore Victor F. Weisskopf, ha tenuto il discorso ufficiale di chiusura, parlando sul tema: "Il privilegio di essere scienziati".

Dopo l'allocuzione di chiusura, che è riprodotta nelle pagine seguenti, il Professor V.F. Weisskopf è passato alla consegna dei premi e delle borse di studio messe a concorso.

Il premio "Best Student" è stato attribuito al Dr. W. Blum del Max-Planck-Institut di Monaco. I due premi per "Best Scientific Secretaries" sono stati attribuiti al Dr. G. Altarelli dell'Istituto di Fisica dell'Università di Firenze ed "ex-aequo" ai Drs. C. Noack del Max-Planck-Institut di Heidelberg e F. Strocchi dell'Università di Parigi.

Due speciali borse di studio erano state messe a concorso dalla Direzione della Scuola per onorare la memoria dei Chiarissimi Professori Antonio Stanghellini ed Alberto Tomasini. Dopo una commossa commemorazione dei due giovani e valorosi fisici, così prematuramente scomparsi, il Professor V.F. Weisskopf ha consegnato ai vincitori le rispettive borse di studio, che sono state così attribuite:

"Antonio Stanghellini" Scholarship, awarded to Professor P.M. Mathews, University of Madras, India; "Alberto Tomasini" Scholarship, awarded to Professor F. Ndili, University of Ibadan, Nigeria.

DISCORSO UFFICIALE DI CHIUSURA DEL CHIARISSIMO PROFESSORE VICTOR F. WEISSKOPF

"The Privilege of Being a Scientist"

In the last two weeks you have tried to understand the secret of the structure of the nucleon. Will we ever be able to understand it? Or is it given to us only to understand atomic structure and a little of nuclear structure, whereas the phenomena in the subnuclear level may never be understood? A biologist friend of mine expressed the view that the human brain is able to understand only those processes that play a role in its functioning -- it reads itself as it were --; hence, we would be able to understand only the quantum mechanics of atoms and that little part of nuclear physics which is closely patterned after it.

I certainly do not share that view. A comparison of the Erice Summer School of 1963 with this year's school shows clearly that our understanding of subnuclear phenomena increases steadily.

We all believe firmly and deeply -- maybe the word is "religiously" -- in the explicability of Nature; that all phenomena, on all levels, follow laws and are subject to systematic regularities which we will be able to find after much work and, even more, after the application of much phantasy.

This belief has an important role to play today, of which I would like to say a few words. We are living in extremely difficult times today. We see around us disorder, suffering, and war, whereas technology based on scientific progress could provide for everybody all the material needs he desires. But the development of this technical progress was so fast that it has rocked mankind out of equilibrium. Traditional thoughts are no longer valid today, the experience of Fathers is not very useful to children. Many people feel deeply disturbed and are looking for ideas and ideals which religion had provided in earlier times. In times like these, you should be grateful to be a scientist. You know better than others what you are here for. You are working on a common aim -- our knowledge of Nature -- which steadily increases. You find and establish law and order where such was not seen before; you find significance and meaning in newly discovered phenomena, which give a lead towards understanding of the essential processes which underlie the workings of the Universe. In this field, the work of our fathers remains useful, since science is a constructive enterprise where you enlarge and improve, but never destory. Newton's theory is not abolished by Einstein's. It remains correct in a given domain of phenomena. In this sense no scientific insight ever becomes obsolete. It makes our work a constructive one, and creates a scientific society of coherence and human value.

What was the new idea, the new twist, which made science so successful in the last centuries? Among many factors, it was that science asked restricted questions, questions which were so limited that a clear answer was possible. Science could develop only when men began to restrain themselves not to ask general questions, such as: what is matter made of? How was the Universe created? What is the essence of life? They had to ask only limited questions, such as: how does an object fall? How does water flow in a tube, etc.? To these limited questions answers could be found, and the answers were general, in the sense that the laws which were found had universal validity.

Instead of asking general questions and receiving limited answers, they asked limited questions and found general answers. The most impressive part of this story, however, is the fact that in the development of science, the answerable questions became gradually more and more general. Today one is allowed to ask what matter is made of, and one is able to give a reasonably definite answer to this question; one begins to get at the essence of life and at the origin of the Universe.

The awareness of this knowledge for us scientists is not only a source of satisfaction, but it also gives our lives a deeper meaning. We are a "happy breed of men" in a world of uncertainty and bewilderment. Could this awareness of absolute laws and order in Nature be of some value also to others in our disturbed times? We do not know. So far, not much effort is spent to transmit our great ideas to the non-scientific public. It seems hopeless and impossible to reach the necessary depth and penetration in non-scientific language. Nevertheless, the great values of the scientific effort have some impact on the world of men, other than the technological applications. Even if people are only vaguely conscious of what is going on in science, they are aware of the great significance of the results in our search for law and order in Nature and in our understanding of the great variety of phenomena and creatures which determine our lives. But, perhaps of equal importance is another human aspect of scientific endeavour: its validity to every man on earth independent of his traditional and national background. Unlike any other cultural activity, science is completely international. Other creative human activities are coloured by their national origin. Here, in Erice, you see an architectorial example: the cathedral of the town. It shows the wonderful result of the blending of Arab and Christian traditions. International collaboration took place in Erice long before the days of the Summer School. There, the results reflect their cultural roots; they show essential features which are characteristic of that origin. The great thing about science is the total absence of any such features. The Maxwell equations are not typically English, relativity theory is not typically Jewish.

Within the restricted field of scientific endeavour, human science has reached general and valid insights into essential problems; these insights transcend fully any national, cultural or political boundary. They are human in the widest sense of the word. Perhaps there is a hope in this thought and it may help to pave the way to a future which is better than the past.

LECTURERS

BELL, J.S.	CERN, Geneva 23, Switzerland.
CABIBBO, N.	CERN, Geneva 23, Switzerland.
COOPER, L.N.	Ecole Normale Supérieure, Laboratoire de Physique, 24, rue Lhomond, Paris V, France.
FOCARDI, S.	Laboratori Nazionali di Frascati, Frascati (Roma), Italy.
FRANZINI, P.	Columbia University, New York, N.Y., U.S.A.
MEYER-BERKHOUT, U.	DESY, Gr. Flottbek, Notkestieg 1 2 Hamburg, Germany.
PAIS, A.	Rockefeller Institute, New York, N.Y., U.S.A.
PRENTKI, J.	CERN, and Collège de France Paris, France.
RADICATI, L.A.	Scuola Normale Superiore di Pisa, Pisa, Italy.
SHARP, D.	Sloan Laboratory of Physics, California Institute of Technology, Pasadena, California, U.S.A.
SNOW, G.A.	Department of Physics and Astronomy, College Park, Maryland, U.S.A.
STEINBERGER, J.	CERN, and Columbia University, New York, N.Y., U.S.A.
WEISSKOPF, V.F.	CERN, Geneva 23, Switzerland.

PARTICIPANTS

ALAM, M.A.
University of Durham, Department of Mathematics,
Durham, England.

ALTARELLI, G.
Istituto di Fisica Teorica,
Via S. Leonardo 71,
Firenze (Arcetri), Italy.

BAGLIN, C.
Laboratoire Leprince-Ringuet, Ecole Polytechnique, 17, rue Descartes
Paris (Ve), France.

BASSOMPIERRE, G.
Université de Strasbourg
Institut de Recherches Nucléaires
Boîte Postale No 16
Strasbourg - Cronenbourg, France.

BELLINI, G.
c/o Dott. G. Occhialini,
Università degli Studi di Milano,
Istituto di Scienze Fisiche,
Via Celoria 16, Milano, Italy.

BERTIN, A.
Istituto di Fisica
dell'Università degli Studi di Bologna,
Via Irnerio 46, Bologna, Italy.

BERTRAM, W.
DESY, Gr. Flottbek, Notkestieg 1
2 Hamburg, Germany.

BLUM, W.
Max-Planck-Institut für Physik
Föhringer Ring 6, 8 München, Germany.

BOOTH, P.S.L.
Nuclear Physics Research Laboratory,
University of Liverpool, Mount Pleasant
Liverpool 3, England.

BUNGERT, J.
Laboratoire de Physique Corpusculaire
à Haute Energie
Saclay, France.

BUSZA, W.
University College London,
Department of Physics
Gower Street, London, W.C.1, England.

de la LANDE de CALAN, C.	Ecole Polytechnique, Centre de Physique Théorique, 17, rue Descartes Paris (Ve), France.
CASTELLI, E.	Istituto Nazionale di Fisica Nucleare Sottosezione di Trieste Trieste, Italy.
CELEGHINI, E.	Istituto di Fisica Teorica Via S. Leonardo 71, Firenze (Arcetri), Italy.
CICOGNA, G.	Scuolo Normale Superiore, Pisa, Italy.
DALPIAZ, P.	CERN, Geneva 23, Switzerland.
DANIELS, J.M.	University of Toronto, Toronto 5, Canada.
DARDO, M.	Istituto di Fisica Generale, Via Pietra Giuria 1, Torino, Italy.
DE BAENST, P.	University of Louvain Louvain, Belgium.
DELOFF, A.	University of Oxford, Department of Theoretical Physics, Clarendon Laboratory, Parks Road, Oxford, England.
DEHNE, H.C.	Physikalisches Staatsinstitut II. Institut für Experimentalphysik Luruper Chaussee 149 Hamburg - Bahrenfeld, Germany.
DI CAPUA, E.	Istituto di Fisica dell'Università Piazzale delle Scienze 5, Roma, Italy.
ELIAS, D.K.	Department of Theoretical Physics 12 Parks Road, Oxford, England.
FERRERO, M.I.	CERN, Geneva 23, Switzerland.

FERRETTI, I.	Istituto di Fisica dell'Università di Torino, Via P. Giuria, 1 Torino, Italy.
FLAMINIO, E.	CERN, Geneva 23, Switzerland.
FREYTAG, D.	CERN, Geneva 23, Switzerland.
FRODESEN, A.G.	Universitetet I Oslo, Institute of Physics, Blindern, Norway.
GRUBER, B.	Università di Napoli, Istituto di Fisica Teorica, Mostra d' Oltremare, Pad. 19, Napoli, Italy.
HARDY, L.M.	Lawrence Radiation Laboratory, University of California, Berkeley, California, U.S.A.
HAMPRECHT, B.	DAMTP, Free School Lane, Cambridge, England.
HERTEL, P.	Institut für Theoretische Physik Philosophenweg 16, 69 Heidelberg, Germany.
HENRI, V.	CERN, Geneva 23, Switzerland.
IWAO, S.	Universität Bern, Institut für Theoretische Physik, Sidlerstrasse 5, Bern, Switzerland.
KIENZLE, W.	CERN, Geneva 23, Switzerland.
KOTANSKI, A.	Uniwersytet Jagielloński, Instytut Fizyki Kraków, ul. Reymonta 4, Poland.
LASSILA, K.E.	University of Helsinki, Research Institute for Theoretical Physics, Siltavuorenpenger 20 B, Helsinki, Finland.

LEE-FRANZINI, J.	c/o Dr. P. Franzini, Department of Physics, Columbia University, 538 West 120th Street, New York 27, U.S.A.
LEVY, N.	Technion - Israel Institute of Technology, Haifa, Israel.
LIPSHUTZ, N.R.	The University of Chicago, The Enrico Fermi Institute for Nuclear Studies, 5630 Ellis Avenue, Chicago, Illinois 50637, U.S.A.
LLEWELLYN SMITH, C.H.	Clarendon Laboratory, Park Road, Oxford, England.
LOVSETH, J.	Institute of Physics, University of Oslo, Oslo 3, Norway.
MAHANTA, P.	Department of Physics, Imperial College, Prince Consort Road, London, S.W.7, England.
MASON, D.	Imperial College of Science and Technology (University of London), Prince Consort Road, South Kensington, London, S.W.7, England.
MATHEWS, P.M.	Brandeis University, Department of Physics, Waltham 54, (Massachusetts), U.S.A.
MÜTTER, K.H.	Deutsches Elektronen-Synchroton DESY, Gr. Flottbek, Notkestieg 1, 2 Hamburg, Germany.
NAGY, T.	Institute of Theoretical Physics, University, Puskin utca 5-7, Budapest VIII, Hungary.
NDILI, F.N.	University of Ibadan, Nigeria Ibadan, Nigeria.
NOACK, C.	Max-Planck-Institut, Saupfercheckweg, 69 Heidelberg, Germany.

OTTOSON, U.I.	Chalmers University of Technology, Department of Theoretical Physics, Göteborg S, Sweden.
OSBORN, H.	University College London, Department of Physics, Gower Street, London, W.C.1, England.
ORZALESI, C.A.	Università de Pavia, Istituto di Fisica Teorica, Via Taramelli 4, Pavia, Italy.
PALMONARI, F.	Istituto di Fisica dell'Università degli Studi di Bologna, Via Irnerio 46, Bologna, Italy.
PELOSI, V.	Università degli Studi di Milano, Istituto di Scienze Fisiche, Via Celoria 16, Milano, Italy.
PERNEGR, J.	CERN, Geneva 23, Switzerland.
POELZ, G.	Max-Planck Institut für Kernphysik, Jahnstrasse 29, Heidelberg, Germany.
PULLIA, A.	Istituto Nazionale di Fisica Nucleare, Sezione di Milano, Gruppo Alte Energie Via Celoria 16, Milano, Italy.
QUERCIGH, E.	CERN, Geneva 23, Switzerland.
REID, G.F.	Tait University of Mathematical Physics, The University, 1 Roxburgh Street Edinburgh 8, Scotland.
REMIDDI, E.	Scuola Normale Superiore, Piazza dei Cavalieri, Pisa, Italy.
REPELLIN, J.P.	Accélérateur Linéaire Orsay (Seine-et-Loire), France.
RIDDIFORD, L.	The University of Birmingham, Department of Physics, The University, Birmingham 15, England.

ROMANO, A.	Istituto di Fisica dell'Università degli Studi di Bari, Via Amendola 173 Bari, Italy.
van ROYEN, R.	Laboratoire de Physique Théorique et Hautes Energies, Bâtiment 211, Facultés des Sciences d'Orsay, Orsay, (Seine-et-Loire), France.
SAEED, A.	Imperial College of Science and Technology, (University of London), Prince Consort Road, South Kensington, London, S.W.7, England.
SAGGION, A.	Università degli Studi di Padova, Istituto di Fisica Galileo Galilei, Via F. Marzolo 8, Padova, Italy.
SANTRONI, A.	Università di Genova, Istituto di Fisica, Viale Benedetto XV, 5, Genova, Italy.
SAUVAGE, G.	Laboratoire de l'Accélérateur Linéaire d'Orsay, Orsay (78), France.
van der SPUY, E.	Atomic Energy Board, Pelindaba, Private Bag 256, Pretoria, Republic of South Africa.
STEINBERGER, C.	CERN, Geneva 23, Switzerland.
STROCCHI, F.	Université de Paris, Faculté des Sciences, Orsay, Laboratoire de Physique Théorique et Hautes Energies, Bâtiment 211 Orsay, (Seine-et-Loire), France.
SUTTORP, L.G.	Instituut voor Theoretische Fysica Universiteit van Amsterdam, Valckenierstraat 65, Amsterdam-C., Holland.
TAK CHIU WONG	Clarendon Laboratory, Oxford, England.
van der WAL, S.G.	Institute for Theoretical Physics of the Physical Laboratory, 34 Westersingel Groningen, Holland.
ZULAUF, M.R.	Institut für Theoretische Physik, Sidlerstrasse 5, 3000 Bern, Switzerland.